Advances in Ring Theory

S. K. JAIN
S. TARIQ RIZVI
Editors

Birkhäuser
Boston • Basel • Berlin

S. K. Jain
Department of Mathematics
Ohio University
Athens, OH 45701

S. Tariq Rizvi
Department of Mathematics
Ohio State University at Lima
Lima, OH 45804

QA
247
.A28
1997

Library of Congress Cataloging-in-Publication Data

Advances in ring theory / S. K. Jain and S. Tariq Rizvi, editors.
    p.   cm. -- (Trends in Mathematics)
  Includes bibliographical references and index.
  ISBN 0-8176-3969-1 (acid-free paper)
  1. Rings (Algebra)   I. Jain, S. K. (Surender Kumar), 1938-
II. Rizvi, S. Tariq.   III. Series.
QA247.A28   1997   97-26397
512'.4--dc21   CIP.

AMS Classifications: 05A10, 13C05, 13F20, 16A08, 16A33, 16D10, 16D50, 16D70, 16D90, 16E50, 16E60, 16L30, 16U20, 16P40, 16N40, 16P50, 16P60, 16P99, 16S99

Printed on acid-free paper

Birkhäuser ®

© Birkhäuser Boston 1997

Copyright is not claimed for works of U.S. Government employees.
All rights reserved. No part of this publication may be reproduced, stored in a retrieval system, or transmitted, in any form or by any means, electronic, mechanical, photocopying, recording, or otherwise, without prior permission of the copyright owner.

Permission to photocopy for internal or personal use of specific clients is granted by Birkhäuser Boston for libraries and other users registered with the Copyright Clearance Center (CCC), provided that the base fee of $6.00 per copy, plus $0.20 per page is paid directly to CCC, 222 Rosewood Drive, Danvers, MA 01923, U.S.A. Special requests should be addressed directly to Birkhäuser Boston, 675 Massachusetts Avenue, Cambridge, MA 02139, U.S.A.

ISBN 0-8176-3969-1
ISBN 3-7643-3969-1
Typeset in TEX by the authors
Printed and bound by Quinn-Woodbine, Woodbine, NJ
Printed in the U.S.A.

9 8 7 6 5 4 3 2 1

CONTENTS

Kasch Modules
T. Albu and R. Wisbauer 1

Compactness in Categories and Interpretations
P. N. Ánh and R. Wiegandt 17

A Ring of Morita Context in Which Each Right Ideal
is Weakly Self-injective
S. Barthwal, S. K. Jain, S. Jhingan, S. R. López-Permouth . . 31

Splitting Theorems and a Problem of Müller
G. F. Birkenmeier, J. Y. Kim and J. K. Park 39

Decompositions of D1 Modules
R. A. Brown and M. H. Wright 49

Right Cones in Groups
H. H. Brungs and G. Törner 65

On Extensions of Regular Rings of Finite Index
by Central Elements
W. D. Burgess and R. M. Raphael 73

Intersections of Modules
J. Dauns 87

Minimal Cogenerators Over Osofsky and Camillo Rings
C. Faith 105

Uniform Modules Over Goldie Prime Serial Rings
F. Guerriero 119

Co-Versus Contravariant Finiteness of Categories
of Representations
B. Huisgen-Zimmermann and S. O. Smalø 129

Monomials and the Lexicographic Order
H. Hulett 145

Rings Over Which Direct Sums of CS Modules Are CS
D. V. Huynh and B. J. Müller 151

Exchange Properties and the Total
F. Kasch and W. Schneider 161

Local Bijective Gabriel Correspondence and
Torsion Theoretic FBN Rings
P. Kim and G. Krause 175

Normalizing Extensions and the Second Layer Condition
K. A. Kosler 191

Generators of Subgroups of Finite Index in $GL_m(\mathbb{Z}G)$
G. T. Lee and S. K. Sehgal 211

Weak Relative Injective M-Subgenerated Modules
S. Malik and N. Vanaja 221

Direct Product and Power Series Formations Over 2-Primal Rings
G. Marks 239

Localization in Noetherian Rings
M. McConnell and F. L. Sandomierski 247

Projective Dimension of Ideals in Von Neumann Regular Rings
B. L. Osofsky 263

Homological Properties of Color Lie Superalgebras
K. L. Price 287

Indecomposable Modules Over Artinian Right Serial Rings
S. Singh 295

Nonsingular Extending Modules
P. F. Smith 305

Right Hereditary, Right Perfect Rings Are Semiprimary
M. L. Teply 313

On the Endomorphism Ring of a Discrete Module:
A Theorem of F. Kasch
J. M. Zelmanowitz 317

Nonsingular Rings with Finite Type Dimension
Y. Zhou 323

PREFACE

This volume is an outcome of invited lectures delivered at the Ring Theory Section of the 23rd Ohio State-Denison Conference in May 1996. It also contains articles by some invited mathematicians who could not attend the conference. These peer-refereed articles showcase the latest developments and trends in classical Ring Theory, highlighting the cross-fertilization of new techniques and ideas with the existing ones. Providing a wide variety of methodologies, this volume should be valuable both to graduate students as well as to specialists in Ring Theory.

We would like to thank our colleagues who invested a lot of their time to make the conference a great success. In particular, our thanks go to Professors Tom Dowling, Dan Sanders, Surinder Sehgal, Ron Solomon and Sergio R. López-Permouth for their help. The financial support for the Conference, provided by the Department of Mathematics, The Ohio State University, and Mathematics Research Institute, Columbus, is gratefully acknowledged. Many thanks go to Dean Violet I. Meek for her commitment to the promotion of research by her continuous encouragement of such efforts and for providing financial support from the Lima campus of The Ohio State University. We have received immense cooperation from all the referees who, meticulously and in a very short time, provided us with their reports in spite of their busy schedules. We express our sincere thanks to all of them. Finally, we thank Ms. Cindy White for her excellent job in typing parts of this volume.

We are pleased to dedicate this volume to Professor Bruno J. Müller on the occasion of his retirement for his many contributions to the Theory of Rings and Modules. As this volume was going to press we have learned that Professor Carl Faith is retiring this year. It is our great pleasure to dedicate this volume also to Carl on his retirement and on his 70th birthday for his outstanding works in the field of Abstract Algebra.

S. K. Jain and S. Tariq Rizvi, editors
March, 1997

KASCH MODULES

Toma Albu and Robert Wisbauer

Abstract

An associative ring R is a *left Kasch ring* if it contains a copy of every simple left R-module. Transferring this notion to modules we call a left R-module M a *Kasch module* if it contains a copy of every simple module in $\sigma[M]$. The aim of this paper is to characterize and investigate this class of modules.

Introduction

Let M be a left R-module over an associative unital ring R, and denote by $\sigma[M]$ the full subcategory of R-Mod consisting of all M-subgenerated R-modules.

In section 1 we collect some basic facts about $\sigma[M]$, torsion theories, and modules of quotients in $\sigma[M]$. In section 2 we introduce the concept of a Kasch module. M is a Kasch module it its M-injective hull $\widehat{M}$ is an (injective) cogenerator in $\sigma[M]$. For ${}_R M = {}_R R$ we regain the classical concept of left Kasch ring. Various characterizations of Kasch modules are provided. In section 3 we present some properties of Kasch modules.

Note that the notion of Kasch module in [10] and [16] is different from ours. Also the notion of Kasch ring used in these papers (R is a Kasch ring if R_R and ${}_R R$ are injective cogenerators in Mod-R and R-Mod respectively) is different from the usual one.

1 Preliminaries

Throughout this paper R will denote an associative ring with nonzero identity, R-Mod the category of all unital left R-modules and M a fixed left R-module. The notation ${}_R N$ will be used to emphasize that N is a left R-module. Module morphisms will be written as acting on the side opposite to scalar multiplication. All other maps will be written as acting on the left.

Any unexplained terminology or notation can be found in [7], [13], [14] and [15].

1.1 M-(co-)generated modules. A left R-module X is said to be M-*generated* (resp. M-*cogenerated*) if there exists a set I and an epimorphism $M^{(I)} \longrightarrow X$ (resp. a monomorphism $X \longrightarrow M^I$). The full subcategory of R-Mod consisting of all M-generated (resp. M-cogenerated) R-modules is denoted by $\text{Gen}(M)$ (resp. $\text{Cog}(M)$).

1.2 The category $\sigma[M]$. A left R-module X is called M-*subgenerated* if X is isomorphic to a submodule of an M-generated module, and the full subcategory of R-Mod consisting of all M-subgenerated R-modules is denoted by $\sigma[M]$. This is a Grothendieck category (see [14]) and it determines a filter of left ideals

$$F_M = \{\, I \leq {}_R R \mid R/I \in \sigma[M]\,\},$$

which is precisely the set of all open left ideals of R in the so called M-*adic topology* on R (see [6]).

For any $X \in \sigma[M]$ we shall denote by $\widehat{X}$ the injective hull of X in $\sigma[M]$, called also the M-injective hull of X. With this terminology, the injective hull of X in R-Mod is the R-injective hull, denoted in the sequel by $E(X)$. It is known (see e.g. [14, 17.9]) that $\widehat{X} = Tr(M, X) = Tr(\sigma[M], X)$, where $Tr(M, X)$ (resp. $Tr(\sigma[M], X)$) denotes the trace of M (resp. $\sigma[M]$) in X.

1.3 Hereditary torsion theories in $\sigma[M]$. The concept of a torsion theory can be defined in any Grothendieck category (cf. [8]), so in particular in $\sigma[M]$. A *hereditary torsion theory* in $\sigma[M]$ is a pair $\tau = (\mathcal{T}, \mathcal{F})$ of nonempty classes of modules in $\sigma[M]$ such that $\mathcal{T}$ is a *hereditary torsion class* or a *localizing subcategory* of $\sigma[M]$ (this means that it is closed under subobjects, factor objects, extensions, and direct sums) and

$$\mathcal{F} = \{\, X \in \sigma[M] \mid \text{Hom}_R(T, X) = 0,\ \forall\, T \in \mathcal{T}\,\}.$$

The objects in $\mathcal{T}$ are called τ-*torsion modules*, and the object in $\mathcal{F}$ are called τ-*torsionfree modules*.

For any $X \in \sigma[M]$ we denote by $\tau(X)$ the τ-*torsion submodule* of X, which is the sum of all submodules of X belonging to $\mathcal{T}$. Clearly, one has $X \in \mathcal{T} \Leftrightarrow \tau(X) = X$, and $X \in \mathcal{F} \Leftrightarrow \tau(X) = 0$.

Note that any hereditary torsion theory $\tau = (\mathcal{T}, \mathcal{F})$ in $\sigma[M]$ if completely determined by its first component $\mathcal{T}$, and so usually the hereditary torsion theories are identified with hereditary torsion classes.

Any injective object $Q \in \sigma[M]$, i.e., any M-injective module belonging to $\sigma[M]$, determines a hereditary torsion theory $\tau_Q = (\mathcal{T}_Q, \mathcal{F}_Q)$, called the *hereditary torsion theory in $\sigma[M]$ cogenerated by Q*:

$$\mathcal{T}_Q = \{\, X \in \sigma[M] \mid \mathrm{Hom}_R(X, Q) = 0 \,\} \text{ and } \mathcal{F}_Q = \mathrm{Cog}(Q) \cap \sigma[M]\,.$$

Note that for any $N \in \sigma[M]$, $\mathrm{Cog}(N) \cap \sigma[M]$ is precisely the class $\mathrm{Cog}_M(N)$ of all objects in $\sigma[M]$ which are cogenerated by N in the category $\sigma[M]$ (i.e., are embeddable in direct products in $\sigma[M]$ of copies of N).

According to [15, 9.4, 9.5], any hereditary torsion theory $\tau = (\mathcal{T}, \mathcal{F})$ in $\sigma[M]$ has this form, i.e., for any such τ there exists an M-injective module Q in $\sigma[M]$ with $\tau = \tau_Q$.

For any M-injective module Q in $\sigma[M]$ we can also consider the hereditary torsion theory $\tau_{E(Q)} = (\mathcal{T}_{E(Q)}, \mathcal{F}_{E(Q)})$ in R-Mod cogenerated by $E(Q)$:

$$\mathcal{T}_{E(Q)} = \{\, {}_R X \mid \mathrm{Hom}_R(X, E(Q)) = 0 \,\} \text{ and } \mathcal{F}_{E(Q)} = \mathrm{Cog}(E(Q))\,.$$

Since for any $X \in \sigma[M]$ and $f \in \mathrm{Hom}_R(X, E(Q))$, one has $\mathrm{Im}(f) \in Tr(\sigma[M], E(Q)) = \widehat{Q} = Q$, we deduce that

$$\mathcal{T}_Q = \mathcal{T}_{E(Q)} \cap \sigma[M] \text{ and } \mathcal{F}_Q = \mathcal{F}_{E(Q)} \cap \sigma[M]\,,$$

that is, any hereditary torsion theory $\tau = (\mathcal{T}, \mathcal{F})$ in $\sigma[M]$ is the "trace" $\tau' \cap \sigma[M]$ of a certain hereditary torsion theory $\tau' = (\mathcal{T}', \mathcal{F}')$ in R-Mod: this means that

$$\mathcal{T} = \mathcal{T}' \cap \sigma[M] \text{ and } \mathcal{F} = \mathcal{F}' \cap \sigma[M]\,.$$

1.4 The Lambek torsion theory in $\sigma[M]$. The M-injective hull $\widehat{M}$ of the module ${}_R M$ cogenerates a hereditary torsion theory $\tau_{\widehat{M}} = (\mathcal{T}_{\widehat{M}}, \mathcal{F}_{\widehat{M}})$ in $\sigma[M]$, namely:

$$\mathcal{T}_{\widehat{M}} = \{\, X \in \sigma[M] \mid \mathrm{Hom}_R(X, \widehat{M}) = 0 \,\}\,,$$

$$\mathcal{F}_{\widehat{M}} = \mathrm{Cog}_M(\widehat{M}) = \sigma[M] \cap \mathrm{Cog}(\widehat{M})\,,$$

called the *Lambek torsion theory* in $\sigma[M]$. Note that this torsion theory depends on the choice of the subgenerator of $\sigma[M]$. If $\sigma[M] = \sigma[N]$ for some ${}_R N$, then in general $\tau_{\widehat{M}} \neq \tau_{\widehat{N}}$.

If $_RM = {_R}R$ then we obtain the torsion theory $\tau_{E(R)}$ on R-Mod, which is precisely the well-known *Lambek* torsion theory in R-Mod. The corresponding Gabriel topology on R is the set

$$D_R = \{\, I \leq {_R}R \,|\, \operatorname{Hom}_R(R/I, E(R)) = 0 \,\}$$

of all *dense* left ideals of R.

In the sequel, we shall denote by D_M the Gabriel topology on R corresponding to the hereditary torsion theory in R-Mod cogenerated by $E(M)$,

$$D_M = \{\, I \leq {_R}R \,|\, \operatorname{Hom}_R(R/I, E(M)) = 0 \,\}.$$

1.5 Modules of quotients in $\sigma[M]$. Let $\tau = (\mathcal{T}, \mathcal{F})$ be a hereditary torsion theory in $\sigma[M]$. For any module $X \in \sigma[M]$ one defines the τ-*injective hull* of X (see [15, 9.10]) as being the submodule $E_\tau(X)$ of the M-injective hull $\widehat{X}$ of X for which

$$E_\tau(X)/X := \tau(\widehat{X}/X).$$

The *module of quotients* $Q_\tau(X)$ of X with respect to τ is defined (see [15, 9.14]) by

$$Q_\tau(X) := E_\tau(X/\tau(X)).$$

In particular one can consider for any $X \in \sigma[M]$ the module of quotients of X with respect to the Lambek torsion theory $\tau_{\widehat{M}}$ in $\sigma[M]$.

The module of quotients of a module $_RX$ with respect to the Lambek torsion theory $\tau_{E(R)}$ in R-Mod is denoted by $Q_{\max}(X)$ and is called the *maximal module of quotients* of X. For $_RX = {_R}R$ one obtains a ring denoted by $Q^\ell_{\max}(R)$ and called the *maximal left ring of quotients* of R.

2 Definition and Characterizations

The following result is well-known (see e.g. [13, Lemma 5.1, p. 235]):

2.1 Proposition. *The following assertions are equivalent for a ring R:*

(1) $D_R = \{R\}$;

(2) $E(R)$ is an injective cogenerator of R-Mod;

(3) Every simple left R–module is isomorphic to a (minimal) left ideal of R;

(4) $\operatorname{Hom}_R(C, R) \neq 0$ for every nonzero cyclic left R-module C;

(5) $\ell(I) \neq 0$ for every left ideal I of R, where $\ell(I) = \{\, r \in R \,|\, rI = 0 \,\}$.

A ring satisfying one of the equivalent conditions above is called a *left Kasch ring*.

The above proposition suggests the following:

2.2 Definition. *A module $_RM$ is called a* Kasch module *if $\widehat{M}$ is an (injective) cogenerator in $\sigma[M]$.*

So, the ring R is a left Kasch ring if and only if $_RR$ is a Kasch module.

2.3 Remarks. (1) Clearly, if M is a Kasch module, then so is $M \oplus N$ for any $N \in \sigma[M]$.

(2) For any $_RN$ there exists $_RK \in \sigma[N]$ such that $\sigma[N] = \sigma[K]$ and K is a Kasch module. Indeed, $\sigma[N]$ has an injective cogenerator, say Q (see e.g. [14, 17.12]). Then $K = N \oplus Q$ is the desired Kasch module.

(3) If M is a Kasch module and $_RN$ is such that $\sigma[M] = \sigma[N]$, then the module N is not necessarily a Kasch module. To see this, take as N a module which is not Kasch and as M the module K considered in (2).

(4) Clearly if $_RM$ is a cogenerator in $\sigma[M]$, then M is a Kasch module. The converse is not true, as the following example shows: let F be a field and denote by R the ring $F[[X]]$ of all formal series in the indeterminate X over F. Then R is a local ring having $P = (X)$ as the only maximal ideal, $R/P \simeq F \leq R$, but R is not a cogenerator of R-Mod since $E(R/P)$ cannot be embedded in R.

2.4 Examples. (1) Any semisimple module M is a Kasch module.

(2) If M is a non-singular module in $\sigma[M]$, i.e. M is *polyform* (see [15]), then M is a Kasch module if and only if M is semisimple. Indeed, one implication is obvious. For the other one, if M is a non-singular module in $\sigma[M]$, then according to [15, 10.2], $N \in \sigma[M]$ is M-singular if and only if $\text{Hom}_R(N, \widehat{M}) = 0$. But, if M is a Kasch module, then such an N must be necessarily zero (see also 2.6). It follows that for any $K \trianglelefteq M$, where "$\trianglelefteq$" means "essential submodule", one obtains an M-singular module M/K, which must be 0. Thus, M has no proper essential submodules, which implies that M is semisimple.

(3) For any nonzero $n \in \mathbb{N}$, the $\mathbb{Z}$-module $\mathbb{Z}/n\mathbb{Z}$ is a Kasch module, which is polyform if and only if n is square-free, i.e., if and only if it is semisimple.

(4) Any torsion abelian group is a Kasch $\mathbb{Z}$-module. More generally, any usual torsion module over a Dedekind domain D is a Kasch module.

Indeed, it is known (see e.g. [3, Proposition 2.2.3]) that the usual torsion modules over a Dedekind domain D which is not a field are precisely the semi-Artinian D-modules, and moreover, any torsion D-module is the direct sum of its U–primary components (see section 3 for definitions). Apply now 3.2 and 3.6.

2.5 Lemma. *The following assertions are equivalent for $X \in \sigma[M]$:*
  (1) $\mathrm{Hom}_R(X, E(M)) = 0$;
  (2) $\mathrm{Hom}_R(X, \widehat{M})) = 0$;
  (3) $\mathrm{Hom}_R(C, M) = 0$ for any (cyclic) submodule C of X.

Proof: Since for any $f \in \mathrm{Hom}_R(X, E(M))$, one has

$$\mathrm{Im}(f) \in \mathrm{Tr}(\sigma[M], E(M)) = \widehat{M},$$

one deduces that (1) $\Leftrightarrow$ (2).

The equivalence (1) $\Leftrightarrow$ (3) is an immediate consequence of [13, Lemma 3.8, p. 142]. □

2.6 Proposition. *The following properties are equivalent for the module $_RM$:*
  (1) M is a Kasch module;
  (2) Any simple module in $\sigma[M]$ can be embedded in M;
  (3) Any simple module in $\sigma[M]$ is cogenerated by M;
  (4) $\mathcal{T}_{\widehat{M}} = \{0\}$;
  (5) $\mathcal{F}_{\widehat{M}} = \sigma[M]$;
  (6) $\{\,_RX \,|\, X \leq\, _RN$ and $\mathrm{Hom}_R(N/X, E(M)) = 0\,\} = \{N\}$ for any $N \leq\, _RM$;
  (7) $F_M \cap D_M = \{R\}$;
  (8) $\mathrm{Hom}_R(C, M) \neq 0$ for any nonzero (cyclic) left R-module C from $\sigma[M]$;
  (9) $N = Q_{\tau_{\widehat{M}}}(N)$ for any $N \in \sigma[M]$, i.e., any module in $\sigma[M]$ is its own module of quotients with respect to the Lambek torsion theory $\tau_{\widehat{M}}$ in $\sigma[M]$.

Proof: (1) $\Rightarrow$ (2) It is known (see [14, 17.12]) that an injective object $_RQ$ in $\sigma[M]$ is a cogenerator of $\sigma[M]$ if and only if it contains a copy of each simple module in $\sigma[M]$. So, (1) implies that for any simple object $U \in \sigma[M]$ there exists a monomorphism $\alpha_U : U \longrightarrow \widehat{M}$. It follows that $\mathrm{Im}(\alpha_U) \cap M \neq 0$, and then $\mathrm{Im}(\alpha_U) \cap M = \mathrm{Im}(\alpha_U)$ because $U \simeq \mathrm{Im}(\alpha_U)$

is a simple module. Thus $U \simeq \text{Im}(\alpha_U) \leq M$, which proves the implication $(1) \Rightarrow (2)$.

$(2) \Rightarrow (3)$ is obvious.

$(3) \Rightarrow (1)$ Let U be an arbitrary simple module in $\sigma[M]$. Since $U \in \text{Cog}(M)$, it follows that there exists a nonzero morphism $f : U \longrightarrow M$, which is necessarily injective because U is a simple module. Thus, any simple module in $\sigma[M]$ can be embedded in M, and so in $\widehat{M}$, showing that $\widehat{M}$ is a cogenerator in $\sigma[M]$.

$(1) \Rightarrow (4)$ If $X \in \mathcal{T}_{\widehat{M}}$ then $\text{Hom}_R(X, \widehat{M}) = 0$. Assume that $X \neq 0$. Then, there exists a nonzero morphism $h : X \longrightarrow \widehat{M}$ because $\widehat{M}$ is a cogenerator in $\sigma[M]$, a contradiction.

$(4) \Leftrightarrow (5)$ and $(4) \Rightarrow (6)$ are clear.

$(6) \Rightarrow (2)$ Assume that (2) is not satisfied. Then, there exists a simple module $U \in \sigma[M]$ such that U cannot be embedded in M. Then $\text{Hom}_R(U, M) = 0$, and so $\text{Hom}_R(U, E(M)) = 0$. But every module in $\sigma[M]$ is an epimorphic image of a submodule of M, i.e., it is a subfactor of M. So, there exists $X \leq N \leq M$ such that $U \simeq N/X$. It follows that $\text{Hom}_R(N/X, E(M)) = 0$, and by assumption, we deduce that $X = N$, a contradiction because $U \neq 0$. This proves the desired implication.

$(4) \Rightarrow (7)$, $(7) \Rightarrow (8)$ and $(8) \Rightarrow (4)$ follow from 2.5.

$(4) \Rightarrow (9)$ Assume that $\mathcal{T}_{\widehat{M}} = \{0\}$. Then $\mathcal{F}_{\widehat{M}} = \sigma[M]$. Let $N \in \sigma[M]$. Then, the module of quotients $Q_{\tau_{\widehat{M}}}(N)$ of N with respect to the Lambek torsion theory $\tau_{\widehat{M}}$ in $\sigma[M]$ is

$$Q_{\tau_{\widehat{M}}}(N) = E_{\tau_{\widehat{M}}}(N/\tau_{\widehat{M}}(N)).$$

But $\tau_{\widehat{M}}(N) = 0$ and

$$E_{\tau_{\widehat{M}}}(N)/N = \tau_{\widehat{M}}(\widehat{N}/N) = 0,$$

by hypothesis. So $E_{\tau_{\widehat{M}}}(N) = N$, and consequently $Q_{\tau_{\widehat{M}}}(N) = N$ for any $N \in \sigma[M]$.

$(9) \Rightarrow (5)$ Suppose that M is such that $Q_{\tau_{\widehat{M}}}(N) = N$ for any $N \in \sigma[M]$. Since $Q_{\tau_{\widehat{M}}}(N) \in \mathcal{F}_{\widehat{M}}$ for all $N \in \sigma[M]$, we deduce that $\mathcal{F}_{\widehat{M}} = \sigma[M]$. $\square$

2.7 Remark. Suppose that $_RM$ is such that any simple module from $\sigma[M]$ is M-cyclic, i.e., isomorphic to a factor module of M. This happens e.g. when $_RM = {}_RR$ or when M is a self-generator. Then, by the proof of

2.6, one deduces that in this case we can add to the equivalent conditions from 2.6 also the following one:

(10) $\{ {}_R X \mid X \leq {}_R M \text{ and } \operatorname{Hom}_R(M/X, E(M)) = 0 \} = \{M\},$

in other words, the only rational submodule of M is M itself.

As an immediate consequene of 2.6 we obtain the following characterization of left Kasch rings:

2.8 Corollary. *The following are equivalent for the ring R:*
(1) R is a left Kasch ring;
(2) Any left R-module X is its own maximal module of quotients.

2.9 Example. The example from [12] we are going to present now provides a module which is its own module of quotients in the Lambek topology, but which is not Kasch. This shows that in 2.6 (resp. 2.8) we need the condition (9) (resp. (2)) to be fulfilled for *all* $X \in \sigma[M]$, and not only for M (resp. for *all* X in R-Mod, and not only for ${}_R R$).

Let R denote the direct product $\prod_{\lambda \in \Lambda} F_\lambda$ of an infinite family $(F_\lambda)_{\lambda \in \Lambda}$ of fields. Then, according to [11, Proposition 9, p. 100], one has

$$Q_{\max}(R) = Q_{\max}(\prod_{\lambda \in \Lambda} F_\lambda) \simeq \prod_{\lambda \in \Lambda} Q_{\max}(F_\lambda) = \prod_{\lambda \in \Lambda} F_\lambda = R,$$

which shows that R is its own maximal ring of quotients. However, R is not a Kasch ring: indeed, if we consider the proper ideal $I = \bigoplus_{\lambda \in \Lambda} F_\lambda$ of R, then clearly $\ell(I) = 0$, and consequently, by 2.1 one deduces that R is not a Kasch ring.

3   PROPERTIES OF KASCH MODULES

Denote by $\mathcal{K}$ the class of all Kasch left R-modules. Consider a module ${}_R N$ which is not a Kasch module, let Q be a cogenerator of $\sigma[N]$, and denote $K = N \oplus Q$. Then N is isomorphic to a submodule, as well as to a factor module of the Kasch module K, which shows that the class $\mathcal{K}$ need not to be closed under subobjects nor under factor objects. The above example shows also that a direct summand of a Kasch module is not necessarily a Kasch module.

We are going now to show that the class $\mathcal{K}$ is closed under direct sums. We need first the following:

3.1 Lemma. *Let $(M_\lambda)_{\lambda \in \Lambda}$ be a nonempty family of nonzero left R-modules. Then, for any simple module $U \in \sigma[\bigoplus_{\lambda \in \Lambda} M_\lambda]$ there exists a $\mu \in \Lambda$ such that $U \in \sigma[M_\mu]$.*

Proof: Denote $M = \bigoplus_{\lambda \in \Lambda} M_\lambda$ and consider the injective hull $\widehat{U}$ of U in $\sigma[M]$. Then, as known, $\widehat{U}$ is M-generated, so there exists a nonzero morphism $h : M \longrightarrow \widehat{U}$. Denote by $\widetilde{U}$ the image of h. It follows that $U \trianglelefteq \widetilde{U} \trianglelefteq \widehat{U}$, and so, we obtain an epimorphism of R-modules

$$g : \bigoplus_{\lambda \in \Lambda} M_\lambda \longrightarrow \widetilde{U}.$$

Denote for each $\lambda \in \Lambda$ by $\varepsilon_\lambda : M_\lambda \longrightarrow M$ the canonical injection. Then, surely there exists a $\mu \in \Lambda$ such that $\varepsilon_\mu g \neq 0$, which produces a nonzero morphism $g_\mu : M_\mu \longrightarrow \widetilde{U}$. Since $U \trianglelefteq \widetilde{U}$ we deduce that U is an epimorphic image of a submodule of M_μ. Thus $U \in \sigma[M_\mu]$. □

3.2 Proposition. *The class $\mathcal{K}$ is closed under arbitrary direct sums and essential submodules.*

Proof: Let $(M_\lambda)_{\lambda \in \Lambda}$ be an arbitrary nonempty family of left R-modules, and $U \in \sigma[\bigoplus_{\lambda \in \Lambda} M_\lambda]$ a simple module. By the previous lemma, there exists a $\mu \in \Lambda$ such that $U \in \sigma[M_\mu]$. Since M_μ is a Kasch module, we deduce that U can be embedded in M_μ, and consequently also in $\bigoplus_{\lambda \in \Lambda} M_\lambda$, proving that $\bigoplus_{\lambda \in \Lambda} M_\lambda$ is a Kasch module.

The last statement of the proposition is obvious. □

We are going now to recall some definitions and results from [1], [2], [4] and [9]. For any full subcategory $\mathcal{C}$ of R-Mod we shall denote by $\mathrm{Sim}(\mathcal{C})$ a representative system of all isomorphism classes of simple modules belonging to $\mathcal{C}$. Clearly, $\mathrm{Sim}(\mathcal{C})$ is a set, possibly empty. For any ${}_R X$ we shall denote

$$\mathrm{Sim}(X) := \mathrm{Sim}(\sigma[X]).$$

So, $\mathrm{Sim}(R)$ denotes $\mathrm{Sim}(R\text{-Mod})$. We always shall assume that $\mathrm{Sim}(\mathcal{C}) \subseteq \mathrm{Sim}(R)$ for any full subcategory $\mathcal{C}$ of R-Mod.

Clearly, for any module ${}_R X$ one has:

$$\mathrm{Sim}(X) = \{\, U \in \mathrm{Sim}(R) \mid \exists\, X' \leq X \text{ and } \exists\, V \leq X/X' \text{ with } V \simeq U \,\}.$$

The next result collects some of the basic properties of "Sim":

3.3 Proposition. *The following assertions hold:*
(1) For an $_RX$ one has $\operatorname{Sim}(X) = \emptyset \Leftrightarrow X = 0$.
(2) If $_RX$ is a module and $Y \in \sigma[X]$, then $\operatorname{Sim}(Y) \subseteq \operatorname{Sim}(X)$.
(3) For any exact sequence in R-Mod:

$$0 \longrightarrow X' \longrightarrow X \longrightarrow X'' \longrightarrow 0,$$

one has

$$\operatorname{Sim}(X) = \operatorname{Sim}(X') \cup \operatorname{Sim}(X'').$$

(4) For any family of $(M_\lambda)_{\lambda \in \Lambda}$ of left R-modules one has

$$\operatorname{Sim}(\bigoplus_{\lambda \in \Lambda} M_\lambda) = \bigcup_{\lambda \in \Lambda} \operatorname{Sim}(M_\lambda).$$

Proof: (1) If $X \neq 0$, then there exists $x \in X$, $x \neq 0$. But, the nonzero cyclic module Rx has a maximal submodule Z, and so, Rx/Z is a simple module in $\sigma[X]$.

(2) is obvious.

(3) Since X', $X'' \in \sigma[X]$ it is clear that $\operatorname{Sim}(X') \cup \operatorname{Sim}(X'') \subseteq \operatorname{Sim}(X)$. Let now $U \in \operatorname{Sim}(X)$. Without loss of generality, we can suppose that $X' \leq X$ and $X'' = X/X'$. There exists a submodule Y of X and an epimorphism $f : Y \longrightarrow U$.

Two cases arise: $(Y \cap X')f = 0$ and $(Y \cap X')f \neq 0$. In the first case f induces an epimorphism $(Y + X')/X' \simeq Y/(Y \cap X') \longrightarrow U$, and so $U \in \operatorname{Sim}(X'')$.

In the second case, $f|_{Y \cap X'}$ yields an epimorphism $Y \cap X' \longrightarrow U$, and then $U \in \operatorname{Sim}(X')$.

(4) is essentially a reformulation of . $\square$

Recall that a module $_RX$ is called a *semi-Artinian* (or *Loewy*) module if any nonzero factor module of X contains a simple submodule.

If $U \in \operatorname{Sim}(R)$, a module $_RX$ is said to be *U-primary* whenever X/X' contains a simple module isomorphic to U for any $X' \leq X$, $X' \neq X$.

The class $\mathcal{L}$ of all semi-Artinian left R-modules is a localizing subcategory of R-Mod, as well as, for each $U \in \operatorname{Sim}(R)$, the class $\mathcal{L}_U$ of all U-primary left R-modules. For any $_RX$ and $U \in \operatorname{Sim}(R)$ we shall denote by X_U the greatest U-primary submodule of X, called the *U-primary component* of X.

If X is a left R-module, then the set

$$\mathcal{S}(X) = \{\, U \in \mathrm{Sim}(R) \mid X_U \neq 0 \,\}$$

is called the *support* of X. One says that X is a module with *finite support* in case $\mathcal{S}(X)$ is a finite set.

It is known that if $X \in \mathcal{L}$, then the sum $\sum_{U \in Sim(R)} X_U$ is a direct sum and $\bigoplus_{U \in Sim(R)} X_U \trianglelefteq X$ (cf. [8]), but in general $X \neq \bigoplus_{U \in Sim(R)} X_U$. Following [2], the module X is said to be *Dickson decomposable* if $X = \bigoplus_{U \in Sim(R)} X_U$.

Following [9] (resp. [1]), the ring R is said to be a left T–ring (resp. a left FT–ring) in case any semi-Artinian module (resp. any semi-Artinian module with finite support) in R-Mod is a Dickson decomposable module. By [1, Corollaire 6], any commutative ring is an FT–ring.

We can extend very naturally these definitions as follows:

3.4 Definitions. *The module $_RM$ is called a T-module (resp. FT-module) in case any semi-Artinian module (resp. any semi-Artinian module with finite support) in $\sigma[M]$ is Dickson decomposable.*

3.5 Lemma. *Let X be a left R-module and $U \in \mathrm{Sim}(R)$. Then*

$$X \in \mathcal{L}_U \Leftrightarrow X \in \mathcal{L} \text{ and } \mathrm{Sim}(X) = \{U\}.$$

Proof: If X is U-primary, then obviously X is semi-Artinian. Let $V \in \mathrm{Sim}(X)$. Then some quotient module X/X' of X contains a simple module W isomorphic to V. Then $V \in \mathcal{L}_U$, and consequently $V = U$. The converse implication is clear. □

3.6 Lemma. *For any $U \in \mathrm{Sim}(R)$, $\mathcal{L}_U \subseteq \mathcal{K}$.*

Proof: If X is a nonzero U-primary module, then the socle $\mathrm{Soc}(X)$ of X contains at least a simple submodule of X isomorphic to U, hence any simple module in $\mathrm{Sim}(X) = \{U\}$ can be embedded in X, showing that X is a Kasch module. □

3.7 Proposition. *Let $(U_\lambda)_{\lambda \in \Lambda}$ be a family of simple modules in $\mathrm{Sim}(R)$ and $X_\lambda \in \mathcal{L}_{U_\lambda}$ for each $\lambda \in \Lambda$. Then $\bigoplus_{\lambda \in \Lambda} X_\lambda \in \mathcal{K}$. In particular any Dickson decomposable module is a Kasch module.*

Proof: Apply 3.6 and 3.2. □

3.8 Corollary. *If $_R M$ is a T-module (resp. an FT-module), then any semi-Artinian module (resp. semi-Artinian module with finite support) in $\sigma[M]$ is a Kasch module.*

Proof: By definition, any semi-Artinian module (resp. semi-Artinian module with finite support) in $\sigma[M]$ is a Dickson decomposable module. Apply now 3.7. □

3.9 Corollary.

If $_R M$ be a semi-Artinian module with finite support, and R is an FT-ring, then any module in $\sigma[M]$ is a Kasch module.

Proof: According to [1, Corollaire 8], for any exact sequence in R-Mod:

$$0 \longrightarrow X' \longrightarrow X \longrightarrow X'' \longrightarrow 0,$$

with X a semi-Artinian module with finite support, one has

$$\mathcal{S}(X) = \mathcal{S}(X') \cup \mathcal{S}(X'').$$

It follows that for each $X \in \sigma[M]$ one has $\mathcal{S}(X) \subseteq \mathcal{S}(M)$, and so X is also with finite support. Note that $\sigma[M] \subseteq \mathcal{L}$ since $M \in \mathcal{L}$. Consequently, any $X \in \sigma[M]$ is Dickson decomposable. Apply now 3.7. □

If R is a commutative ring, then $\mathrm{Ass}(X)$ will denote the "Assasin" of X (see [13]).

3.10 Corollary. *Let M be a semi-Artinian module over the commutative ring R. If $\mathrm{Ass}(M)$ is a finite set, then any module in $\sigma[M]$ is a Kasch module.*

Proof: As noted above, any commutative ring is an FT-ring. Since M has finite support if and only if $\mathrm{Ass}(M)$ is a finite set, the result follows now from 3.3. □

3.11 Corollary. *If M be a semi-Artinian module over the commutative semi-local ring R, then any module in $\sigma[M]$ is a Kasch module.*

Proof: By [2, Proposition 1], any $P \in \mathrm{Ass}(M)$ is a maximal ideal of R. Apply now 3.10. □

3.12 Corollary. *If R is a commutative semi-local semi-Artinian ring, then any R-module is a Kasch module.*

3.13 Corollary. *Any module over a commutative Artinian ring is a Kasch module.*

3.14 Remarks. (1) The observation in 3.13 does not hold for noncommutative Artinian rings. For this let R be the ring of upper triangular $(2,2)$-matrices over a field F. The left R-module

$$M = \begin{pmatrix} 0 & F \\ 0 & F \end{pmatrix} \text{ has socle } S = \begin{pmatrix} 0 & F \\ 0 & 0 \end{pmatrix},$$

and M/S is not isomorphic to S. Hence M is not a Kasch module.

(2) In case any factor module of M is a Kasch module, then M must be necessarily a semi-Artinian module, as this can be shown by considering the ascending Loewy series of M.

(3) The result in 3.10 fails if $\mathrm{Ass}(M)$ is an infinite set. To see this, consider the following example given in [4, 3.34]:

Let F a field and Λ an infinite set. Denote by B the direct product $\prod_\Lambda F_\lambda$, where $F_\lambda = F$ for all $\lambda \in \Lambda$, and by A the subring $\bigoplus_\Lambda F_\lambda + Fe$ of B, where e is the identity element of B. Denote for each $\lambda \in \Lambda$ by $\varepsilon_\lambda : F_\lambda \longrightarrow \bigoplus_{\mu \in \Lambda} F_\mu$ the canonical injection, and $U_\lambda = \varepsilon_\lambda(F_\lambda)$. Then

$$\sum_{\lambda \in \Lambda} U_\lambda = \bigoplus_{\lambda \in \Lambda} U_\lambda = \bigoplus_{\lambda \in \Lambda} F_\lambda,$$

is precisely the socle $\mathrm{Soc}(A)$ of A, this is a maximal ideal of A, U_λ's are mutually nonisomorphic simple A-modules, and $U_\lambda \not\cong U_0$ for all $\lambda \in \Lambda$, where $U_0 = A/\mathrm{Soc}(A)$. The ring A is a semi-Artinian regular ring with the Loewy length 2 which is not semi-simple, the A-module A is not Dickson decomposable, $\mathrm{Ass}(A)$ is an infinite set, and A is not a Kasch ring.

The exact sequence

$$0 \longrightarrow \mathrm{Soc}(A) \longrightarrow A \longrightarrow U_0 \longrightarrow 0$$

of A modules shows also that $\mathcal{K}$ need not to be closed under extensions.

(4) The example considered in 2.9 shows that $\mathcal{K}$ is not closed under direct products. Let R denote the direct product $\prod_{\lambda \in \Lambda} F_\lambda$ of an infinite family $(F_\lambda)_{\lambda \in \Lambda}$ of fields. Each F_λ is a simple R-module in a canonical way, but their product is R itself, which as we have already seen in 2.9, is not a Kasch module.

(5) We are going now to show that a direct sum of two modules which both are not Kasch could be a Kasch module. For this, consider the example due to P.M. Cohn, exhibited in [9]:

Let F be any field possesing an endomorphism $\varphi : F \longrightarrow F$ which is not onto, and denote by A the skew polynomial ring $F[X, \varphi]$ consisting of all polynomials $\sum_{0 \leq i \leq n} X^i a_i$, where $a_i \in F$, with the multiplication $aX = X\varphi(a)$ for any $a \in F$. Then A is a principal right ideal domain.

Let $\beta \in F \setminus \varphi(F)$ and consider the elements $a = X$, $b = X + \beta$. If we denote $U = A/aA$ and $V = A/bA$ then $V \simeq aA/abA$ and $U \simeq bA/baA$, U and V are simple right A-modules which are not isomorphic, and the canonical exact sequences

$$0 \longrightarrow aA/abA \longrightarrow A/abA \longrightarrow A/aA \longrightarrow 0$$

$$0 \longrightarrow bA/baA \longrightarrow A/baA \longrightarrow A/aA \longrightarrow 0$$

are not splitting. This shows that both the right A-modules A/abA and A/baA are not Kasch modules, but their direct sum is a Kasch module.

3.15 Proposition. *Any faithful left R-module over a left Kasch ring R is a Kasch module.*

Proof: If N is a faithful module over the Kach ring R, then the module $_RR$ can be embedded in N^N, hence any simple left R-module is cogenerated by N, proving that N is a Kasch module. □

We have proved so far that $\mathcal{K}$ is closed under direct sums and under essential subobjects, but need not to be closed under subobjects, nor factor objects, nor extensions and nor direct products.

Some natural questions arise:

Question 1. *For which rings R is the class $\mathcal{K}$ of all left Kasch R-modules $\mathcal{K}$ closed under extensions resp. direct products?*

Question 2. *Let M be a Kasch module. When is any submodule (resp. factor module) of M again a Kasch module?*

Question 3. *For which modules M is any module in $\sigma[M]$ a Kasch module? In particular, for which rings R are all left R-modules Kasch modules?*

If all modules in $\sigma[M]$ are homo-serial then they are all Kasch modules (see [14, 56.7, 56.8]). As a special case all left (and right) R-modules are Kasch provided R is left and right an artinian and principal ideal ring (see

[14, 56.9]). Moreover all modules over commutative (semi-) local (semi-) Artinian rings are Kasch (by 3.12).

Acknowledgements

This paper was written while the first author was a Humboldt Fellow at the Mathematical Institute of the Heinrich-Heine-University in Düsseldorf (May – June 1996). In the final stage of its preparation he was partially supported by the Grant 281/1996 of CNSCU (Romania).

He would like to thank the Alexander von Humboldt Foundation and CNSCU for financial support and the Heinrich-Heine-University Düsseldorf for hospitality.

The authors are grateful to Friedrich Kasch for drawing to their attention the titles [10] and [16], and to John Clark for helpful comments.

References

[1] T. Albu, *Modules de torsion à support fini*, C. R. Acad. Sci. Paris **273**, Série A (1971), 335-338.

[2] T. Albu, *Modules décomposables de Dickson*, C. R. Acad. Sci. Paris **273**, Série A (1971), 369-372.

[3] T. Albu, *On some classes of modules I* (in Romanian), Stud. Cerc. Mat. **24** (1972), 1329-1392.

[4] T. Albu and C. Năstăsescu, *Décompositions primaires dans les catégories de Grothendieck commutatives I*, J. Reine Angew. Math. **280** (1976), 172-194.

[5] T. Albu and C. Năstăsescu, *"Relative Finiteness in Module Theory"*, Marcel Dekker, Inc., New York and Basel, 1984.

[6] T. Albu and R. Wisbauer, *M-density, M-adic completion and M-subgeneration*, Rend. Sem. Mat. Univ. Padova **98** (1997), to appear.

[7] F.W. Anderson and K.R. Fuller, *"Rings and Categories of Modules"*, Springer-Verlag, New York Heidelberg Berlin (1992).

[8] S.E. Dickson, *A torsion theory for abelian categories*, Trans. Amer. Math. Soc. **121** (1966), 223-235.

[9] S.E. Dickson, *Decomposition of modules II. Rings without chain conditions*, Math. Z. **104** (1969), 349-357.

[10] D.N. Dikranjan, E. Gregorio and A. Orsatti, *Kasch bimodules*, Rend. Sem. Mat. Univ. Padova **85** (1991), 147-160.

[11] J. Lambek, *"Lectures on Rings and Modules"*, Blaisdell Publishing Company (1966).

[12] J. Lambek, *"Torsion Theories, Additive Semantics, and Rings of Quotients"*, Springer-Verlag, New York Heidelberg Berlin (1971).

[13] B. Stenström, *"Rings of Quotients"*, Springer-Verlag, Berlin Heidelberg New York (1975).

[14] R. Wisbauer, *"Foundations of Module and Ring Theory"*, Gordon and Breach, Reading (1991).

[15] R. Wisbauer, *"Modules and Algebras: Bimodule Structure and Group Actions on Algebras"*, Pitman Monographs 81, Longman (1996).

[16] W. Xue, *On Kasch duality*, Algebra Colloq. **1** (1994), 257-266.

FACULTATEA DE MATEMATICĂ, UNIVERSITATEA BUCUREŞTI, 70109 BUCUREŞTI 1, ROMANIA
E-mail address: talbu@roimar.imar.ro

MATHEMATISCHES INSTITUT DER HEINRICH-HEINE-UNIVERSITÄT, 40225 DÜSSELDORF, GERMANY
E-mail address: wisbauer@math.uni-duesseldorf.de

COMPACTNESS IN CATEGORIES AND INTERPRETATIONS

P. N. ÁNH AND R. WIEGANDT

ABSTRACT. It is the purpose of this note to give a definition for compactness of objects in a category in terms of a covariant functor $\mathcal{F}$. Choosing the category and the functor $\mathcal{F}$ appropriately, we rediscover many kinds of compactnesses, so for instance, we get the usual compactness for T_1-topological spaces and linear compactness of rings and modules. Generalizing a module-theoretic theorem of Leptin [6] we prove that if an object X of a locally small complete abelian category endowed with a linearly compact topology, has a dense semisimple subobject then X is a product of simple objects. We prove also a decomposition theorem for linearly compact rings into the product of simple rings, in particular, for linearly compact Brown–McCoy semisimple rings. Also a connection between compactness and sheaves is shown.

1. DEFINITION AND BASIC PROPERTIES

The notion of compactness is of fundamental importance in topology. Working, however, in module categories, or more generally, in abelian categories, some related notions such as (discrete) linear compactness and algebraic linear compactness turned out to be useful, in particular in the theory of Morita duality as seen from Müller's paper [8].

The objective of the present note is to give a general definition for compactness of objects of a category in terms of a covariant functor, and to discuss its theory. Specifying the category and the functor we shall prove that our notion of compactness provides all sorts of compactnesses. These statements will be called examples, not merely from modesty but rather to emphasize the novelty and generality of our categorical definition of compactness.

Definition. Let $\mathcal{F}\colon \mathcal{C} \longrightarrow \mathcal{D}$ be a covariant functor between two categories $\mathcal{C}$ and $\mathcal{D}$. An object $X \in \mathrm{Ob}\,\mathcal{C}$ is said to be *compact*, or more precisely

---

1991 *Mathematics Subject Classification.* Primary 18A35; Secondary 16D90, 16W80, 18B30, 54D30.

Key words and phrases. Compactness, linear compactness, semisimple object, dense subobject, Brown–McCoy semisimple ring.

Research carried out under the auspices of the Hungarian National Foundation for Scientific Research Grant # T16432. The first author gratefully acknowledges the financial support of the Alexander von Humboldt Foundation.

$\mathcal{F}$-compact, if for any inverse system $\{X_\alpha, f_\beta^\alpha \colon X_\alpha \longrightarrow X_\beta \mid \alpha > \beta\}$ with compatible morphisms $f_\alpha \colon X \longrightarrow X_\alpha$ in $\mathcal{C}$ such that every morphism $\mathcal{F}f_\alpha$ is an epimorphism in $\mathcal{D}$, the canonical morphism $\mathcal{F}X \longrightarrow \varprojlim\{\mathcal{F}X_\alpha, \mathcal{F}f_\beta^\alpha\}$ is an epimorphism, whenever it exists in $\mathcal{D}$.

As is well-known, categorically Grothendieck's condition $AB\,5^*$ means that the inverse limit functor is right exact. According to our definition compactness means that for a fixed object $\mathcal{F}X$ in $\mathcal{D}$ the inverse limit functor is right exact and this is just (discrete) linear compactness of the object X, if $\mathcal{D}$ is a complete abelian category (cf. Examples 8 and also [2]). Thus, according to the Definition we are going to investigate a kind of local right exactness of the inverse limit functor at certain objects. In the interesting cases $\mathcal{D}$ is always complete.

Dualizing the Definition, an object X may be called $\mathcal{F}$-cocompact, if for any direct system $\{X_\alpha, f_\beta^\alpha \colon X_\alpha \to X_\beta\}$ with compatible morphisms $f_\alpha \colon X_\alpha \to X$ in $\mathcal{C}$ such that every morphism $\mathcal{F}f_\alpha$ is a monomorphism, the canonical morphism $\varinjlim\{\mathcal{F}X_\alpha, \mathcal{F}f_\beta^\alpha\} \to \mathcal{F}X$ is a monomorphism. Choosing $\mathcal{F}$ to be the identical functor, this means exactly that direct limits of subobjects are subobjects. Demanding cocompactness for every object is exactly Grothendieck's axiom $AB5$, that is, the direct limit functor is left exact. Examples for categories satisfying $AB5$ are the categories of algebraic structures as it is well known.

Proposition 1. *Assume that the functor $\mathcal{F}$ preserves epimorphisms. If $X \in \mathrm{Ob}\,\mathcal{C}$ is compact and $f \colon X \to Y$ is an epimorphism, then also Y is compact.*

Proof. Let $\{Y_\alpha, f_\beta^\alpha \colon Y_\alpha \to Y_\beta\}$ be an inverse system with compatible morphisms $f_\alpha \colon Y \to Y_\alpha$ in $\mathcal{C}$ such that every $\mathcal{F}f_\alpha$ is an epimorphism in $\mathcal{D}$. Then we have also an inverse system $\{Y_\alpha, f_\beta^\alpha\}$ with compatible morphisms $ff_\alpha \colon X \to Y_\alpha$. Since $\mathcal{F}$ preserves epimorphisms, each morphism $\mathcal{F}(ff_\alpha) = \mathcal{F}f\mathcal{F}f_\alpha$ is an epimorphism. Hence taking into account that X is compact, the canonical morphism

$$g_\mathcal{D} \colon \mathcal{F}X \longrightarrow \varprojlim\{\mathcal{F}Y_\alpha, \mathcal{F}f_\beta^\alpha\} \quad \text{is an epimorphism.}$$

Putting $h_\mathcal{D} \colon \mathcal{F}Y \to \varprojlim\{\mathcal{F}Y_\alpha, \mathcal{F}f_\beta^\alpha\}$ for the canonical morphism, the diagram

$$\begin{array}{ccc} \mathcal{F}X & \xrightarrow{\mathcal{F}f} & \mathcal{F}Y \\ {}_{g_\mathcal{D}}\searrow & & \swarrow {}_{h_\mathcal{D}} \\ & \varprojlim\{\mathcal{F}Y_\alpha, \mathcal{F}f_\beta^\alpha\} & \end{array}$$

commutes. Hence also $h_\mathcal{D}$ must be an epimorphism, proving that also Y is a compact object.

Proposition 2. *Suppose that the category $\mathcal{C}$ has pushouts, $\mathcal{F}$ is pushout preserving, $\mathcal{D}$ has inverse limits, in $\mathcal{D}$ inverse limits and pushouts commute, further, if*

$$\begin{array}{ccc} \cdot & \longrightarrow & \cdot \\ g\downarrow & & \downarrow f \\ \cdot & \longrightarrow & \cdot \end{array}$$

is a pushout diagram in $\mathcal{D}$ and f is an epimorphism, then also g is an epimorphism. If X is a compact object in $\mathcal{C}$, then every subobject Y of X is compact.

Proof. Let $h\colon Y \to X$ be any monomorphism. Consider an inverse system $\{Y_\alpha, g_\beta^\alpha\}$ with compatible morphisms $g_\alpha\colon Y \to Y_\alpha$ such that each $\mathcal{F}g_\alpha$ is an epimorphism. For every index α we have a pushout diagram

$$\begin{array}{ccc} Y & \xrightarrow{h} & X \\ g_\alpha\downarrow & & \downarrow f_\alpha \\ Y_\alpha & \xrightarrow{h_\alpha} & X_\alpha \end{array}$$

Since $\{Y_\alpha, g_\beta^\alpha\}$ is an inverse system, these pushout diagrams induce an inverse system $\{X_\alpha, f_\beta^\alpha\}$ with compatible morphisms $f_\alpha\colon X \to X_\alpha$. By assumption the functor $\mathcal{F}$ preserves pushouts, so also

$$\begin{array}{ccc} \mathcal{F}Y & \xrightarrow{\mathcal{F}h} & \mathcal{F}X \\ \mathcal{F}g_\alpha\downarrow & & \downarrow \mathcal{F}f_\alpha \\ \mathcal{F}Y_\alpha & \xrightarrow{\mathcal{F}h_\alpha} & \mathcal{F}X_\alpha \end{array}$$

is a pushout diagram for every index α. Since each $\mathcal{F}g_\alpha$ is an epimorphism, so too is every $\mathcal{F}f_\alpha$. Since $\mathcal{D}$ has inverse limits and pushouts commute in $\mathcal{D}$, we have a pushout diagram

$$\begin{array}{ccc} \mathcal{F}Y & \xrightarrow{\mathcal{F}h} & \mathcal{F}X \\ g_\mathcal{D}\downarrow & & \downarrow f_\mathcal{D} \\ \varprojlim\{\mathcal{F}Y_\alpha, \mathcal{F}g_\beta^\alpha\} & \longrightarrow & \varprojlim\{\mathcal{F}X_\alpha, \mathcal{F}f_\beta^\alpha\} \end{array}$$

The compactness of X yields that $f_\mathcal{D}$ is an epimorphism and therefore by the assumption on $\mathcal{D}$ also $g_\mathcal{D}$ is an epimorphism.

As is well-known from topology, a subspace of a compact space need not be compact. We shall see, however, examples for $\mathcal{F}$-compactness such that every subobject of any $\mathcal{F}$-compact object is again $\mathcal{F}$-compact (Examples 1, 3, 4(ii), 5).

One would guess that at least finite products of compact objects would be compact. This is not so. Normak [10] investigated compactness of S-acts and his compactness can be interpreted as $\mathcal{F}$-compactness (cf. Example 13). He proved that finite products of compact S-acts need not be compact, but finite coproducts of compact S-acts are compact.

In the rest of this section we indicate a connection between compactness and sheaves. Let $\mathcal{C} = \mathcal{D}$, and assume that $\mathcal{C}$ is small, has pullbacks and for every non-isomorphic monomorphism f there exist morphisms g and h such that $g \neq h$ and $gf = hf$. For every object $X \in \mathcal{C}$ we denote by h_X the functor

$$h_X \colon \mathcal{C}^{\mathrm{op}} \to \mathrm{Set}, \qquad Y \in \mathcal{C}^{\mathrm{op}}, \ \mathrm{Hom}_\mathcal{C}(Y, X) \in \mathrm{Set}.$$

Following [4] Exercise 0.8 on p. 20 we call a family $\{f_i \colon X_i \to X, i \in I\}$ *epimorphic on* X, if each f_i is a monomorphism, they form a directed set of subobjects of X and for every $X \underset{h}{\overset{g}{\rightrightarrows}} Z$ in $\mathcal{C}$ such that $g \neq h$, there exists an f_i with $gf_i = hf_i$. Note that our notion of epimorphic family is somewhat stronger than the one in [4] Exercise 0.8 on p.20, where it is not assumed that the considered morphisms are monomorphisms. For every morphism $g \colon Y \to X$ the pullback diagrams

$$\begin{array}{ccc} Y_i & \xrightarrow{h_i} & X_i \\ {\scriptstyle g_i}\downarrow & & \downarrow{\scriptstyle f_i} \\ Y & \xrightarrow{g} & X \end{array}$$

show that also $\{g_i \colon Y_i \to Y, i \in I\}$ forms a directed set of subobjects of Y. If the family $\{g_i \colon Y_i \to Y\}$ is epimorphic for every $g \colon Y \to X$, then the family $\{f_i \colon X_i \to X\}$ is said to be *universally epimorphic on* X. Therefore universally epimorphic families define a Grothendieck pre-topology on $\mathcal{C}$ for which each representable functor $h_X \colon \mathcal{C}^{\mathrm{op}} \to \mathrm{Set}$ is a separated presheaf for every $X \in \mathcal{C}$. Assume now that every directed set of subobjects of an object $X \in \mathcal{C}$ has a colimit (which is in general not a subobject of X). If X is, in addition, cocompact, that is, X is compact in $\mathcal{C}^{\mathrm{op}}$ with respect to the identity functor $\mathcal{F} \colon \mathcal{C}^{\mathrm{op}} \to \mathcal{C}^{\mathrm{op}}$ then h_Y is a sheaf for every monomorphism

$Y \to X$. For, let $\{f_i : Y_i \to Y\}$ be a universally epimorphic family on Y and $\mathcal{R}_Y$ the corresponding presheaf defined by

$$\mathcal{R}_Y(Z) = \{f : Z \to Y \mid f \text{ factors through some } f_i\}.$$

The cocompactness of X implies that Y is the colimit of the Y_i. If $\mathcal{G}$ is any presheaf with a morphism $g : \mathcal{R}_Y \to \mathcal{G}$, then the Yoneda Lemma implies the existence of the compatible family of morphisms $g_i : Y_i \to \mathcal{G}(Y)$. Hence there is a unique morphism $Y \to \mathcal{G}(Y)$ because Y is the colimit of the Y_i. This shows that h_Y is a sheaf for every subobject Y of X.

Conversely, let $\mathcal{C}$ be a small category with pullbacks such that for every non-isomorphic $f : X \to Z$ there are morphisms $Z \xrightarrow[h]{g} W$ that $g \neq h$ and $gf = hf$. Endow $\mathcal{C}$ with the Grothendieck pretopology defined by universally epimorphic families. Let $X \in \mathcal{C}$ be an object such that every epimorphic family on a subobject Y of X is universally epimorphic, the lattice of subobjects of X is complete, and for every directed set of subobjects Y_i of X the family $\{f_i : Y_i \to Y\}$ is universally epimorphic where $Y = \cup(Y_i \mid i \in I)$. Therefore every compatible system of morphisms $\{g_i : Y_i :\to Z\}$ induces a morphism g^* from the presheaf $\mathcal{R}_Y$ to the representable functor H_Z. If now h_Y is a sheaf for every subobject Y, then there is the unique extension of g^* to h_Y and hence it is representable by a morphism $g : Y \to Z$ by Yoneda Lemma and $g_i = gf_i$ for every i. Therefore Y is in fact the colimit of the Y_i and hence X is cocompact and also every subobject Y of X is cocompact. In this case every directed set of subobjects has a colimit which is a subobject of X, too.

This observation led to the strong embedding of some abelian categories in Grothendieck categories (see [1]).

2. Applications to Sets and Topological Spaces

Example 1. $\mathcal{C} = \mathcal{D} = \text{Set}$, $\mathcal{F} = $ identical functor. *A set $X \in \mathcal{C}$ is compact if and only if X is finite.*

Proof. If X is finite, then it is trivially compact.

Let $|X| \geq \aleph_0$ and let us choose an element $p \in X$. For all subsets I of X not containing p, the sets $I \cup \{p\}$ form an inverse system along with the mappings

$$f_K^I(x) = \begin{cases} x & \text{if } x \in K \\ p & \text{if } x \in (I \cup \{p\}) \setminus K \end{cases}$$

for every subset K of I. Obviously

$$Y = \varprojlim \{I \cup \{p\}, f_K^I\}$$

is the set of all characteristic functions of X. Since $|X| \geq \aleph_0$, it follows $|X| < |Y|$, and hence the canonical morphism $X \longrightarrow \varprojlim\{I \cup \{p\}, f_K^I\}$ cannot be an epimorphism. Thus X is not compact.

Example 2. Let $\mathcal{C}$ denote the category of finite sets and $\mathcal{F}$ the identical functor. The category $\mathcal{C} = \mathcal{D}$ is not complete, though by definition *every object of $\mathcal{C}$ is compact*.

Example 3. Let $\mathcal{C}$ be the category of (undirected) graphs with edge preserving mappings, and $\mathcal{F}\colon \mathcal{C} \longrightarrow$ Set the forgetful functor. *A graph X is compact if and only if X is finite.* The proof is the same as in Example 1, since the inverse systems are chosen in $\mathcal{D} =$ Set.

Examples 4.
 i) Let $\mathcal{C}$ be the category of T_1-spaces with continuous mappings and $\mathcal{F}$ the forgetful functor. *A T_1-space X is $\mathcal{F}$-compact if and only if X is a compact topological space.*

Proof. Let X be a compact T_1-space, and $\{X_\alpha, f_\beta^\alpha\}$ an inverse system of mappings as required in the Definition. Since $\mathcal{F}$ is the forgetful functor, $\{\mathcal{F}X_\alpha, \mathcal{F}f_\beta^\alpha\}$ is an inverse system with epimorphism $\mathcal{F}f_\alpha$ if and only if $\{X_\alpha, f_\beta^\alpha\}$ is an inverse system with continuous surjective mappings f_α, moreover, the canonical mapping $f\colon X \longrightarrow \varprojlim\{X_\alpha, f_\beta^\alpha\}$ is surjective if and only if $\mathcal{F}f$ is an epimorphism. An element

$$\overline{x} = (\ldots, x_\alpha, \ldots) \in \prod X_\alpha$$

is contained in $\varprojlim\{X_\alpha, f_\beta^\alpha\}$ if and only if $\{f_\alpha^{-1}(x_\alpha)\}$ is a filter of closed subspaces in X, because X is a T_1-space. Since X is compact, this filter has a nonempty intersection, and for $x \in \cap_\alpha f_\alpha^{-1}(x_\alpha)$ clearly $f(x) = \overline{x}$ holds true. Thus f is surjective, and hence $\mathcal{F}f$ is an epimorphism in Set.

Suppose that X is an $\mathcal{F}$-compact object in $\mathcal{C}$, and let $\{C_\alpha\}$ be a filter in X consisting of closed subspaces. This filter defines the system $\{X_\alpha\}$ of factor spaces of X where $X_\alpha = (X \setminus C_\alpha) \cup \{C_\alpha\}$ is endowed with the factor topology. Since $\{C_\alpha\}$ is a filter, the system $\{X_\alpha\}$ induces an inverse system $\{X_\alpha, f_\beta^\alpha\}$, and since X is $\mathcal{F}$-compact, the canonical morphism $\mathcal{F}f\colon \mathcal{F}X \longrightarrow \varprojlim\{\mathcal{F}X_\alpha, \mathcal{F}f_\beta^\alpha\}$ is an epimorphism. Hence the canonical mapping $f\colon X \longrightarrow \varprojlim\{X_\alpha, f_\beta^\alpha\}$ is surjective, and therefore there exists an element $x \in X$ such that $f(x) = (\ldots, \{C_\alpha\}, \ldots)$. This means exactly that $x \in \cap_\alpha \{C_\alpha\}$, and so the space X is compact.

ii) In the category $\mathcal{C}$ of all T_1-spaces with closed continuous mappings every epimorphism is surjective. Thus, if $\mathcal{F}$ is chosen to be the identical functor, then a T_1-space is $\mathcal{F}$-compact if and only if it is compact in the usual sense.

Remark 1. The assertion that an $\mathcal{F}$-compact object is a compact space, holds true also for the category of all topological spaces with continuous mappings. The converse, however, is not true: let $\mathcal{F}$ be the forgetful functor, and let us consider the ring $\mathbf{Z}$ of integers endowed with the indiscrete topology, and the surjective continuous mappings $f_n : \mathbf{Z} \longrightarrow \mathbf{Z}/(p^n)$ for a fixed prime p and $n = 1, 2, \ldots$. These mappings f_n induce an inverse system $\{\mathbf{Z}/(p^n), f_m^n\}$ and $\varprojlim\{\mathbf{Z}/(p^n), f_m^n\}$ is the ring of p-adic integers with the indiscrete topology. Thus the canonical mapping $f : \mathbf{Z} \longrightarrow \varprojlim\{\mathbf{Z}/(p^n), f_m^n\}$ is not an epimorphism, and so $\mathbf{Z}$ is not $\mathcal{F}$-compact, though compact in the usual sense.

Example 5. Let $\mathcal{C}$ be the category of T_2-topological spaces with continuous mappings and $\mathcal{F}$ the identical functor. *Every space $X \in \mathcal{C}$ is $\mathcal{F}$-compact.** As we shall see, this strange state of affairs is due to the fact that in $\mathcal{C}$ epimorphisms are the dense continuous mappings. Let X be a T_2-space and $Y = \varprojlim\{X_\alpha, \pi_{\alpha\beta}\}$ such that for every $\alpha > \beta$

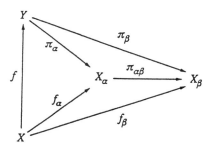

is a commutative diagram and every f_α is an epimorphism, that is a dense continuous mapping. In order to prove that X is $\mathcal{F}$-compact, we have to show that f is a dense mapping. Let $\langle y_\alpha \rangle$ be any element of Y. A neighborhood U of $\langle y_\alpha \rangle$ has the form

$$U = Y \cap \left(\bigcap_{i=1}^{n} \pi_{\alpha_i}^{-1}(U_{\alpha_i})\right)$$

---

*Sincere thanks are due to Dr. E. Makai for the proof.

where U_{α_i} is a neighborhood of y_{α_i} for $i = 1, \ldots, n$. Let us choose an index α_0 such that $\alpha_0 \geq \alpha_i$ for $i = 1, \ldots, n$, and let U_{α_0} be a neighborhood of y_{α_0} such that $\pi_{\alpha_0 \alpha_i}(U_{\alpha_0}) \subseteq U_{\alpha_i}$, $i = 1, \ldots, n$. Now

$$Y \cap \pi_{\alpha_0}^{-1}(U_{\alpha_0}) \subseteq Y \cap \left(\bigcap_{i=1}^{n} \pi_{\alpha_i}^{-1}(U_{\alpha_i})\right)$$

is a neighborhood of $\langle y_\alpha \rangle$. Further, we have $\pi_{\alpha_0} f(X) = f_{\alpha_0}(X)$ and $f_{\alpha_0}(X)$ is dense in X_{α_0}. Hence $\pi_{\alpha_0} f(X) \cap U_{\alpha_0} \neq \emptyset$, and so also

$$f(X) \cap (Y \cap \pi_{\alpha_0}^{-1}(U_{\alpha_0})) = f(X) \cap \pi_{\alpha_0}^{-1}(U_{\alpha_0}) \neq \emptyset.$$

Thus every neighborhood of $\langle y_\alpha \rangle$ intersects $f(X)$, which means that $f(X)$ is dense in Y, that is, f is an epimorphism in $\mathcal{C}$.

Remark 2. In the subcategories of T_3- and T_4-spaces with continuous mappings the same result holds.

3. APPLICATIONS TO ALGEBRAIC STRUCTURES

Example 6. Let us consider the forgetful functor $\mathcal{F} \colon R - \mathrm{Mod} \longrightarrow \mathrm{Set}$. An R-module X is said to be *discrete linearly compact*, if every filter of cosets of X has a nonempty intersection. *An R-module X is $\mathcal{F}$-compact if and only if X is discrete linearly compact.* The proof is straightforward. Considering the identical functor $\mathcal{F} \colon R - \mathrm{Mod} \longrightarrow R - \mathrm{Mod}$, we get again discrete linear compactness. It was B. J. Müller who has first shown the importance of this notion in the study of Morita duality ([8]).

Examples 7. Let $\mathcal{C}$ be the category of linearly topological Hausdorff R-modules with continuous mappings.

i) Let $\mathcal{F}_1$ be the forgetful functor $\mathcal{F}_1 \colon \mathcal{C} \longrightarrow \mathrm{Set}$. It is straightforward to see that *a module $X \in \mathcal{C}$ is $\mathcal{F}_1$-compact if and only if X is linearly compact* (that is, every filter of closed cosets of X has a nonempty intersection).

ii) Let $\mathcal{F}_2$ (or $\mathcal{F}_3$) be the functor which forgets topology (or forgets topology and multiplication by R, respectively). Now *$\mathcal{F}_2$- and $\mathcal{F}_3$-compactness mean again linear compactness*.

iii) Let $\mathcal{F}_4 \colon \mathcal{C} \longrightarrow \mathcal{C}$ be the identical functor. Now the epimorphisms in $\mathcal{F}\mathcal{C} = \mathcal{C}$ are dense mappings, and as in Example 5 we get that *every module of $\mathcal{C}$ is $\mathcal{F}_4$-compact*.

Examples 8. Let $\mathcal{A}$ be a complete abelian category.

i) Choosing $\mathcal{F}$ to be the identical functor of $\mathcal{A}$, $\mathcal{F}$-*compactness means discrete linear compactness* in the sense of [2] (an object $X \in \text{Ob } \mathcal{A}$ is discrete linearly compact, if for every filter $\{X_\alpha\}$ consisting of subobjects of X the canonical morphism $X \longrightarrow \varprojlim\{X/X_\alpha\}$ is an epimorphism).

ii) Starting from the category $\mathcal{A}$ we construct a category $\mathcal{C}$ by topologizing the objects of $\mathcal{A}$. A topology on an object $X \in \text{Ob } \mathcal{A}$ is a filter $\{X_\alpha\}$ of subobjects X_α. The filter $\{X_\alpha\}$ induces a canonical morphism $\tau\colon X \longrightarrow \varprojlim X/X_\alpha$ and τ determines uniquely the filter $\{X_\alpha\}$. Thus we may say that the object X is *endowed with topology* τ. The topology τ is said to be Hausdorff, if $\cap X_\alpha = 0$, that is, τ is a monomorphism. τ is complete if it is an epimorphism. A subobject Y of X endowed is *closed* (with respect to τ), if $Y = \cap(Y + X_\alpha)$ holds. Y is *open* if it belongs to the filter generated by τ, i.e., if it contains at least one of the X_α. Y is said to be *dense* if $Y + X_\alpha = X$ for every X_α in the topology τ induced by $\{X_\alpha\}$. Let Z be another object topologized by a filter base $\{Z_\beta\}$. A morphism $\varphi\colon X \longrightarrow Z$ of Mor $\mathcal{A}$ is said to be *continuous*, if for every Z_β there exists an X_α such that $\varphi X_\alpha \leq Z_\beta$. Now, let us define a category $\mathcal{C}$ as follows.

Ob $\mathcal{C} = \{(X, \tau) \mid X \in \text{Ob } \mathcal{A}, \tau \text{ is a Hausdorff topology on } X\}$

Mor $\mathcal{C} = \{\varphi \in \text{Mor } \mathcal{A} \mid \varphi \text{ is continuous}\}$.

Obviously, also $\mathcal{C}$ is complete, additive, has kernels and cokernels. If, in addition, $\mathcal{A}$ is locally small, then so is $\mathcal{C}$.

Choosing the forgetful functor $\mathcal{F}\colon \mathcal{C} \longrightarrow \mathcal{A}$ (which forgets topology), $\mathcal{F}$-*compactness means exactly linear compactness* in the sense of [2], i.e., an object (X, τ) is linearly compact, if for every filter base $\{Y_\beta\}$ of closed subobjects of (X, τ) the canonical morphism $X \longrightarrow \varprojlim\{X/Y_\beta\}$ is an epimorphism. This shows that the linear compactness of (X, τ) depends only on the set of closed subobjects of (X, τ). This observation justifies the following definition. Two topologies τ and κ on object X is said to be *equivalent* if they have the same subobjects closed. In view of this notion the linear compactness of (X, τ) depends only on the equivalence class of τ. An alternative definition of linear compactness on (X, τ) is that it is complete with respect to every topology $\{Y_\beta\}$ where Y_β are closed subobjects of X. Example 6 in [2] shows that infinite direct products of linearly compact objects or even objects of finite length are in general not linearly compact. In contrast to the module categories, infinite direct sums of discrete linearly compact objects may be discrete linearly compact even in Grothendieck categories. For example the product of cyclic groups of order p^{n_p} where n_p are arbitrary natural integers and p runs over all prime integers, is discrete linearly compact in the category of all abelian torsion groups with respect to the identity functor. However, we have the following statement.

Proposition 3. *Let X be a coproduct of X_α, $\alpha \in A$. If X is discrete linearly compact, then it is also a product of X_α.*

Proof. For every finite subset I of A put $X_I = \prod(X_\alpha \mid \alpha \in A)$. The discrete linear compactness of X implies obviously that the canonical morphism $X \longrightarrow \varprojlim\{X/X_I\}$ is an isomorphism.

Remark 3. In view of [2] Theorem 1.1 the product topology on X is equivalent to the discrete topology, although its set of open subobjects does not contain all subobject of X if the direct sum is infinite.

In the rest of this example we are going to prove a generalization of Leptin's [6], II. Satz 2. Let us recall that an object $X \in \mathcal{A}$ is called a *semisimple object*, if X is a coproduct of simple objects.

Theorem 1. *Let $\mathcal{A}$ be a locally small complete abelian category and $\mathcal{C}$ the corresponding category of topologized objects and continuous morphisms. If an object (X, τ) of $\mathcal{C}$ is linearly compact and has a dense subobject (Y, ϑ) such that Y is semisimple, then X is a product of simple objects and τ is equivalent to the product (Tychonoff) topology.*

Proof. Let (X, τ) be a linearly compact object in $\mathcal{C}$ where the topology τ is given by a filter base $\{X_\alpha\}$, and let $Y = \sum(Y_\beta \mid \beta \in B)$ be a semisimple subobject of X which is dense in X, where every Y_β is simple. Since (X, τ) is linearly compact and τ is Hausdorff, we have $X = \varprojlim\{X/X_\alpha\}$ also as topologized objects. Moreover, as one easily sees, the linear compactness of (X, τ) implies that for every $X_\alpha \in \{X_\alpha\}$ the factor object X/X_α is discrete linearly compact in $\mathcal{A}$. Since Y is dense in (X, τ), we get

$$X/X_\alpha \cong (Y + X_\alpha)/X_\alpha \cong Y/(Y \cap X_\alpha) \cong \sum(Y_\beta \mid \beta \in C \subseteq B).$$

Therefore also X/X_α is semisimple. Thus in X/X_α there are maximal subobjects having zero intersection. Consequently every X_α in $\{X_\alpha\}$ is the intersection of maximal open subobjects of X. Hence, the intersection of open maximal subobjects of X is 0. Among the maximal open subobjects X_β we choose an independent system $\{X_\beta\}$ which means that for any $X_{\beta_1}, \ldots, X_{\beta_n}$ the factor object $X/\bigcap_{i=1}^{n} X_{\beta_i}$ is a finite product of simple objects X/X_{β_i}, $i = 1, \ldots, n$. Since $\mathcal{A}$ is locally small, Zorn's Lemma yields the existence of a maximal independent system $\{Z_\gamma\}$ of maximal subobjects of X. We claim that $\cap Z_\gamma = 0$. Suppose that $Z = \cap Z_\gamma \neq 0$. Since the

intersection of maximal open subobjects is 0, there is a maximal open subobject Z_0 such that $Z \not\subseteq Z_0$. It is straightforward to see that $\{Z_\gamma\} \cup \{Z_0\}$ is an independent system of maximal open subobject, a contradiction. Hence $\cap Z_\gamma = 0$ as claimed. Obviously $\{X/\underset{\text{finite}}{\cap} Z_\gamma\}$ forms an inverse system such that
$$X = \varprojlim\{X/X_\alpha\} = \varprojlim\{X/\underset{\text{finite}}{\cap} Z_\gamma\} = \prod X/Z_\gamma.$$
The fact that the topology τ is equivalent to the product topology, follows immediately from the linear compactness of X and [2] Theorem 1.1. Thus the Theorem is proved.

For related results we refer to [6], I. Satz 15 and [15] Theorem 6. It is an interesting question to decide whether the product of simple objects is linearly compact with respect to the product topology.

Examples 9. Let R be a ring with local units and $\mathcal{C}$ the category of all R-modules X with $RX = X$.

Taking $\mathcal{C} = \mathcal{D}$ and $\mathcal{F}_1$ = identical functor, $\mathcal{F}_1$-*compactness of an R-module X means that every filter of the form $\{ex_\alpha + X_\alpha\}$ with an arbitrary but fixed idempotent $e \in R$ and submodules X_α, has a non-empty intersection.*

Choosing $\mathcal{D}$ = Set (or **Z**-modules), $\mathcal{F}_2$ = forgetful functor (functor forgetting the multiplication by elements of R, respectively), $\mathcal{F}_2$-*compactness means exactly discrete linear compactness in the sense of Example 6.*

These two kinds of compactnesses are different, in fact $\mathcal{F}_1$-compactness is weaker. For details we refer to Menini and del Rio Mateos [7].

Example 10. Let $\mathcal{C}$ be the category R-mod with mappings f such that each ker f is a finitely generated submodule, and let $\mathcal{F}$ be the forgetful functor. In this case $\mathcal{F}$-compactness means algebraic linear compactness in the sense of Oberst [11] Definitions 4.8 and 4.9 and Lemma 4.10.

Example 11. Let $\mathcal{C}$ be the category of linearly topologized rings having a filter base of two-sided ideals and of continuous mappings, and let $\mathcal{F}$ be the forgetful functor. *A topological ring $X \in \mathcal{C}$ is $\mathcal{F}$-compact if and only if every filter of cosets by closed ideals has a nonempty intersection.*

Let **M** be a class of simple idempotent rings. Clearly, each simple ring endowed with the discrete topology in $\mathcal{F}$-compact. Let us consider a ring X and its structure M-space
$$S(X) = \{Y \triangleleft X \mid X/Y \in \mathbf{M}\}.$$

The finite intersections of ideals from $S(X)$ form a filter which defines a topology $\vartheta = \vartheta(\mathbf{M})$ on X. Thus X endowed with the topology ϑ is an object of the category $\mathcal{C}$. The topology ϑ will be referred to as the *structure space topology* of X. Moreover, taking into account that $\mathbf{M}$ consists of idempotent simple rings, by Zelinsky [17] Lemma 2 we have a ring-direct decomposition

$$X/(Y_1 \cap \ldots \cap Y_n) \cong X/Y_1 \boxplus \ldots \boxplus X/Y_n$$

for every finite subset $\{Y_1, \ldots, Y_n\}$ of $S(X)$. These factor rings form an inverse system. Obviously, the structure space topology ϑ is Hausdorff if and only if $\cap(Y \mid Y \in S(X)) = 0$. Hence, using the fact that all rings from $\mathbf{M}$ as well as their finite direct sums are $\mathcal{F}$-compact in the discrete topology and using standard methods we arrive at

Theorem 2. *Let X be a ring endowed with the structure space topology $\vartheta(\mathbf{M})$. $(X, \vartheta) \in \mathcal{C}$ is $\mathcal{F}$-compact and ϑ is Hausdorff if and only if $X = \prod_\alpha (X_\alpha \mid X_\alpha \in \mathbf{M})$ and ϑ is the product topology.*

Assigning the ideal

$$\rho(X) = \cap(Y \triangleleft X \mid Y \in S(X))$$

to each ring X, a Hoehnke radical ρ is defined in the category of rings. This Hoehnke radical is determined by its semisimple class

$$\mathbf{H} = \{X \mid X = \sum_{\text{subdirect}} (X_\alpha \mid X_\alpha \in \mathbf{M})\},$$

and $X \in \mathbf{H}$, that is, $\rho(X) = 0$ means exactly that the structure space topology on X is Hausdorff. Thus Theorem 2 characterizes certain rings which are semisimple with respect to the radical ρ.

This radical ρ need not be a Kurosh–Amitsur radical. More precisely, the Hoehnke semisimple class may be properly contained in the Kurosh–Amitsur semisimple class $\mathcal{SU}\mathbf{M}$ of the upper radical $\mathcal{U}\mathbf{M}$. (Let us recall that

$$\mathcal{U}\mathbf{M} = \{X \mid S(X) = \emptyset \text{ or } X = 0\}$$

and

$$\mathcal{SU}\mathbf{M} = \{X \mid X \text{ has no nonzero ideal in } \mathcal{U}\mathbf{M}\}.)$$

In fact, Leavitt [5] proved that $\mathbf{H} = \mathcal{SU}\mathbf{M}$ for a class $\mathbf{M}$ of simple idempotent rings if and only if each ring from $\mathbf{M}$ has a unity. Choosing $\mathbf{M}$ to be *the class of all simple rings with unity*, the upper radical $\mathcal{U}\mathbf{M}$ is known as the Brown–McCoy radical. Hence in view of Theorem 2 we have got

Theorem 3. *A ring X endowed with the structure space topology is $\mathcal{F}$-compact and Brown–McCoy semisimple if and only if*

$$X = \prod_\alpha (X_\alpha \mid X_\alpha \text{ is a simple ring with } 1)$$

endowed with the product topology.

Comparing Theorem 1 with Theorems 2 and 3, we make the following observation.

The linear compactness of products of idempotent simple rings as well as the uniqueness of the product topology in Theorems 2 and 3 can be prove in the same way as the assertion that products of compact topological spaces are compact. An alternative proof is based on the following observation. In contrast to semisimple modules of finite length, every ideal M of a finite direct sum $R = \prod (R_i \mid i \in I)$ of simple idempotent ring R_i is also a ring-theoretic direct summand, hence it is a direct sum of those R_i which are contained in M. Therefore it determines uniquely its complement which is exactly a direct sum of those R_i which are not contained in M and hence have the intersection 0 with M. Taking now the product R of simple idempotent rings R_i as the inverse limit of finite direct sums $R_J = \prod (R_i \mid i \in J)$ where J runs over all finte subset of I, every ideal M of R induces the following two inverse systems consisting of the canonical images of M and their complements in R_J, respectively. Hence, if M is closed then M is an inverse limit of its images in R_J, consequently R is a direct sum of the rings M and N where N is exactly the direct product of those R_i which are not contained in M. This implies obviously that R is linearly compact in the sense of Theorems 2 and 3. This reasoning is however not applicable to the case of abelian categories, because the complements of a submodule of semisimple module (even of finite length) are in general not unique.

For related results on Brown–McCoy semisimple rings (and objects in categories) and structure spaces we refer to [12] and [14] (and to [13] [15] and [16], respectively).

Example 12. Considering $\mathcal{C} =$ Ring and $\mathcal{F} =$ forgetful functor, *a ring X is $\mathcal{F}$-compact if and only if every filter of cosets by ideals has a nonempty intersection.* Similarly as in Example 11 one gets the corresponding assertions of Theorems 2 and 3 but *with finite products.*

Example 13. In the case when $\mathcal{C}$ is the category of Hausdorff topological universal algebras of a given type with continuous homomorphisms, and $\mathcal{F}$ is the forgetful functor, *$\mathcal{F}$-compactness is just linear compactness in the sense of Hannák* [3]. In the case of unary algebras (that is, S-acts) Normak [9], [10] calls this linear compactness as *congruence compactness.*

The authors are indebted to the referee for pointing out some shortcomings.

REFERENCES

1. Ánh, P. N., *An Embedding Theorem for Abelian Categories*, J. Algebra **167** (1994), 627–633.
2. Ánh, P. N., Wiegandt, R., *Morita Duality for Grothendieck Categories*, J. Algebra **168** (1994), 273–293.
3. Hannák, L., *Linearly Compact Algebras*, Annales Univ. Sci. Budapest. **21** (1978), 129–137.
4. Johnstone, P. T., *Topos Theory*, Academic Press, 1977.
5. Leavitt, W. G., *A Minimally Embeddable Ring*, Periodica Math. Hungar. **12** (1981), 129–140.
6. Leptin, H., *Linear kompakte Moduln und Ringe, I, II*, Math. Zeitschr. **62** (1955), 241–267 and **66** (1957), 289–327.
7. Menini, C., Rio, A. del, *Morita Dualities and Graded Rings*, Comm. in Algebra **19** (1991), 1743–1794.
8. Müller, B. J., *Linear Compactness and Morita Duality*, J. Algebra **16** (1970), 60–66.
9. Normak, P., *Topological S-acts: Preliminaries and Problems, Transformation Semigroups*, Proc. Intern. Conf. Colchester 1993 (editor: P. Higgins), Univ. Essex 1994, pp. 60–69.
10. Normak, P., *Congruence Compact Acts*, Semigroup Forum, to appear.
11. Oberst, U., *Duality Theory for Grothendieck Categories and Linearly Compact Rings*, J. Algebra **15** (1970), 473–542.
12. Suliński, A., *On Subdirect Sums of Simple Rings with Unity*, Bull. Acad. Polon. Sci. **8** (1960), 223–228.
13. Suliński, A., *The Brown–McCoy Radical in Categories*, Fund. Math. **59** (1966), 23–41.
14. Szász, F. A., *Radicals of Rings*, John Wiley & Sons, 1981.
15. Wiegandt, R., *Radical and Semisimplicity in Categories*, Acta Math. Acad. Sci. Hungar. **19** (1968), 345–364.
16. Wiegandt, R., *On Compact Objects in Categories*, Publ. Math. Debrecen **15** (1968), 267–281.
17. Zelinsky, D., *Linearly Compact Modules and Rings*, Amer. J. Math. **76** (1953), 79–90.

MATHEMATICAL INSTITUTE, HUNGARIAN ACADEMY OF SCIENCES, P. O. BOX 127, H–1364 BUDAPEST, WIEGANDT@MATH-INST.HU; FAKULTÄT FÜR MATHEMATIK, UNIVERSITÄT BIELEFELD, POSTFACH 100131, D-33501 BIELEFELD, GERMANY, ANH@MATHEMATIK.UNI-BIELEFELD.DE

A RING OF MORITA CONTEXT IN WHICH EACH RIGHT IDEAL IS WEAKLY SELF-INJECTIVE *

S. BARTHWAL, S.K. JAIN, S. JHINGAN, AND SERGIO R. LÓPEZ-PERMOUTH

ABSTRACT. In this paper, among others, an example of a noetherian ring of Morita Context in which each right ideal is weakly self-injective, has been studied.

1. INTRODUCTION

A right R-module M is said to be self-injective if for every R-homomorphism $\varphi : M \to E(M)$, from M into its injective hull $E(M)$, $\varphi(M) \subseteq M$. Similarly, M is weakly self-injective if for every homomorphism $\varphi : M \to E(M)$, there exists $X \subseteq M$ such that $\varphi(M) \subseteq X \cong M$. If K is a right Ore domain then each right ideal is clearly weakly self-injective. Let K be a right and left noetherian domain. We consider the ring

$$R = \begin{pmatrix} K & K^* & 0 & \cdots & 0 & 0 \\ 0 & K & K^* & \cdots & 0 & 0 \\ \cdots & \cdots & \cdots & \cdots & \cdots \\ \cdots & \cdots & \cdots & \cdots & \cdots \\ 0 & 0 & 0 & \cdots & K & K^* \\ K^* & 0 & 0 & \cdots & 0 & K \end{pmatrix}$$

of all $n \times n$ matrices whose only nonzero possible entries are on the main diagonal and at $(1,2), (2,3), ..., (n-1,n), (n,1)$ places. The entries on the main diagonal belong to K and on the $(i, i+1)$th places, for all $i, 1 \leq i \leq n$, belong to a null K-algebra K^* of rank 1. We prove that each right and left ideal in R is weakly self-injective and address some related questions by considering K-algebra K^* of rank greater than 1. Rings whose right ideals are self-injective have been studied in great detail by many authors. It is interesting to note that the injective hull of R is a ring with the property that each right ideal is self-injective([1]).

---
*1991 Mathematics Subject Classification. 16D50.

Throughout this paper, by a module we mean a right module. K will denote a right noetherian ring. Q, the right maximal ring of quotients of K. K^* is a two-sided K-module isomorphic to K itself such that $xy = 0$ for all x, y in K^*. We fix an isomorphism ${}_K K_K \to_K K_K^*$ and for each element a of K we denote by a^* the corresponding element of K^*. Then we have $ab^* = (ab)^* = a^*b$ for all a, b in K, and in particular we have $a1^* = a^* = 1^*a$ for all a. For any module M, $E(M)$ will denote the injective hull of M. For any two sets I and J, $(I, J)_i$ will denote the set of $n \times n$ matrices with (i,i)th element from I, $(i,i+1)$th element from J^* and all other elements zero. For all $i \neq j$, e_{ij}^* will denote the $n \times n$ matrix whose (i,j)th entry is 1^* and other entries are all zero. Note that in our notation, $R = \sum_{i=1}^{n}(K, K)_i$. The fact that R is a ring under usual matrix operations is clear, because $K^*K^* = 0$.

2. MAIN RESULT

We begin with a lemma whose proof is straightforward.

Lemma 2.1 *If K is right noetherian then R is right noetherian.*

Proof. Writing R as

$$R = \begin{pmatrix} T & P \\ S & K \end{pmatrix}$$

where

$$T = \begin{pmatrix} K & K^* & 0 & \cdots & 0 & 0 \\ 0 & K & K^* & \cdots & 0 & 0 \\ \cdots & \cdots & \cdots & \cdots & \cdots & \cdots \\ \cdots & \cdots & \cdots & \cdots & \cdots & \cdots \\ 0 & 0 & 0 & \cdots & K & K^* \\ 0 & 0 & 0 & \cdots & & K \end{pmatrix},$$

$$P = \begin{pmatrix} 0 \\ 0 \\ 0 \\ \cdots \\ \cdots \\ K^* \end{pmatrix}, \quad S = \begin{pmatrix} K^* & 0 & 0 & \cdots & \cdots & 0 \end{pmatrix}.$$

It is easy to check that T is a right noetherian ring or one may prove this by invoking proposition 1.7 in ([4], page 12). Furthermore, because both P_K and S_T are cyclic modules, it follows again by appealing to the same proposition that R is right noetherian. ∎

Lemma 2.2 *Let K be a right noetherian ring. Then the right ideals of the ring R are precisely $\sum_{k=1}^{n}(I_k, J_k)_k$, where*

$$\sum_{k=1}^{n} (I_k, J_k)_k = \begin{pmatrix} I_1 & J_1^* & 0 & \ldots & 0 \\ 0 & I_2 & J_2^* & \ldots & 0 \\ \ldots & \ldots & \ldots & \ldots & \ldots \\ J_n^* & 0 & 0 & \ldots & I_n \end{pmatrix}$$

and for each k, $1 \leq k \leq n$, I_k and J_k are right ideals of K such that $I_k \subseteq J_k$.

Proof. Clearly, $\sum_{k=1}^{n}(I_k, J_k)_k$ is a right ideal of R whenever I_k and J_k are right ideals of K such that $I_k \subseteq J_k$. Conversely, if I is a right ideal of R, then since R is right noetherian, I is finitely generated. We shall prove that I is of the form $\sum_{k=1}^{n}(I_k, J_k)_k$, by induction on the number m of generators of I. For $m = 1$, suppose that I is generated by $a = \sum_{k=1}^{n}\left(a_{ii}e_{ii} + a_{ii+1}e_{ii+1}^*\right)$. Then,

$$\begin{aligned}
I &= aR \\
&= \sum_{i=1}^{n}\left(a_{ii}e_{ii+1} + a_{ii+1}e_{ii+1}^*\right)R \\
&= \begin{pmatrix} a_{11}K & a_{11}K^* + a_{12}^*K & 0 & 0 & \ldots & 0 \\ 0 & a_{22}K & a_{22}K^* + a_{23}^*K & 0 & \ldots & 0 \\ \ldots & \ldots & \ldots & \ldots & \ldots & \ldots \\ \ldots & \ldots & \ldots & \ldots & \ldots & \ldots \\ \ldots & \ldots & \ldots & \ldots & \ldots & \ldots \\ a_{n1}^*K + a_{nn}K^* & 0 & \ldots & \ldots & 0 & \ldots & a_{nn}K \end{pmatrix}
\end{aligned}$$

which is clearly of the form $\sum_{k=1}^{n}(I_k, J_k)_k$ with $I_k \subseteq J_k$ for all $1 \leq k \leq n$.

Suppose the result holds for $m-1$ generators. Let I be generated by m generators $a_1, a_2, \ldots, a_m$. Then $I = a_1R + a_2R + \ldots + a_{m-1}R + a_mR$. By induction hypothesis, $a_1R + a_2R + \ldots + a_{m-1}R = \sum_{k=1}^{n}(I_k, J_k)_k$ and $a_mR = \sum_{k=1}^{n}(I_k', J_k')_k$ with $I_k \subseteq J_k$ and $I_k' \subseteq J_k'$ for all $1 \leq k \leq n$. So $I = \sum_{k=1}^{n}(I_k, J_k)_k + \sum_{k=1}^{n}(I_k', J_k')_k = \sum_{k=1}^{n}((I_k + I_k')(J_k + J_k'))_k$. Since $I_k \subseteq J_k$ and $I_k' \subseteq J_k'$, $I_k + I_k' \subseteq J_k + J_k'$. So I has the desired form. Hence the proof is complete by induction.

Lemma 2.3 *If I, J are right ideals of K with $I \subset J$ then for $1 \leq i,j \leq n$, $j \neq i, j \neq i+1 (\text{mod } n)$, $Hom_R((I,J)_i, (Q,Q)_j) = 0$, where Q is the right maximal quotient ring of K.*

Proof. Let $\phi \in Hom_R((I,J)_i, (Q,Q)_j)$ and $\phi(ae_{ii} + be_{ii+1}^*) = pe_{jj} + qe_{jj+1}^*$. Then $(ae_{ii} + be_{ii+1}^*)e_{j+1j+1} = 0$ and $(ae_{ii} + be_{ii+1}^*)e_{jj+1}^* = 0$, yields $q = 0$ and $p = 0$ respectively.

Next we compute the injective hull of R_R.

Theorem 2.4 *Let K be a noetherian ring. Then the injective hull of R_R*

is

$$\sum_{i=1}^{n}(Q,Q)_i = \begin{pmatrix} Q & Q^* & 0 & \dots & 0 & 0 \\ 0 & Q & Q^* & \dots & 0 & 0 \\ \dots & \dots & \dots & \dots & \dots & \dots \\ \dots & \dots & \dots & \dots & \dots & \dots \\ 0 & 0 & 0 & \dots & Q & Q^* \\ Q^* & 0 & 0 & \dots & 0 & Q \end{pmatrix}.$$

Proof. Suppose $0 \neq x = \sum_{i=1}^{n}(p_{ii}e_{ii} + q_{ii+1}e_{ii+1}^*) \in \sum_{i=1}^{n}(Q,Q)_i$. To be definite, let $p_{11} \neq 0$. For each i, $1 \leq i \leq n$, there exists $r_i \in K$ such that $p_{ii}r_i \in K, q_{ii+1}r_i \in K$ and $p_{11}r_1 \neq 0$. But then $0 \neq x\sum_{i=1}^{n}r_ie_{ii} = \sum_{i=1}^{n}(p_{ii}r_ie_{ii} + q_{ii+1}r_{i+1}e_{ii+1}^*)$ is in R. Hence $\sum_{i=1}^{n}(Q,Q)_i$ is an essential extension of R.

To prove injectivity of $\sum_{i=1}^{n}(Q,Q)_i$, it is sufficient to prove that for all i, $(Q,Q)_i$ is injective as an R-module. To prove this, let $\phi : \sum_{k=1}^{n}(I_k, J_k)_k \to (Q,Q)_i$ be an R-homomorphism. Then $\phi = \sum_{i=1}^{n}\phi_k$, where $\phi_k : (I_k, J_k)_k \to (Q,Q)_i$. By Lemma , $\phi_k = 0$ for $k \neq i, i+1$, so that $\phi = \phi_i + \phi_{i+1}$. Observe, now, that if $\phi_{i+1}(ae_{i+1i+1} + be_{i+1i+2}^*) = pe_{ii} + qe_{ii+1}^*$ then, as $(ae_{i+1i+1} + be_{i+1i+2}^*)e_{ii+1}^* = 0$, we have $p = 0$. If $I_{i+1} = 0$, then since $(ae_{i+1i+1} + be_{i+1i+2}^*)e_{i+1i+1} = 0$ we have $\phi_{i+1} = 0$. Thus, in this case, $\phi = \phi_i$ and consequently

$$\begin{aligned}\phi(\sum_{k=1}^{n}(a_ke_{kk} + b_ke_{kk+1}^*)) &= \phi_i(ae_{ii} + be_{ii+1}^*) \\ &= \alpha\phi_i(ae_{ii} + be_{ii+1}^*) \\ &= \alpha e_{ii}(\sum_{k=1}^{n}(a_ke_{kk} + b_ke_{kk+1}^*)).\end{aligned}$$

Assume, now, that $I_{i+1} \neq 0$. Define $\phi_i' : J_i \to Q$ as follows : if $b \in J_i$ and $\phi_i(ae_{ii} + be_{ii+1}^*) = pe_{ii} + qe_{ii+1}^*$ for some $a \in I_i$, then set $\phi_i'(b) = q$. The map ϕ_i' is well defined, for if, $ae_{ii} \in (I, J)_i$ and $\phi(ae_{ii}) = pe_{ii} + qe_{ii+1}^*$ then, as $ae_{ii}e_{i+1i+1} = 0$, we have $q = 0$. ϕ_i' is clearly a K-homomorphism. Since Q is injective as K-module, there exist $\alpha \in Q$ such that $\phi_i'(b) = \alpha b$ for all $b \in J_i$. If $\phi_i(ae_{ii} + be_{ii+1}^*) = pe_{ii} + qe_{ii+1}^*$ then $pe_{ii+1}^* = (pe_{ii} + qe_{ii+1}^*)e_{ii+1}^* = \phi_i(ae_{ii+1}^*)$, so that $\phi_i'(a) = p$. Thus $p = \alpha a$, $q = \alpha b$. It follows that $\phi_i(ae_{ii} + be_{ii+1}^*) = \alpha ae_{ii} + \alpha be_{ii+1}^* = \alpha e_{ii}(ae_{ii} + be_{ii+1}^*)$.

Define $\phi_{i+1}' : I_{i+1} \to Q$ as follows: if $\phi_{i+1}(ae_{i+1i+1} + be_{i+1i+2}^*) = qe_{ii+1}^*$, set $\phi_{i+1}'(a) = q$. The map ϕ_{i+1}' is well defined, for if $\phi_{i+1}(be_{i+1i+2}^*) = qe_{ii+1}^*$, then, as $be_{i+1i+2}^*e_{i+1i+1} = 0$, we have $q = 0$. Clearly ϕ_{i+1}' is a K-homomorphism. As above, there exists $\beta \in Q$ such that $\phi_{i+1}'(a) = \beta a$ for all $a \in I_{i+1}$, that is, if $\phi_{i+1}(ae_{i+1i+1} + be_{i+1i+2}^*) = qe_{ii+1}^*$ then $q = \phi_{i+1}'(a) = \beta a$. Thus,

$$\phi_{i+1}(ae_{i+1i+1} + be^*_{i+1i+2}) = \beta a e^*_{ii+1} = \beta e^*_{ii+1}(ae_{i+1i+1} + be^*_{i+1i+2}).$$

Hence,

$$\phi(\sum_{k=1}^{n}(a_k e_{kk} + b_k e^*_{kk+1})) = \alpha e_{ii}(a_i e_{ii} + b_i e^*_{ii+1})$$
$$+ \beta e^*_{ii+1}(a_{i+1} e_{i+1i+1} + b_{i+1} e^*_{i+1i+2})$$
$$= \alpha e_{ii}(\sum_{k=1}^{n}(a_k e_{kk} + b_k e^*_{kk+1}))$$
$$+ \beta e^*_{ii+1}(\sum_{k=1}^{n}(a_k e_{kk} + b_k e^*_{kk+1}))$$
$$= (\alpha e_{ii} + \beta e^*_{ii+1})(\sum_{k=1}^{n}(a_k e_{kk} + b_k e^*_{kk+1})).$$

Thus, in any case, there exist $\lambda \in (Q,Q)_i$ such that $\phi(x) = \lambda x$ for all $x \in I$. Consequently, for all $i, 1 \le i \le n$, $(Q,Q)_i$ is injective as R-module and the proof of the theorem is complete. ∎

Remark 2.1 Müller [5] and Sakano [6] obtained the injective hull of generalized matrix. One may obtain the above result by using their method. Our construction is direct and less technical. A slight modification of the above argument yields the following theorem:

Theorem2.5 *Let K be a noetherian ring. If I and J are right ideals of K such that $I \subset J$ and J is non-zero then for $1 \le i \le n$, the injective hull of the right ideal $(I,J)_i$ is $(Q,Q)_i$.*

It is known that a domain is right weakly self-injective if and only if it is a right Ore domain (Example 1.11(ii), [3]). For the following discussion we assume that K is right Ore. Then for $n > 1$, we have the following:

Theorem 2.6 *Let K be a right Ore domain and $n > 1$. Then the ring $R = \sum_{i=1}^{n}(K,K)_i$ is right weakly self-injective if and only if K is left Ore.*

Proof. First assume R is right weakly self-injective. Let a and b be any two non-zero elements of K, and let $x = a^{-1} e_{11} + b^{-1} e^*_{12}$. Then x is a non-zero element of $E(R_R)$. Since R is weakly self-injective, by Remark 1.5 in [3], there exists $0 \ne y = \sum_{i=1}^{n}(a_i e_{ii} + b_i e^*_{ii+1})$ such that $x \in yR$ and $r.ann_R(y) = 0$. If $a_i = 0$ for some i, say $i = k$, then $ye^*_{kk+1} = 0$, a contradiction to the fact that $r.ann_R(y) = 0$. Thus $a_i \ne 0$ for all i. But then y is invertible and $y^{-1} = \sum_{i=1}^{n}(a_i^{-1} e_{ii} - a_i^{-1} b_i a_{i+1}^{-1} e_{ii+1})$. Since $x \in yR$, we get $a_1^{-1} a^{-1} = \alpha$ and $a_1^{-1} b^{-1} = \beta$ for some $\alpha, \beta \in K$. Thus $\alpha a = \beta b$ for some $\alpha, \beta \in K$ and hence K is left Ore.

Conversely, assume that K is left Ore and let $0 \neq x = \sum_{i=1}^{n}(a_i e_{ii} + b_i e_{ii+1}^*)$ be an arbitrary element of $E(R_R) = \sum_{i=1}^{n}(Q,Q)_i$. Since ${}_K K$ is essential in Q, there exists $0 \neq r \in K$ such that for $1 \leq i \leq n$, $ra_i, rb_i \in K$. Let $ra_i = s_i$, $rb_i = t_i$ and $y = \sum_{i=1}^{n} r^{-1} e_{ii}$. Then y is a non-zero element of $E(R_R)$ and $s_i, t_i \in K$ for $1 \leq i \leq n$. Also $x = \sum_{i=1}^{n}(r^{-1} s_i e_{ii} + r^{-1} t_i e_{ii+1}^*) = \sum_{i=1}^{n} r^{-1} e_{ii} \sum_{i=1}^{n}(s_i e_{ii} + t_i e_{ii+1}^*) \in yR$. Moreover, if $\sum_{i=1}^{n}(x_i e_{ii} + y_i e_{ii+1}^*) \in r.ann_R(y)$ then $r^{-1} x_i = 0 = r^{-1} y_i$ for all i, $1 \leq i \leq n$. It follows that $x_i = 0 = y_i$ for all i. Consequently, $r.ann_R(y) = 0$ and therefore R is weakly self-injective.

In view of the above theorem, it is clear that if K is a right noetherian domain then $R = \sum_{i=1}^{n}(K,K)_i$ is a weakly self injective ring.

Proposition 2.7 *Let K be a noetherian domain and*

$$R = \sum_{i=1}^{n}(K,K)_i = \begin{pmatrix} K & K^* & 0 & \ldots & 0 & 0 \\ 0 & K & K^* & \ldots & 0 & 0 \\ \ldots & \ldots & \ldots & \ldots & \ldots & \ldots \\ \ldots & \ldots & \ldots & \ldots & \ldots & \ldots \\ 0 & 0 & 0 & \ldots & K & K^* \\ K^* & 0 & 0 & \ldots & 0 & K \end{pmatrix}.$$

Then every right ideal as well as every left ideal of R is weakly self-injective.

Proof. Suppose $0 \neq I = \sum_{k=1}^{n}(I_k, J_k)_k$ be a right ideal of R and let Λ be the set of k, $1 \leq k \leq n$, for which I_k and J_k are not both zero. Then $I = \bigoplus \sum_{k \in \Lambda}(I_k, J_k)_k$ so that $E(I) = \bigoplus \sum_{k \in \Lambda}(Q,Q)_k$. Let $0 \neq \phi : I \to E(I)$ be any R-homomorphism and let for $k \in \Lambda$, $\phi_k = \pi_k \phi$, where π_k is the projection of $\bigoplus \sum_{k \in \Lambda}(Q,Q)_k$ onto $(Q,Q)_k$. Then $\phi = \sum_{k \in \Lambda} \phi_k$. Proceeding, now, as in Theorem , for all $k \in \Lambda$, we can find $\alpha_k, \beta_k \in Q$ such that $\phi_k(x) = (\alpha_k e_{kk} + \beta_{k+1} e_{kk+1}^*) x$ for all $x \in I$. Observe that, if for some $k \in \Lambda$, $k - 1 \notin \Lambda$, then $\beta_k = 0$. Consequently, for all $x \in I$, $\phi(x) = \left[\sum_{k \in \Lambda}(\alpha_k e_{kk} + \beta_{k+1} e_{kk+1}^*)\right] x$. Since $\phi \neq 0$, at least one of α_k's and β_k's is non-zero. Since Q is left maximal ring of quotients of K, there exist $0 \neq r \in K$ such that for all $k \in \Lambda$, $\alpha_k, \beta_k \in K$. Let $r\alpha_k = x_k$, $r\beta_k = y_k$. Now, if for some $k \in \Lambda$, $I_k \neq 0$, pick $0 \neq \lambda_k \in I_k$, and if $I_k = 0$, pick $0 \neq \lambda_k \in J_k$. Then, for all $k \in \Lambda$, $\lambda_k x_k \in I_k$, and $\lambda_k y_{k+1} \in J_k$. Consequently, for all $x \in I$,

$$\phi(x) = \left(\sum_{k \in \Lambda} r^{-1} \lambda_k^{-1} e_{kk}\right) \left(\sum_{k \in \Lambda}(\lambda_k x_k e_{kk} + \lambda_k y_{k+1} e_{kk+1}^*) x\right)$$
$$\in \left(\sum_{k \in \Lambda} r^{-1} \lambda_k^{-1} e_{kk}\right) I$$

Thus $\phi(I) \subset yI$, where $y = \sum_{k \in \Lambda} r^{-1} \lambda_k^{-1} e_{kk}$. It is easy to see that

$yI \simeq I$. It follows that I is weakly self-injective. The proof is similar for left ideals of R. ∎

A slight modification of the above argument yields our next result. Before we can state it we need to mention the following definition:

Definition 2.1 A module M is said to be weakly $N-$ injective if for every homomorphism $\varphi : N \to E(M)$, there exist $X \subseteq E(M)$ such that $\varphi(N) \subseteq X \cong M$ and weakly-injective if it is weakly N-injective for every finitely generated module N.

Theorem 2.8 *For every n, every right ideal I of R is weakly I^n-self-injective.*

Corollary 2.9 *R is weakly-injective.*

We now give an example to show that if the null algebra K^* is of rank more than 1 then the result may not be true.

Example 2.1 Let K be a right Ore domain. Then, neither of the rings

$$R_1 = \begin{pmatrix} K & K^* \times K^* \\ K^* \times K^* & K \end{pmatrix} \text{ and } R_2 = \begin{pmatrix} K & K^* \times K^* \\ K^* & K \end{pmatrix}$$

is weakly self-injective.

Proof. We prove the result for R_2. The argument for R_1 is similar. It is easy to see that $R_2 \simeq R_2'$, where

$$R_2' = \left\{ \begin{pmatrix} a & 0 & b^* \\ 0 & a & c^* \\ d^* & 0 & e \end{pmatrix} : a, b, c, d, e \in K \right\}$$

and that

$$E(R_2) \simeq \begin{pmatrix} Q & Q & Q^* \\ Q & Q & Q^* \\ Q^* & 0 & Q \end{pmatrix}$$

where Q is the right maximal quotient ring of K.

Suppose R_2 is weakly self-injective. Then for $e_{12} \in E(R_2)$, there exist

$$y = \begin{pmatrix} y_{11} & y_{12} & y_{13}^* \\ y_{21} & y_{22} & y_{23}^* \\ y_{31}^* & 0 & y_{33} \end{pmatrix} \in E(R_2) \text{ such that } e_{12} \in yR_2 \text{ and } r.ann_{R_2}(y) = 0.$$

Since $e_{12} \in yR_2$, therefore,

$$\begin{pmatrix} 0 & 1 & 0 \\ 0 & 0 & 0 \\ 0 & 0 & 0 \end{pmatrix} = \begin{pmatrix} y_{11} & y_{12} & y_{13}^* \\ y_{21} & y_{22} & y_{23}^* \\ y_{31}^* & 0 & y_{33} \end{pmatrix} \begin{pmatrix} a & 0 & b^* \\ 0 & a & c^* \\ d^* & 0 & e \end{pmatrix}$$

for some $a, b, c, d, e \in K$. It follows that $a \neq 0$, $y_{11} = 0 = y_{21} = y_{22}$. But then $ye_{13}^* = 0$, contradicting the fact that $r.ann_{R_2}(y) = 0$. Hence R_2 is not weakly self-injective. ∎

Remark 2.2 A similar argument can be used to show that the ring $R' = \begin{pmatrix} D & D^2 \\ 0 & D \end{pmatrix}$, where D is a division ring, is not weakly self-injective. It follows that every right ideal in a non-singular, artinian ring need not be weakly self-injective. However, for the ring of upper triangular matrices over a division ring, it is not hard to see that every right ideal is weakly self-injective.

References

[1] G. Ivanov, *Non-local rings whose ideals are all quasi-injective*, Bull. Austral. Math. Soc. **6** (1972), 45-52.

[2] G. Ivanov, *Non-local rings whose ideals are all quasi-injective: addendum*, Bull. Austral. Math. Soc. **12** (1975), 159-160.

[3] S. K. Jain and S. R. López-Permouth, *A survey on the theory of weakly-injective modules*, Computational Algebra, Marcel Dekker, NY (1994), 205-232.

[4] J. C. McConnell and J. C. Robson, *Noncommutative Noetherian Rings*, Wiley, NY, 1987.

[5] Marianne Müller, *Rings of quotients of generalized matrix rings*, Comm. Alg. **15**(10), 1987, 1991-2015.

[6] Kazunori Sakano, *Maximal quotient rings of generalized matrix rings*, Comm. Alg. **12**(16), 1984, 2055-2065.

Department of Mathematics, Ohio University, Athens, OH 45701
email: sbarthwa@oucsace.cs.ohiou.edu

Department of Mathematics, Ohio University, Athens, OH 45701
email: jain@oucsace.cs.ohiou.edu

Department of Mathematics, Ohio University, Athens, OH 45701
email: sjhingan@oucsace.cs.ohiou.edu

Department of Mathematics, Ohio University, Athens, OH 45701
email: slopez@bing.math.ohiou.edu

SPLITTING THEOREMS AND A PROBLEM OF MÜLLER

Gary F. Birkenmeier, Jin Yong Kim, and Jae Keol Park

Dedicated to Professor B. J. Müller on his retirement

ABSTRACT. In this paper we introduce and investigate a condition (FI) which encompases a large class of rings including duo rings, FPF rings, and GFC rings. This condition is used in our main results to generalize a splitting theoren of C. Faith, and it is also used to provide a large class of self-injective rings on which a question of B. J. Müller has an affirmative answer.

0. Introduction

The motivation for this paper is provided by the following theorem of Faith [5, Theorem 7] and a problem of Müller [9]:

1. (Faith) The maximal regular ideal $M(R)$ splits off any continuous ring, in particular, of any left and right self-injective ring.

2. (Müller) Let R be a right and left self-injective ring (so $J(R) = Z(R_R) = Z(_RR)$). Show $Z_2(R_R) = Z_2(_RR)$, or provide a counterexample.

In this paper we introduce and investigate a condition (FI) which encompasses a large class of rings including duo rings, FPF rings, and GFC rings. This condition is used in our main results to obtain a splitting theorem (Theorem 2.1) similar to (1) and a corollary (Corollary 2.3) which provides a large class of self-injective rings for which $Z_2(R_R) = Z_2(_RR)$. In the last section we focus our investigation on the class of GFC rings which is an important subclass of the class of FI rings. We show that if a ring R is GFC, then $Z_2(R_R) = Z_2(_RR)$. Moreover, a ring R is right FPF if and only if the full $n \times n$ matrix ring over R is right GFC for every positive integer n. These results allow us to construct a class of non-GFC rings R for which $Z_2(R_R) = Z_2(_RR)$.

Throughout this paper R denotes an associative ring with unity and Mod-R is the category of right R-modules. For a nonempty subset X of R, $\ell_R(X)$ and $r_R(X)$ ($\ell(X)$ and $r(X)$ if the context is clear) denote the left and right annihilators of X in R, respectively. A right R-module M is said to be *faithful* if $\text{Ann}_R(M) = \{r \in R \mid Mr = 0\} = 0$. R is called a right *GFC* (resp. *FPF*) ring, if every faithful cyclic (resp. finitely generated) right R-module

---

1991 *Mathematics Subject Classification.* Primary 16D50, 16D70.

Typeset by $\mathcal{A}_{\mathcal{M}}\mathcal{S}$-TEX

generates Mod-R. For examples and basic results for right GFC rings see [2] and for right FPF rings see [6]. $J(R), \mathcal{P}(R), Soc(_RR), Soc(R_R), Z(_RR),$ and $Z(R_R)$ denote the Jacobson radical, prime radical, left socle, right socle, left singular ideal, and right singular ideal of R, respectively. An ideal I of R is called *ideal essential* in an ideal H of R if I has nonzero intersection with every nonzero ideal of R contained in H. A right (resp. left) ideal X of R is called *right* (resp. *left*) *essential* in a right (resp. left) ideal Y of R if X has nonzero intersection with every nonzero right (resp. left) ideal of R contained in Y. $\overline{J}(R)$ and $\overline{\mathcal{P}}(R)$ denote the sum of all ideals of R which are ideal essential extensions of $J(R)$-radical and $\mathcal{P}(R)$-radical ideals, respectively [4]. $M(R)$ symbolizes the maximal regular ideal of R [5]. An $e \in R$ is called a *right* (resp. *left*) *semicentral idempotent* of R if $e = e^2$ and $ex = exe$ (resp. $xe = exe$), for all $x \in R$. For basic properties of right semicentral idempotents see [1]. $Z_2(R_R)$ is defined by $Z_2(R_R)/Z(R_R) = Z(R/Z(R_R))$, the singular submodule of the right R-module $R/Z(R_R)$. We note that $Z_2(R_R)$ is right closed in R and a right essential extension of $Z(R_R)$. $Z_2(_RR)$ is defined similarly. If S is a ring, then $\operatorname{Mat}_n(S)$ denotes the full $n \times n$ matrix ring over S.

1. Preliminaries

We start with the following definition.

Definition 1.1. We say R is a *right* (resp. *left*) *FI (faithful idempotents) ring* if $e \in R$ such that $e = e^2 \neq 1$ and $r(eR) = 0$ (resp. $\ell(Re) = 0$) implies $\operatorname{Hom}_R(eR, (1-e)R) \neq 0$ (resp. $\operatorname{Hom}_R(Re, R(1-e)) \neq 0$). If R is both a left and a right FI ring, then we say R is an FI ring.

Lemma 1.2. *The following conditions are equivalent:*
  (i) R *is a right FI ring.*
  (ii) *If* $e = e^2 \in R$ *such that* $r(eR) = 0$ *and* $\operatorname{Hom}_R(eR, (1-e)R) = 0$, *then* $e = 1$.
  (iii) *If* $0 \neq e = e^2 \in R$ *such that* $(1-e)R$ *contains no nonzero ideals, then either* $e = 1$ *or* $\operatorname{Hom}_R(eR, (1-e)R) \neq 0$.
  (iv) *Let* P *be a faithful cyclic projective module generated by* x *(i.e., $P = xR$). If* $\operatorname{Hom}_R(P, \operatorname{Ann}_R(x)) = 0$, *then* $\operatorname{Ann}_R(x) = 0$.
  (v) *If* $R = \begin{pmatrix} A & X \\ 0 & B \end{pmatrix}$ *with rings A, B, and a left A-right B-bimodule X, then the right B-module X is not faithful.*

Proof. The proof of the equivalence (i)–(iv) is straightforward. For (i)$\Rightarrow$(v), assume R is right FI. Since $R = \begin{pmatrix} A & X \\ 0 & B \end{pmatrix}$, there exists $1 \neq e = e^2 \in R$ such that $A = eRe$, $X = eR(1-e)$, and $B = (1-e)R(1-e)$. In this situation, $(1-e)Re = 0$ and so $\operatorname{Hom}_R(eR, (1-e)R) = 0$. Since R is right FI, we have $r(eR) \neq 0$. So there is $0 \neq b \in R$ such that $eRb = 0$. In particular, $eb = 0$ and thus $0 \neq b = (1-e)b(1-e)$. So $0 = eRb = [eR(1-e)](1-e)b(1-e) = X(1-e)b(1-e)$ with $0 \neq (1-e)b(1-e) \in B$. Thus $0 \neq \text{Ann}_B(X)$ and hence X is not faithful as a right B-module.

Conversely, assume $\text{Hom}_R(eR, (1-e)R) = 0$ with $1 \neq e = e^2 \in R$. Then $(1-e)Re = 0$; and so $R = \begin{pmatrix} A & X \\ 0 & B \end{pmatrix}$ with $A = eRe$, $B = (1-e)R(1-e)$, and $X = eR(1-e)$. By assumption, X is not faithful as a right B-module and hence $\text{Ann}_B(X) \neq 0$. Therefore there exists $0 \neq (1-e)r(1-e) \in B$ such that $0 = X(1-e)r(1-e) = [eR(1-e)](1-e)r(1-e)$. So $eR(1-e)r(1-e) = 0$ and thus $r(eR) \neq 0$. Therefore R is right FI.

Note that if R is the 2×2 upper triangular matrix ring over a ring with unity, then R is *not* a right FI ring by Lemma 1.2(v). Also, take $e = \begin{pmatrix} 1 & 0 \\ 0 & 0 \end{pmatrix}$. Then $r(eR) = 0$, but $\text{Hom}_R(eR, (1-e)R) = 0$.

Lemma 1.3. *If $e \in R$ is an idempotent such that eR is faithful and $\text{Hom}_R(eR, (1-e)R) = 0$, then eR is an ideal of R which is left essential in R.*

Proof. Since $\text{Hom}_R(eR, (1-e)R) \cong (1-e)Re$ (as abelian groups), then $(1-e)Re = 0$. Hence $Re = eRe$. So eR is an ideal of R. Let $0 \neq x \in R$. Then since $r(eR) = 0$, it follows that $eRx \neq 0$ and so there exists $s \in R$ such that $0 \neq esx \in Rx \cap eR$. So eR is left essential in R.

Note that in general if I is an ideal of R such that $r(I) = 0$, then I is left essential in R.

Lemma 1.4. *If R satisfies any of the following conditions, then R is a right FI ring:*

(i) *Every right semicentral idempotent is central;*

(ii) *Every left semicentral idempotent is central;*

(iii) *R is semiprime;*

(iv) *Every faithful idempotent generated right ideal is a generator in Mod-R (e.g., R is right GFC or right FPF);*

(v) *Every principal ideal X with $r(X) = 0$ is right essential;*

(vi) *Every principal ideal which is left essential in R is also right essential in R.*

Proof. Let $e = e^2 \in R$ such that eR is faithful and $\text{Hom}_R(eR, (1-e)R) = 0$. By Lemma 1.3, eR is an ideal. Hence $1-e$ is a right semicentral idempotent [1, Lemma 1].

(i) Hence $1-e$ is central. Since eR is faithful, $1-e = 0$. Thus $e = 1$. By Lemma 1.2, R is a right FI ring.

(ii) Note that an idempotent e is right semicentral if and only if $1-e$ is left semicentral. Hence (ii) is equivalent to (i).

(iii) If R is semiprime, then $eR(1-e) = 0$ [1, Lemma 1]. From the proof of Lemma 1.3, $(1-e)Re = 0$. Hence $1-e$ is central. Thus as in the proof of part (i), R is a right FI ring.

(iv) Since eR is a generator in the category Mod-R of right R-modules, if $1 - e \neq 0$, then $\text{Hom}_R(eR, (1-e)R) \neq 0$. So $1 - e = 0$. Hence R is a right FI ring.

(v) This part is an immediate consequence of the definition of a right FI ring.

(vi) This part follows from Lemma 1.3.

Proposition 1.5. *Let c be a right semicentral idempotent. If R is a right FI ring, then cRc is a right FI ring.*

Proof. Let $cR = C$ and $(1-c)R = B$. Observe that $C = cRc$. If $e = e^2 \in C$ such that $e \neq c$ and eC is faithful in C, then $eC \oplus B$ is faithful in R. Now $eC = eR$ and $eR \oplus B = (1 - (c - e))R$. Hence there exists a nonzero homomorphism $f : (1 - (c - e))R \to (c - e)R$. So there exists $x \in R$ such that $f(1 - (c - e)) = (c - e)x \neq 0$. But $(c - e)x = (c - e)xc$. Hence $f(1 - (c - e)) = f(1 - (c - e))c = f(ec) = f(e) \neq 0$. Therefore $\text{Hom}_R(eC, (c-e)C) \neq 0$, so C is a right FI ring.

2. Splitting Theorems

In this section, we present our main results. In the following theorem we replace the left continuous condition in Faith's theorem (see (1) of the Introduction) with a variety of conditions to obtain a decomposition of a right continuous ring R in terms of $Z_2(R_R)$ and a right essential extension of $M(R)$. A consequence of this result shows that the class of self-injective FI rings satisfies the condition $Z_2(R_R) = Z_2(_RR)$.

Theorem 2.1. *Let R be a right continuous ring which satisfies at least one of the following conditions:*

(i) *R is a right FI ring;*

(ii) *$R = \oplus_{i=1}^{n} e_i R$, where $\{e_i \mid i = 1, 2, \ldots, n\}$ is a complete set of orthogonal primitive idempotents;*

(iii) *$\text{Soc}(R_R)$ is right essential in R.*

Then we have the following:

(a) *$R = Z_2(R_R) \oplus K$ (right ideal direct sum), where K is a right continuous regular ring with unity;*

(b) *$Z_2(R_R) = \overline{J}(R)$;*

(c) *$M(R)$ is right essential in K;*

(d) *For conditions (ii) or (iii), K is a semisimple Artinian ring.*

Proof. (a) This part follows from Theorem 3.9(i) of [4].

(b)–(d) From Theorem 3.9 of [4], $R = Z_2(R_R) \oplus A \oplus \overline{S}$ (right ideal direct sum), $\overline{J}(R) = Z_2(R_R) \oplus A$, $K = A \oplus \overline{S}$ (ring direct sum), $M(R)$ is right essential in $\overline{S}$, and A contains no nonzero ideals of R. There exists $e = e^2 \in R$ such that $Z_2(R_R) \oplus \overline{S} = eR$ and $A = (1-e)R$. Observe that $r(eR) = 0$.

Assume condition (i). Let $f \in \text{Hom}_R(eR, (1-e)R)$. Since A is a nonsingular right R-module, it follows that $f(Z_2(R_R)) = 0$. Moreover, $f(\overline{S}) = 0$ because $K = A \oplus \overline{S}$ is a ring decomposition. So $\text{Hom}_R(eR, (1-e)R) = 0$. By Lemma 1.2(ii), $A = 0$. Therefore $\overline{J}(R) = Z_2(R_R)$ and $K = \overline{S}$.

Next observe that conditions (ii) or (iii) implies that K is a semisimple Artinian ring. If $A \neq 0$, there exists a minimal right ideal X of R such that $X \subseteq A$. Hence there exists $x = x^2$ such that $X = xR$. By Theorem 3.2 of [3], X is R-isomorphic to a nilpotent right ideal of R. But since R is right continuous, direct summands are only R-isomorphic to direct summands, a contradiction to the fact that $X \neq 0$. Hence $A = 0$. Therefore $\overline{J}(R) = Z_2(R_R)$ and $K = \overline{S}$.

Note that in the proof of Theorem 2.1, conditions (ii) and (iii) can be replaced with any condition which makes K a semisimple Artinian ring.

Corollary 2.2. *If R is right continuous and $Z_2(R_R) = cR$ for a central idempotent c, then $Z_2(R_R) = \overline{J}(R)$ and $(1-c)R = M(R)$.*

Proof. From the proof of Theorem 2.1, we have $K \oplus \overline{S}$ is a ring decomposition. Hence A is an ideal of R. But A contains no nonzero ideals of R. Hence $A = 0$. So $Z_2(R_R) = \overline{J}(R)$ and $K = M(R)$.

Observe that a semiprime right continuous ring satisfies the hypothesis of Corollary 2.2. Hence if there exists a semiprime right self-injective ring which is not regular, then there exists a semiprime right self-injective ring R such that $R = Z_2(R_R)$.

Corollary 2.3. *Let R be a continuous ring. If R satisfies any of the following conditions, then $R = \overline{J}(R) \oplus M(R)$ (ring direct sum) and $\overline{J}(R) = Z_2(R_R) = Z_2(_RR)$:*

(i) *R is an FI ring.*
(ii) *R is semiprime.*
(iii) *Every right semicentral idempotent is central.*
(iv) *Every left semicentral idempotent is central.*
(v) *R is GFC.*
(vi) *$R = \oplus_{i=1}^n e_i R$, where $\{e_i \mid i = 1, 2, \ldots, n\}$ is a complete set of orthogonal primitive idempotents.*
(vii) *$\text{Soc}(_RR)$ is left essential in R and $\text{Soc}(R_R)$ is right essential in R.*

Proof. Parts (i), (vi) and (vii) are consequences of Corollary 3.10 of [4] and Theorem 2.1. From Lemma 1.4, parts (ii), (iii), (iv) and (v) imply that R is an FI ring.

Proposition 2.4. *Let R be a right self-injective ring. If $e = e^2 \in R$ such that $r(eR) = 0$ and $Hom_R(eR, (1-e)R) = 0$, then $(1-e)R$ contains no nonzero minimal right ideals of R.*

Proof. Assume that $0 \neq X$ is a minimal right ideal of R with $X \subseteq (1-e)R$. Suppose there exists a nonzero homomorphism $f : X \to eR$. Let $E(X)$ denote the injective hull of X in $(1-e)R$. Now f can be extended to $\bar{f} : E(X) \to eR$. Since X is a minimal right ideal of R, $\bar{f}$ is a monomorphism. Hence there exists $c = c^2 \in eR$ such that $E(X) \cong cR$. But this contradicts the hypothesis, $Hom_R(eR, (1-e)R) = 0$. So $eR \cap H = 0$, where H is the homogeneous component of X. Since H is an ideal, it follows that $(eR)H = 0$. This contradicts $r(eR) = 0$. Therefore $(1-e)R$ contains no nonzero minimal right ideals of R.

Corollary 2.5. *If R is a right self-injective ring with $Soc(R_R)$ right essential in R, then R is a right FI ring.*

Proof. This result is a consequence of Proposition 2.4 and Lemma 1.2(ii).

Observe that the "right self-injective" condition in Proposition 2.4 and Corollary 2.5 can be relaxed to "right p-injective and right CS" with virtually the same proof.

3. GFC Rings

From Lemma 1.4(iv), we see that the class of GFC rings (which includes the strongly bounded rings and the FPF rings) forms an important subclass of the class of FI rings. In this section we investigate the condition $Z_2(R_R) = Z_2(_RR)$ in the class of GFC rings.

Proposition 3.1. *If R is a GFC ring, then $\overline{\mathcal{P}}(R) = Z_2(R_R) = Z_2(_RR)$.*

Proof. This result is a consequence of Proposition 3.11 of [4].

Proposition 3.2. *If R is a right GFC ring, then there exists an ideal I of R such that $I^2 = 0$, I is right essential in $\mathcal{P}(R)$ and $\mathcal{P}(R)$ is right essential in $Z_2(R_R) = \overline{\mathcal{P}}(R)$.*

Proof. Let $\{X_\alpha \mid \alpha \in \Lambda\}$ be the family of all nilpotent ideals of index 2. Using techniques of Fisher in [7], there is a maximal direct sum I of elements of this family. Then $I^2 = 0$. Assume that there exists a right ideal X of R such that $X \cap I = 0$ and $X \oplus I$ is right essential in $\mathcal{P}(R)$.

Observe that X contains no nonzero ideals of R. By Proposition 3.11 of [4], there exists an ideal S such that $(I \oplus X) \cap S = 0$ and $I \oplus X \oplus S$ is right essential in R. Since X contains no nonzero ideals of R, it follows that R/X is faithful. Hence R/X is a generator. By Lemma 2 of [10], $I \oplus S$ is right essential in R. Hence $X = 0$. So I is right essential in $\mathcal{P}(R)$. By Proposition 3.11 of [4], $\mathcal{P}(R)$ is right essential in $Z_2(R_R) = \overline{\mathcal{P}}(R)$.

Note that the right FPF condition is Morita invariant by [6, Theorem 1.2D].

Theorem 3.3. *R is a right FPF ring if and only if $\mathrm{Mat}_n(R)$ is right GFC for every positive integer n.*

Proof. Assume R is right FPF. Then since the right FPF condition is Morita invariant, $\mathrm{Mat}_n(R)$ is right FPF and hence it is right GFC for every positive integer n.

Conversely, to show that R is right FPF, let $M = m_1 R + m_2 R + \cdots + m_n R$ be a finitely generated faithful right R-module with generators $m_1, m_2, \ldots, m_n$. Now let $N = M \times M \times \cdots \times M$ (n-times). For $x = (x_1, x_2, \ldots, x_n) \in N$ and $\alpha = (a_{ij}) \in \mathrm{Mat}_n(R)$, define

$$x \cdot \alpha = (\sum_i x_i a_{i1}, \sum_i x_i a_{i2}, \ldots, \sum_i x_i a_{in})$$

as an element in N. Then N is a right module over the ring $\mathrm{Mat}_n(R)$ under this scalar multiplication. In this situation, since $M = m_1 R + m_2 R + \cdots + m_n R$, it follows that N is a cyclic module over the ring $\mathrm{Mat}_n(R)$ generated by the element $(m_1, m_2, \ldots, m_n)$ in N. Also it can be easily checked that N is a faithful right $\mathrm{Mat}_n(R)$-module. Since $\mathrm{Mat}_n(R)$ is right GFC, N is a generator for the category Mod-$\mathrm{Mat}_n(R)$. Therefore for a set Λ such that there is a $\mathrm{Mat}_n(R)$-module epimorphism f from $N^{(\Lambda)}$ to $\mathrm{Mat}_n(R)$, where $N^{(\Lambda)}$ denotes the direct sum of the $|\Lambda|$ copies of N. So f is an R-module epimorphism. Now since $N \cong M \oplus \cdots \oplus M$ (n-times) as R-modules, there is a set A such that $M^{(A)} \cong N^{(\Lambda)}$ as R-modules and f is an R-module epimorphism from $M^{(A)}$ to $\mathrm{Mat}_n(R)$. Thus M is a generator for the category of Mod-R. Consequently R is right FPF.

We note that it was shown in [8] and [11] that a regular ring R is right FPF if and only if $\mathrm{Mat}_n(R)$ is right GFC for some $n \geq 2$.

The following fact may be known, but for the sake of completeness we provide its proof.

Lemma 3.4. *For a ring R and a positive integer n, let $S = \mathrm{Mat}_n(R)$. Then we have that $Z_2(S_S) = \mathrm{Mat}_n(Z_2(R_R))$ and $Z_2(_SS) = \mathrm{Mat}_n(Z_2(_RR))$.*

Proof. Firstly we show that $Z(S_S) = \mathrm{Mat}_n(Z(R_R))$. For $a \in R$, aE_{ij} denotes the matrix in $\mathrm{Mat}_n(R)$ with a in (i, j)-position and 0 otherwise.

Let $a \in Z(R_R)$. Then $\text{Mat}_n(r_R(a)) \subseteq r_S(aE_{ij})$ for all i, j. Then it can be easily checked that $\text{Mat}_n(r_R(a))$ is a right essential right ideal of S and so is $r_S(aE_{ij})$. Thus $aE_{ij} \in Z(S_S)$ for all i, j. Therefore $\text{Mat}_n(Z(R_R)) \subseteq Z(S_S)$. Conversely, assume that $(a_{ij}) \in Z(S_S)$. Since $Z(S_S)$ is an ideal, $E_{1i}(a_{ij})E_{j1} = a_{ij}E_{11} \in Z(S_S)$. Then

$$r_S(a_{ij}E_{11}) = \begin{pmatrix} r_R(a_{ij}) & r_R(a_{ij}) & \cdots & r_R(a_{ij}) \\ R & R & \cdots & R \\ \vdots & \vdots & \ddots & \vdots \\ R & R & \cdots & R \end{pmatrix}$$

is a right essential right ideal of S. Let $0 \neq x \in R$. Then there exists $(b_{ij}) \in S$ such that $0 \neq xE_{11}(b_{ij}) \in r_S(a_{ij}E_{11})$. Hence there is a positive integer k such that $0 \neq xb_{1k} \in r_R(a_{ij})$. Thus $r_R(a_{ij})$ is right essential in R. So $a_{ij} \in Z(R_R)$ and hence $(a_{ij}) \in \text{Mat}_n(Z(R_R))$. Therefore $Z(S_S) = \text{Mat}_n(Z(R_R))$.

Now for proving that $Z_2(S_S) = \text{Mat}_n(Z_2(R_R))$, let (a_{ij}) be a matrix in $Z_2(S_S)$. Since $Z_2(S_S)$ is an ideal, $E_{1i}(a_{ij})E_{j1} = a_{ij}E_{11} \in Z_2(S_S)$. Hence there exists a right essential right ideal K of S such that $a_{ij}E_{11}K \subseteq Z(S_S)$. Let K_{ij} be the set of (i,j)-components of K. Then $K_{1j} = K_{1k}$ for all j, k and K_{1j} is right essential in R. So $a_{ij}K_{1j} \subseteq Z(R_R)$. Thus $a_{ij} \in Z_2(R_R)$ and so $Z_2(S_S) \subseteq \text{Mat}_n(Z_2(R_R))$. Conversely, let $a \in Z_2(R_R)$. Then there is a right essential right ideal K of R such that $aK \subseteq Z(R_R)$. In this case $\text{Mat}_n(K)$ is a right essential right ideal of S and also note that $aE_{ij}\text{Mat}_n(K) \subseteq \text{Mat}_n(Z(R_R)) = Z(S_S)$. So $aE_{ij} \in Z_2(S_S)$ and hence $\text{Mat}_n(Z_2(R_R)) \subseteq Z_2(S_S)$. Consequently, $Z_2(S_S) = \text{Mat}_n(Z_2(R_R))$.

Similarly we also can prove that $Z_2(_SS) = \text{Mat}_n(Z_2(_RR))$.

Observe that Proposition 3.1, Theorem 3.3 and Lemma 3.4 allow us to construct non-GFC rings R for which $Z_2(R_R) = Z_2(_RR)$ in the following example.

Example 3.5. Let A be a GFC ring which is not FPF (e.g., any commutative non-Prüfer domain). By Theorem 3.3 there exists a positive integer $n > 1$ such that $R = \text{Mat}_n(A)$ is not GFC. By Proposition 3.1, $Z_2(A_A) = Z_2(_AA)$. Therefore by Lemma 3.4, $Z_2(R_R) = Z_2(_RR)$.

Acknowledgments

The authors wish to thank the referee for suggestions and especially for the equivalence of conditions (i) and (v) of Lemma 1.2. The second author was partially supported by KOSEF Research Grant 96-k3-0101 (RCAA) and the Basic Science Research Institute Program, Ministry of Education, Korea in 1996, Project No. BSRI-96-1432, while the third author was supported in part by KOSEF Research Grant 96-k3-0101 (RCAA) and the

Basic Science Research Institute Program, Ministry of Education in 1996, Project No. BSRI-96-1402.

References

1. G. F. Birkenmeier, *Idempotents and completely semiprime ideals*, Comm. Algebra **11** (1983), 567–580.
2. G. F. Birkenmeier, *A generalization of FPF rings*, Comm. Algebra **17** (1989), 855–884.
3. G. F. Birkenmeier, *Rings which are essentially supernilpotent*, Comm. Algebra **22** (1994), 1063–1082.
4. G. F. Birkenmeier, *When does a supernilpotent radical essentially split off?*, J. Algebra **172** (1995), 49–60.
5. C. Faith, *The maximal regular ideal of self-injective and continuous rings splits off*, Arch. Math. **44** (1985), 511–521.
6. C. Faith and S. Page, *"FPF Ring Theory: Faithful Modules and Generators of Mod-R"*, London Math. Soc. Lecture Notes Series, Vol. **88**, Cambridge Univ. Press, Cambridge, 1984.
7. J. W. Fisher, *On the nilpotency of nil subrings*, Canad. J. Math. **22** (1970), 1211–1216.
8. S. Kobayashi, *On regular rings whose cyclic faithful modules are generators*, Math. J. Okayama Univ. **30** (1988), 45–52.
9. B. J. Müller, *Problem 1*, Ring Theory, Proceedings of the Biennial Ohio State–Denison Conference 1992, eds. S. K. Jain and S. T. Rizvi, World Scientific, Singapore, 1993.
10. H. Yoshimura, *On finitely pseudo-Frobenius rings*, Osaka J. Math. **28** (1991), 285–294.
11. H. Yoshimura, *On rings whose cyclic faithful modules are generators*, Osaka J. Math. **32** (1995), 591–611.

DEPARTMENT OF MATHEMATICS, UNIVERSITY OF SOUTHWESTERN LOUISIANA, LAFAYETTE, LOUISIANA 70504, U. S. A.

DEPARTMENT OF MATHEMATICS, KYUNG HEE UNIVERSITY, SUWON 449-701, SOUTH KOREA

DEPARTMENT OF MATHEMATICS, BUSAN NATIONAL UNIVERSITY, BUSAN 609-735, SOUTH KOREA

DECOMPOSITIONS OF D1 MODULES

ROBERT A. BROWN AND MARY H. WRIGHT

ABSTRACT. Continuity and quasicontinuity for modules may be viewed as generalizations of quasi-injectivity. A key property of quasi-continuous modules is that complements are summands. Modules with this special property are called extending modules or C1 modules. We investigate decomposition properties of dual-extending (D1) modules, those modules which are supplemented and for which each supplement is a summand. The notions of hollowness and dual Goldie dimension play a prominent role. Our results are analogous to results for extending modules developed by Camillo and Yousif.

Generalizing results known for continuous and quasicontinuous modules, Camillo and Yousif showed in [1] that if M is a C1 module whose socle has finite Goldie dimension then M may be decomposed as the direct sum of a finite dimensional module and a semisimple module. Much of their work depends on the properties of essentially closed modules. In section 1 we develop a dual notion, the concept of the interior of a submodule. This in turn leads to the definition of open submodules. In section 3 we show that the concept of dual Goldie dimension developed by Grzeszczuk and Puczylowski allows us to formulate a decomposition criterion involving the Jacobson radical of the module.

All rings are assumed to have unity. Unless stated otherwise, all modules are assumed to be unitary right R-modules. We follow the convention that the intersection of the empty family of submodules of a module M is M itself.

The symbol $<$ denotes proper inclusion; where equality is permitted we write $\leq$. The set of all elements belonging to set A but not to B is denoted $A - B$. If $a \in A$ we write $A - a$ rather than $A - \{a\}$.

1. Supplements and Interiors

In this section we use the concept of smallness, which is dual to essentiality, to develop a concept dual to essential closure. We begin with some basic definitions. See [8].

1991 *Mathematics Subject Classification*. Primary 16D70.

Typeset by $\mathcal{A}_{\mathcal{M}}\mathcal{S}$-TEX

Definition 1.1. $A \leq M$ is <u>small in M</u>, denoted $A \stackrel{\circ}{\hookrightarrow} M$, provided that, for each $B \leq M$, $A + B = M$ implies that $B = M$. M is <u>hollow</u> if every proper submodule of M is small.

Definition 1.2. Let $A \leq M$. B is a <u>supplement</u> of A in M provided that B is minimal among submodules C of M such that $A + C = M$.

Lemma 1.3. *Suppose that $B + B^\bullet = M$. Then $B^\bullet$ is a supplement of B in M if and only if $B \cap B^\bullet \stackrel{\circ}{\hookrightarrow} B^\bullet$.*

Proof. $\Rightarrow$: Suppose that $B \cap B^\bullet + C = B^\bullet$. Then $M = B + B^\bullet = B + B \cap B^\bullet + C = B + C$. Since $C \leq B^\bullet$ and $B^\bullet$ is a supplement of B in M, we have $C = B^\bullet$.

$\Leftarrow$: Suppose that $C \leq B^\bullet$ and $B + C = M$. Then $B^\bullet \leq B \cap B^\bullet + C \leq B^\bullet$. Since $B \cap B^\bullet \stackrel{\circ}{\hookrightarrow} B^\bullet$, $C = B^\bullet$.

Recall that $C \leq M$ is a <u>complement</u> of $A \leq M$ if C is maximal among submodules $B \leq M$ such that $A \cap B = 0$. A Zorn's lemma argument shows that every submodule of a given module has a complement in that module.

Definition 1.4. A module M is <u>supplemented</u> if for any two submodules A and B with $A + B = M$, B contains a supplement of A in M.

The notion of supplement is dual to that of complement. Note that not every submodule of M need have a supplement. Hence, not all modules are supplemented.

Definition 1.5. A module M is <u>semiprimitive</u> provided $\operatorname{Rad} M = 0$. If every factor of M is semiprimitive, we say that M is <u>completely semiprimitive</u>.

Proposition 1.6. *Let M be a supplemented module. If M is semiprimitive then M is semisimple.*

Proof. Let $B \leq M$. Since M is supplemented, there exists $B^\bullet \leq M$ such that $B + B^\bullet = M$ and $B \cap B^\bullet \stackrel{\circ}{\hookrightarrow} B^\bullet$. Since M is semiprimitive, it has no nonzero small submodule. Hence $M = B \oplus B^\bullet$.

Thus every submodule of M is a summand.

Since any semisimple module is semiprimitive, the proposition above shows that for supplemented modules the notions of semiprimitivity and semisimplicity are equivalent.

The following proposition, whose proof is straightforward, gathers some relevant facts about supplemented modules.

Proposition 1.7. *Any factor of a supplemented module is supplemented. Any summand of a supplemented module is supplemented. If M is a supplemented module then $\frac{M}{\operatorname{Rad} M}$ is semisimple.*

If $A \leq X \leq M$ then the notation $A \stackrel{*}{\hookrightarrow} X$ denotes that A is an essential submodule of X, and we say that X is an <u>essential extension</u> of A in M. If

$A < X$ then the extension is proper. Recall that if A has no proper essential extension in M, we say that A is (essentially) <u>closed</u> in M. If $A \stackrel{*}{\hookrightarrow} C \leq M$ and C is closed in M, then C is an essential <u>closure</u> of A in M. A submodule is closed if and only if it is a complement. These ideas are discussed in [8]. For supplemented modules we have the following concepts, which are dual to these notions.

Definition 1.8. Let M_R be a supplemented module. Given $B \leq M$, let $B^\bullet$ be a supplement of B in M, and let $B^{\bullet\bullet}$ be a supplement of $B^\bullet$ in M such that $B^{\bullet\bullet} \leq B$. Any $B^{\bullet\bullet}$ so constructed is called an <u>interior of B in M</u>.

When no confusion is likely we shall omit reference to the larger module M.

The closure of a submodule $A \leq M$ is maximal among essential extensions of A in M. The duality between the notions of submodule and factor module and between smallness and essentiality suggests the following minimality property for interiors:

Proposition 1.9. *Let M_R be a supplemented module and let $B \leq M$. Let $B^{\bullet\bullet}$ be an interior of B in M. Then*

$$\frac{B}{B^{\bullet\bullet}} \stackrel{\circ}{\hookrightarrow} \frac{M}{B^{\bullet\bullet}}, \tag{1}$$

If $Y \leq B^{\bullet\bullet}$ is such that

$$\frac{B}{Y} \stackrel{\circ}{\hookrightarrow} \frac{M}{Y}, \tag{2}$$

then $Y = B^{\bullet\bullet}$.

Proof. (1): Suppose $B^{\bullet\bullet} \leq U \leq M$ and $\frac{B}{B^{\bullet\bullet}} + \frac{U}{B^{\bullet\bullet}} = \frac{M}{B^{\bullet\bullet}}$. Then $M = B + U$. Since $M = B^\bullet + B^{\bullet\bullet}$, we have $U = B^{\bullet\bullet} + (B^\bullet \cap U)$, so $M = B + (B^\bullet \cap U)$. Since $B^\bullet$ is a supplement of B in M, $B^\bullet \cap U = B^\bullet$, i.e. $B^\bullet \leq U$. Hence $M = B^\bullet + B^{\bullet\bullet} \leq U$ so $U = M$.

(2): $\frac{B}{Y} + \frac{B^\bullet + Y}{Y} = \frac{M}{Y}$ so $\frac{B^\bullet + Y}{Y} = \frac{M}{Y}$. Hence $B^\bullet + Y = M$. Since $B^{\bullet\bullet}$ is a supplement of $B^\bullet$ in M, $Y = B^{\bullet\bullet}$.

Closed submodules are equal to their closures. We define the notion of an open submodule and show that a module is open if and only if it equals any of its interiors.

Definition 1.10. Let $B \leq M$. We say that B is <u>open in M</u> if there exists no proper submodule C of B such that

$$\frac{B}{C} \stackrel{\circ}{\hookrightarrow} \frac{M}{C}.$$

Proposition 1.11. Let $B \leq M$ and suppose that $B^\bullet$ is a supplement of B in M. Then $B^\bullet$ is open in M.

Proof. Suppose $C \leq B^\bullet$ with $\dfrac{B^\bullet}{C} \overset{\circ}{\hookrightarrow} \dfrac{M}{C}$. Then $\dfrac{B^\bullet}{C} + \dfrac{B+C}{C} = \dfrac{M}{C}$ so $\dfrac{B+C}{C} = \dfrac{M}{C}$. Hence $B+C = M$, so $C = B^\bullet$ since $B^\bullet$ is a supplement of B in M.

Corollary 1.12. Let M_R be a supplemented module and let $B \leq M$. Then B is open in M if and only if: Given any supplement $B^\bullet$ of B in M, if we take any supplement $B^{\bullet\bullet}$ of $B^\bullet$ in M with $B^{\bullet\bullet} \leq B$, then $B = B^{\bullet\bullet}$.

Proof. $\Rightarrow$: Let $B^{\bullet\bullet}$ be an interior of B. By Proposition 1.9, $\dfrac{B}{B^{\bullet\bullet}} \overset{\circ}{\hookrightarrow} \dfrac{M}{B^{\bullet\bullet}}$. Since B is open, $B = B^{\bullet\bullet}$.
$\Leftarrow$: Immediate from the proposition.

2. Meet-Independence and Dual Dimension

In this section we define the notion of dual Goldie dimension, which is based on the concept of meet-independent families of submodules of a module. This dual notion was developed by Grzeszczuk and Puczylowski in [4]. Their work extended and modified ideas of Fleury [2] and Varadarajan [9]. Recall that a module M is said to have <u>finite Goldie dimension</u> provided that it does not contain an infinite direct sum of submodules. For any such module M there exists a unique nonnegative integer n such that M contains a direct sum of n submodules but no direct sum of $n+1$ submodules. This number n is called the Goldie dimension of M. Equivalently, a module M has Goldie dimension n if and only if there exist submodules $A_1, A_2, \ldots, A_n$ such that $\sum_{i=1}^n A_i$ is direct, $\oplus_{i=1}^n A_i \overset{*}{\hookrightarrow} M$, and each A_i is a uniform submodule of M.

A family $\{X_1, \ldots, X_n\}$ of submodules of a module M is <u>independent</u> provided that the sum $\sum_{i=1}^n X_i$ is direct. Equivalently, for each $1 \leq j \leq n$, $X_j \cap \left(\sum_{i \neq j} X_i\right) = 0$. The following definition dualizes this concept.

Definition 2.1. A finite family $\{A_1, A_2, \ldots, A_n\}$ of proper submodules of M is <u>meet-independent</u> provided that for each $1 \leq k \leq n$ we have

$$A_k + \bigcap_{i \neq k} A_i = M.$$

An arbitrary family of submodules of M is meet-independent if every finite subfamily is meet-independent in M.

The notion of meet-independence may be extended to any modular lattice. This approach is taken in [4]. In particular, the following is a consequence of observation B of [4, p. 148].

Lemma 2.2. *Suppose* $\{A_i\}_{i \in I}$ *is a meet-independent family of proper submodules of* M *and* $B < M$ *is such that* $B + \bigcap_{i \in F} A_i = M$ *for each finite* $F \leq I$. *Then* $\{B\} \cup \{A_i\}_{i \in I}$ *is also meet-independent.*

Proposition 2.3. *The family* $\{A_1, A_2, \ldots, A_n\}$ *of submodules of* M *is meet-independent if and only if the canonical monomorphism* $\dfrac{M}{\bigcap_{i=1}^n A_i} \xrightarrow{\alpha} \bigoplus_{i=1}^n \dfrac{M}{A_i}$ *is an isomorphism.*

Proof. The result clearly holds for families of size 1. Consider a family $\{A, B\}$ of size 2. If $\{A, B\}$ is meet-independent, then $M = A + B$. Hence, given any $m_1, m_2 \in M$, we can write $m_1 = a_1 + b_1$ and $m_2 = a_2 + b_2$. Letting $m = a_2 + b_1$, we see that $m - m_1 \in A$ and $m - m_2 \in B$. Thus the canonical monomorphism in question is surjective. Conversely, if the monomorphism is surjective then given $m \in M$ there exists $m' \in B$ such that $m - m' \in A$. Hence $M = A + B$.

Now assume that the proposition holds for all families of cardinality k where $k \geq 2$ is some fixed integer. Suppose that $\{A_1, A_2, \ldots, A_{k+1}\}$ is a meet-independent family. Since $\left\{\bigcap_{i=1}^k A_i, A_{k+1}\right\}$ is meet-independent, we have a canonical isomorphism $\dfrac{M}{\bigcap_{i=1}^{k+1} A_i} \xrightarrow{\alpha} \dfrac{M}{\bigcap_{i=1}^k A_i} \oplus \dfrac{M}{A_{k+1}}$. Since $\{A_1, \ldots, A_k\}$ is meet-independent, by assumption we have a canonical isomorphism $\dfrac{M}{\bigcap_{i=1}^k A_i} \xrightarrow{\beta} \bigoplus_{i=1}^k \dfrac{M}{A_i}$. Letting ι denote the identity map on $\dfrac{M}{A_{k+1}}$, the canonical monomorphism from $\dfrac{M}{\bigcap_{i=1}^{k+1} A_i}$ to $\bigoplus_{i=1}^{k+1} \dfrac{M}{A_i}$ may be expressed as $\eta = (\beta \oplus \iota)\alpha$. Hence η is surjective.

Conversely, suppose that the canonical monomorphism $\dfrac{M}{\bigcap_{i=1}^{k+1} A_i} \xrightarrow{\gamma} \bigoplus_{i=1}^{k+1} \dfrac{M}{A_i}$ is surjective. Hence the map β described above is also surjective. Thus, by our induction assumption, $\{A_1, A_2, \ldots, A_k\}$ is meet-independent.

Next let $m_1, m_2 \in M$. Consider $(m_1 + A_1, \ldots, m_1 + A_k, m_2 + A_{k+1})$. Since β is surjective, there exists $m \in M$ such that $m - m_1 \in \bigcap_{i=1}^k A_i$ and $m - m_2 \in A_{k+1}$. Thus $\{\bigcap_{i=1}^k A_i, A_{k+1}\}$ is meet-independent. Therefore $A_{k+1} + \bigcap_{i=1}^k A_i = M$, and $\{A_1, \ldots, A_{k+1}\}$ is meet-independent.

The notion of meet-independence motivates a definition of dual-Goldie dimension. The following was proved in a lattice-theoretic setting by Grzeszczuk and Puczylowski. See [4, Thm. 9 and Cor. 13].

Proposition 2.4. *For any module M the following are equivalent:*
(1) *There exists no infinite meet-independent family of submodules of M.*
(2) *There exists a positive integer n and proper submodules $A_1, A_2, \ldots, A_n$ such that the following hold:*
   (a) *$\{A_1, A_2, \ldots, A_n\}$ is meet-independent.*
   (b) *$\bigcap_{i=1}^{n} A_i \stackrel{\circ}{\hookrightarrow} M$.*
   (c) *$\dfrac{M}{A_i}$ is hollow for each $1 \leq i \leq n$.*
(3) *$\sup\{k : \text{there exists a meet-independent family of cardinality } k \text{ consisting of proper submodules of } M\} = n < \infty$.*
(4) *For any sequence*
$$X_1 \geq X_2 \geq \cdots$$
of submodules of M, there exists j such that
$$\frac{X_j}{X_k} \stackrel{\circ}{\hookrightarrow} \frac{M}{X_k} \text{ for all } k \geq j.$$

Definition 2.5. A module M satisfying conditions (2) or (3) of the proposition above is said to have <u>finite dual Goldie dimension n.</u>

The next proposition shows that taking interiors of modules comprising a meet-independent family maintains the meet-independence.

Proposition 2.6. *Suppose $\{A_i\}_{i \in I}$ is a meet-independent family of submodules of M, and let $IntA_i$ denote some interior of A_i in M. Then $\{IntA_i\}_{i \in I}$ is also a meet-independent family of submodules of M.*

Proof. Let $\{i_1, i_2, \ldots, i_k\}$ be a finite subset of I. Consider the following commutative diagram:

$$\begin{array}{ccc}
\dfrac{M}{\bigcap_{j=1}^{k} IntA_{i_j}} & \stackrel{\alpha}{\longrightarrow} & \dfrac{M}{IntA_{i_1}} \times \cdots \times \dfrac{M}{IntA_{i_k}} \\
\sigma \downarrow & & \downarrow \tau \\
\dfrac{M}{\bigcap_{j=1}^{k} A_{i_j}} & \stackrel{\eta}{\longrightarrow} & \dfrac{M}{A_{i_1}} \times \cdots \times \dfrac{M}{A_{i_k}}
\end{array}$$

The maps α and η are the canonical monomorphisms, while σ and τ are epimorphisms defined by

$$\sigma(m + \bigcap_{j=1}^{k} IntA_{i_j}) = m + \bigcap_{j=1}^{k} A_{i_j}$$

and

$$\tau(m_1 + IntA_{i_1}, \ldots, m_k + IntA_{i_k}) = (m_1 + A_{i_1}, \ldots, m_k + A_{i_k}).$$

It is easily checked that $\tau\alpha = \eta\sigma$ and $Ker\,\tau = \dfrac{A_{i_1}}{IntA_{i_1}} \times \cdots \times \dfrac{A_{i_k}}{IntA_{i_k}}$. Now for each $j = 1,\ldots,k$, $\dfrac{A_{i_j}}{IntA_{i_j}} \overset{\circ}{\hookrightarrow} \dfrac{M}{IntA_{i_j}}$. Hence $Ker\,\tau \overset{\circ}{\hookrightarrow} \dfrac{M}{IntA_{i_1}} \times \cdots \times \dfrac{M}{IntA_{i_k}}$. Since $\{A_i\}_{i\in I}$ is a meet-independent family, η is an isomorphism and $\tau\alpha = \eta\sigma$ is surjective. Since $Ker\,\tau$ is small, α is also surjective.

Thus every finite subfamily of $\{IntA_i\}_{i\in I}$ is meet-independent, so the family is itself meet-independent.

Various formulas involving dual Goldie dimension were developed by Haack in [5]. We shall need only the following proposition due to Varadarajan (cf. Prop. 4 of [5]).

Proposition 2.7. *Suppose that X and Y are modules having dual Goldie dimensions m and n respectively. Then $X \oplus Y$ has dual dimension $m + n$.*

3. Decomposition of D1 Modules

The results of this section are motivated by results of Camillo and Yousif in [1] concerning decompositions of extending modules.

Definition 3.1. *M is a D1 module provided that M is supplemented and every supplement in M is a summand of M.*

Definition 3.2. *M is eventually completely semiprimitive provided that given any countably infinite meet-independent family $\{X_1, X_2, \ldots\}$ there exists a positive integer n such that, for each $k \geq n$, $\dfrac{M}{X_k}$ is completely semiprimitive.*

In the remainder of this section, the module M is assumed to be supplemented.

Theorem 3.3. *Let M be a D1 module which is eventually completely semiprimitive. Then M may be decomposed as $M = K \oplus S$ where K has finite dual Goldie dimension and S is semisimple.*

The proof of this theorem begins with two reduction arguments which show that it suffices to prove the theorem for special cases in which more information about the structure of M is available.

Reduction 1. *Suppose that Theorem 3.3 holds with the additional assumption that $Rad\,M \overset{\circ}{\hookrightarrow} M$. Then it holds as stated.*

Proof. Let $S = Rad\,M$ and let K be a supplement of S in M. Now $Rad\,K \leq K \cap S \overset{\circ}{\hookrightarrow} K$. Moreover, since M is $D1$, K, being a supplement, is in fact a summand of M. Hence K is a $D1$ module which is eventually completely semiprimitive.

Say $M = K \oplus T$. We may assume without loss of generality that $T \leq S$. Then T has finite dual dimension. For suppose that $\{T_1, T_2, \dots\}$ is a meet-independent family of submodules of T. Since M is eventually completely semiprimitive, there exists a positive integer n such that $\dfrac{M}{T_k}$ is completely semiprimitive for all $k \geq n$. In particular, $\text{Rad}\,(\dfrac{M}{T_k}) = 0$ for all $k \geq n$. Thus, for each such k, $T \geq T_k \geq \text{Rad}\,M = S \geq T$, so $T_k = T$. Hence there is no infinite meet-independent family of proper submodules of T.

Since $\text{Rad}\,K \overset{\circ}{\hookrightarrow} K$, by hypothesis K admits a decomposition $K = A \oplus B$ where A has finite dual dimension and B is semisimple. Then $M = K \oplus T \cong A \oplus T \oplus B$. Since $A \oplus T$ has finite dual dimension, the proof is complete.

Lemma 3.4. *Suppose $G \leq M$ and $\dfrac{M}{I(G)}$ is semiprimitive for some interior $I(G) = \text{Int}\,G$ of G in M. Then G is open in M.*

Proof. Since $\dfrac{G}{I(G)} \overset{\circ}{\hookrightarrow} \dfrac{M}{I(G)}$, $\dfrac{G}{I(G)} \leq \text{Rad}\,\dfrac{M}{I(G)} = 0$. Hence $G = I(G)$.

Lemma 3.5. *Suppose G is a non-open maximal submodule of M and $I(G)$ is an interior of G in M. Then $\dfrac{M}{I(G)}$ is hollow.*

Proof. Suppose $I(G) < X, Y < M$ and $X + Y = M$. Since $\dfrac{M}{G}$ is simple, either $X + G = M$ or $Y + G = M$. Suppose the former holds. Since $\dfrac{G}{I(G)} \overset{\circ}{\hookrightarrow} \dfrac{M}{I(G)}$, we have $X = M$, a contradiction.

Reduction 2. *Suppose that Theorem 3.3 holds with the additional assumptions that $\text{Rad}\,M \overset{\circ}{\hookrightarrow} M$ and every maximal submodule of M is open. Then it holds as stated.*

Proof. We assume, in light of reduction 1, that M has small radical. If every maximal submodule of M is open, then M has a decomposition of the desired form. If not, let G_1 be a non-open maximal submodule of M, and let I_1 be an interior of G_1 in M. Since I_1 is open in M, we may write $M = H_1 \oplus I_1$. Now if I_1 has a non-open maximal submodule G_2, let I_2 be an interior of G_2 in I_1, and write $I_1 = H_2 \oplus I_2$. We continue in this manner, stopping only when we reach a stage at which every maximal submodule of I_n is open. We claim that such a stage must in fact be reached.

For if not, then, as shown in the next paragraph, $\{I_1, H_1 \oplus I_2, H_1 \oplus H_2 \oplus I_3, \dots\}$ is an infinite meet-independent family of submodules of M. Since M is eventually completely semiprimitive, there exists m such that for all $k \geq m$, $H_k \cong \dfrac{M}{H_1 \oplus \cdots \oplus H_{k-1} \oplus I_k}$ is semiprimitive. Now $H_k \cong \dfrac{I_{k-1}}{I_k}$, so, by the lemma, G_k is open in I_{k-1}, hence in M. This is a contradiction since by construction G_k is not open in M.

To prove the assertion of meet-independence made in the previous paragraph, first observe that, by a straightforward induction argument, $I_k = I_{k-1} \cap [H_1 \oplus H_2 \oplus \cdots \oplus H_{k-1} \oplus I_k]$ for all $k \geq 2$. Since $I_1 + (H_1 \oplus I_2) = M$, $\{I_1, H_1 \oplus I_2\}$ is meet-independent. Assume $\{I_1, H_1 \oplus I_2, \ldots, H_1 \oplus \cdots \oplus H_{k-1} \oplus I_k\}$ is meet- independent for some $k \geq 2$. We have

$$(H_1 \oplus \cdots \oplus H_k \oplus I_{k+1}) + I_1 \cap (H_1 \oplus I_2) \cap \cdots \cap (H_1 \oplus \cdots \oplus H_{k-1} \oplus I_k)$$
$$\geq H_1 \oplus \cdots \oplus H_k \oplus I_k = M.$$

Thus, by induction, we see that the infinite family is meet-independent.

Hence there does indeed exist a positive integer n such that $M = H_1 \oplus H_2 \oplus \cdots \oplus H_n \oplus I_n$ where every maximal submodule of I_n is open. Since I_n is D1 and eventually completely semiprimitive and has small radical, we may write $I_n = X \oplus Y$, where X has finite dual Goldie dimension and Y is semisimple. Now, by Lemma 3.5, each H_i is hollow, hence of dual dimension 1. Thus M has a decomposition of the type asserted by the theorem.

We proceed with the proof of Theorem 3.3 in the presence of the two reductions just completed. First we make several observations concerning the supplemented module M. Since $\dfrac{M}{\operatorname{Rad} M}$ is semisimple it admits a decomposition $\dfrac{M}{\operatorname{Rad} M} = \bigoplus_{i \in I} \dfrac{B_i}{\operatorname{Rad} M}$ where each summand is simple. The next proposition follows immediately from the decomposition of $\dfrac{M}{\operatorname{Rad} M}$.

Proposition 3.6. *With the notation above, for any $K \leq I$,*

(1) $\sum_{i \in K} B_i + \sum_{i \in I-K} B_i = M$
(2) $(\sum_{i \in K} B_i) \cap (\sum_{i \in I-K} B_i) = \operatorname{Rad} M$.

To simplify our notation, we define $X_K = \sum_{i \in I-K} B_i$. If $K = \{i\}$ is a singleton, we write simply X_i. By (2) above, $X_K \geq \operatorname{Rad} M$ for each $K \leq I$. Note also that for each $i \in I$, $\dfrac{M}{X_i} \cong \dfrac{B_i}{\operatorname{Rad} M}$, which is simple. Thus X_i is a maximal submodule of M.

Observe that if $K \leq K' \leq I$ then $I - K' \leq I - K$ so $X_{K'} \leq X_K$. With this notation, we have

Proposition 3.7.

$$X_K = \bigcap_{i \in K} X_i \text{ for each } K \leq I.$$

Proof. Let $m \in \bigcap_{i \in K} X_i$. Express $m + \operatorname{Rad} M$ as $m + \operatorname{Rad} M = b_{i_1} + \cdots + b_{i_n} + \operatorname{Rad} M$ where each $b_{i_k} \in B_{i_k}$. For any $i_0 \in K$ we have

$$\sum_{k=1}^{n} b_{i_k} + \operatorname{Rad} M \in \dfrac{X_{i_0}}{\operatorname{Rad} M} = \bigoplus_{i \neq i_0} \dfrac{B_i}{\operatorname{Rad} M}.$$

By uniqueness of representation, $i_k \neq i_0$ for $k = 1, \ldots, n$. Since this holds for each $i_0 \in K$, we see that for each $k = 1, \ldots, n$, we have $i_k \notin K$. Hence $b_{i_1}, \ldots, b_{i_n}$ are all in $\sum_{i \in I-K} B_i = X_K$. Therefore $m \in X_K + \operatorname{Rad} M = X_K$ since $X_K \geq \operatorname{Rad} M$.

Thus $\bigcap_{i \in K} X_i \leq X_K$. The reverse inclusion is clear.

Using the notation already introduced, we have the following result:

Proposition 3.8. *For any partition $\{I_\gamma\}_{\gamma \in \Gamma}$ of I, the associated family $\{X_{I_\gamma}\}_{\gamma \in \Gamma}$ is a meet-independent family of submodules of M.*

Proof. We must show that any finite subfamily is meet-independent. We proceed by induction on the size n of the subfamily.

Let $\gamma, \delta \in \Gamma$. We show that the canonical monomorphism

$$\frac{M}{X_{I_\gamma} \cap X_{I_\delta}} \to \frac{M}{X_{I_\gamma}} \times \frac{M}{X_{I_\delta}}$$

is surjective.

Let $m_1, m_2 \in M$. Now $M = X_{I_\gamma} + X_{I-I_\gamma} \leq X_{I_\gamma} + X_{I_\delta}$. Write $m_1 = x + y$ and $m_2 = x' + y'$ where x and x' belong to I_γ and y and y' belong to I_δ. Let $m = x' + y$. Then $m_1 - m \in I_\gamma$ and $m_2 - m \in I_\delta$. This completes the base step $n = 2$.

For the induction step assume that every subfamily of size $n - 1$ is meet-independent, and let $\{X_{I_{\gamma_1}}, \ldots, X_{I_{\gamma_n}}\}$ be a subfamily of size n. By the induction hypothesis and the case $n = 2$, we have the two canonical isomorphisms $\dfrac{M}{\bigcap_{i=1}^{n-1} X_{I_{\gamma_i}}} \cong \prod_{i=1}^{n-1} \dfrac{M}{X_{I_{\gamma_i}}}$ and $\dfrac{M}{\bigcap_{i=1}^{n} X_{I_{\gamma_i}}} \cong \dfrac{M}{\bigcap_{i=1}^{n-1} X_{I_{\gamma_i}}} \times \dfrac{M}{X_{I_{\gamma_n}}}$ Composing these two isomorphisms establishes the required meet-independence.

In summary, for any supplemented module M with $\dfrac{M}{\operatorname{Rad} M} = \bigoplus_{i \in I} \dfrac{B_i}{\operatorname{Rad} M}$, as above, and I infinite, we may partition I into countably many infinite subsets: $I = \bigcup_{n=1}^{\infty} I_n$. Then $\operatorname{Rad} M = \bigcap_{n=1}^{\infty} X_{I_n}$. By the last proposition, $\{X_{I_n}\}_{n=1}^{\infty}$ is a meet-independent family of submodules of M. Hence, by Proposition 2.6, $\{Int X_{I_n}\}$ is also meet-independent.

Definition 3.9. $D \leq M_R$ is <u>completely meet-irreducible in M</u> if whenever

$$D = \bigcap_{i \in I} C_i,$$

where each $C_i \leq M$, then $D = C_j$ for some $j \in I$.

Proposition 3.10. *Suppose M is a module with the property that every completely meet-irreducible submodule is maximal. Then M is semiprimitive.*

Proof. Suppose, by way of contradiction, that $0 \neq x \in \text{Rad}\, M$. Since x belongs to every maximal submodule of M, $x \in D$ for each completely meet-irreducible submodule $D \leq M$. By Zorn's lemma there exists a submodule $D' \leq M$ which is maximal among submodules of M not containing x.

We claim that D' is itself completely meet-irreducible. For suppose not; say $D' = \bigcap A_i{}_{i \in I}$ with $D' < A_i \leq M$ for each $i \in I$. Then $x \in A_i$ for each $i \in I$ by the maximality property of D'. But then $x \in D'$, a contradiction.

Since x belongs to every completely meet-irreducible submodule of M, in particular it belongs to D'. This contradicts the construction of D' and shows that in fact $\text{Rad}\, M = 0$.

Now recall our assumptions that M is eventually completely semiprimitive, has small radical, and has no non-open maximal submodule. We show that every completely meet-irreducible submodule $D < M$ is maximal.

Let $D^\bullet$ be a supplement of D in M. Thus $M = D + D^\bullet$ and $D \cap D^\bullet \overset{\circ}{\hookrightarrow} D^\bullet$. Moreover, $\dfrac{M}{D} = \dfrac{D + D^\bullet}{D} \cong \dfrac{D^\bullet}{D \cap D^\bullet}$.

Since M is D1, $D^\bullet$ is a summand of M. As such it inherits various properties from M. In particular, $D^\bullet$ is D1 and eventually completely semiprimitive, has small radical, and has no non-open maximal submodules.

Since D is completely meet-irreducible in M, $\dfrac{M}{D}$ has a unique smallest submodule, namely the intersection of all of its nonzero submodules. Therefore $\dfrac{D^\bullet}{D \cap D^\bullet}$ also has a unique smallest submodule. Hence $D \cap D^\bullet$ is completely meet-irreducible in $D^\bullet$.

We apply the results summarized above to $D^\bullet$, retaining the notation introduced earlier. Suppose that $\dfrac{D^\bullet}{\text{Rad}\, D^\bullet} = \bigoplus_{i \in I} \dfrac{B_i}{\text{Rad}\, D^\bullet}$ where the summands are simple and I is infinite. We have $\text{Rad}\, D^\bullet = \bigcap_{n=1}^\infty X_{I_n}$ where the I_ns form a partition of I into infinite subsets.

Now since $D^\bullet$ is eventually completely semiprimitive, for all j sufficiently large we have $\text{Rad}\left(\dfrac{D^\bullet}{Int X_{I_j}}\right) = 0$. But $\dfrac{X_{I_j}}{Int X_{I_j}}$ is a small submodule of $\dfrac{D^\bullet}{Int X_{I_j}}$, so $X_{I_j} = Int X_{I_j}$ for all j sufficiently large, say for $j \geq k$.

Write $D^\bullet = X_{I_k} \oplus Y_{I_k}$. Since $D \cap D^\bullet$ is small in $D^\bullet$, $D \cap D^\bullet \leq \text{Rad}\, D^\bullet \leq X_{I_k}$. Suppose that $x = d + y$ where $x \in X_{I_k}$, $d \in D \cap D^\bullet$, and $y \in Y_{I_k}$. Then $y = x - d \in X_{I_k} \cap Y_{I_k} = 0$, so $x = d \in D \cap D^\bullet$. Therefore $D \cap D^\bullet = X_{I_k} \cap (D \cap D^\bullet + Y_{I_k})$. Since $D \cap D^\bullet$ is completely meet-irreducible in $D^\bullet$, one of the following two cases must occur:

Case 1: $D \cap D^\bullet = D \cap D^\bullet + Y_{I_k}$. Then $Y_{I_k} \leq D \cap D^\bullet \leq X_{I_k}$, so $Y_{I_k} = 0$ and $D^\bullet = X_{I_k}$. Since X_{I_k} is a maximal submodule of $D^\bullet$, this gives a contradiction, showing that case 1 cannot occur.

Case 2: $D \cap D^\bullet = X_{I_k} = \bigcap_{i \in I_k} X_i$. Again using the complete meet-irreducibility of $D \cap D^\bullet$ in $D^\bullet$, we conclude that $D \cap D^\bullet = X_{i_0}$ for some $i_0 \in I_k$. Hence $X_{i_0} \leq \operatorname{Rad} D^\bullet$. But the maximality of X_{i_0} implies that the reverse inclusion also holds. Being maximal, X_{i_0}, is open in $D^\bullet$, and therefore a summand. Since $X_{i_0} = \operatorname{Rad} D^\bullet$ is small in $D^\bullet$, $\operatorname{Rad} D^\bullet = 0$. Hence $D^\bullet$ is simple and D is a maximal submodule of M.

Thus if $\dfrac{D^\bullet}{\operatorname{Rad} D^\bullet} = \bigoplus_{i \in I} \dfrac{B_i}{\operatorname{Rad} D^\bullet}$ where I is infinite, then D is maximal. It remains to consider the case where I is finite, say $I = \{1, 2, \ldots, n\}$.

Consider $\dfrac{M}{D} \cong \dfrac{D^\bullet}{D \cap D^\bullet}$. Since $D \cap D^\bullet \leq \operatorname{Rad} D^\bullet$, we have $\operatorname{Rad}\left(\dfrac{D^\bullet}{D \cap D^\bullet}\right) = \dfrac{\operatorname{Rad} D^\bullet}{D \cap D^\bullet}$. Since $D^\bullet$ has small radical, so also does $\dfrac{D^\bullet}{D \cap D^\bullet}$. It follows that $\operatorname{Rad}\left(\dfrac{M}{D}\right) \hookrightarrow \dfrac{M}{D}$.

$\dfrac{\frac{M}{D}}{\operatorname{Rad}\left(\frac{M}{D}\right)} \cong \dfrac{D^\bullet}{\operatorname{Rad}(D^\bullet)}$, so $\dfrac{\frac{M}{D}}{\operatorname{Rad}\left(\frac{M}{D}\right)}$ is a direct sum of finitely many simple modules. Therefore $\operatorname{Rad}\left(\dfrac{M}{D}\right)$ is a finite meet-independent intersection of maximal submodules:

$$\operatorname{Rad}\left(\dfrac{M}{D}\right) = \dfrac{C_1}{D} \cap \cdots \cap \dfrac{C_n}{D}.$$

Now maximal submodules of M are open in M and are therefore summands of M. Write $M = C_i \oplus T_i$ for $i = 1, \ldots, n$. Then $\dfrac{M}{D} = \dfrac{C_i}{D} \oplus \dfrac{T_i + D}{D}$. Hence $C_i \cap (T_i + D) = D$. Since D is completely meet-irreducible, either $D = C_i$, which is a maximal submodule of M, or $D = T_i + D$. In the latter case, $T_i \leq D \leq C_i$, so $T_i = T_i \cap C_i = 0$. But then $C_i = M$, a contradiction.

Thus D is maximal whether I is finite or infinite. Since every completely meet-irreducible submodule of M is maximal, M is semiprimitive. This completes the proof of Theorem 3.3. ∎

The next proposition gives a condition sufficient for M to be eventually completely semiprimitive.

Proposition 3.11. *Suppose that M is supplemented and $\operatorname{Rad} M$ has finite dual Goldie dimension. Then M is eventually completely semiprimitive.*

Proof. Let n be the dual Goldie dimension of $\operatorname{Rad} M$. Suppose $\{M_1, M_2, \ldots\}$ is a meet-independent family of submodules of M. We claim that at most n of these submodules fail to contain $\operatorname{Rad} M$. For suppose not, i.e. suppose, re-numbering if necessary, that, for each $i = 1, 2, \ldots, n+1$, M_i does not contain $\mathrm{Rad}\, M$. We show that the family $\{M_1 \cap \mathrm{Rad}\, M, \ldots, M_{n+1} \cap \mathrm{Rad}\, M\}$ is meet-independent in $\mathrm{Rad}\, M$. This gives a contradiction.

We show that for each $j = 1, 2, \ldots, n$, $A_j \cap \mathrm{Rad}\, M + M_{j+1} \cap \mathrm{Rad}\, M = \mathrm{Rad}\, M$, where $A_j = \bigcap_{i=1}^{j} M_i$. Let $U_j = (A_j \cap \mathrm{Rad}\, M) + M_{j+1}$. By the modular law, $U_j \cap \mathrm{Rad}\, M = A_j \cap \mathrm{Rad}\, M + M_{j+1} \cap \mathrm{Rad}\, M$. It therefore suffices to show that each $U_j \geq \mathrm{Rad}\, M$.

Since M is supplemented, $\dfrac{M}{\mathrm{Rad}\, M}$ is semisimple. Writing $\dfrac{M}{\mathrm{Rad}\, M} = \dfrac{A_j + \mathrm{Rad}\, M}{\mathrm{Rad}\, M} \oplus \dfrac{T_j}{\mathrm{Rad}\, M}$, we have $A_j \cap \mathrm{Rad}\, M = A_j \cap (A_j + \mathrm{Rad}\, M) \cap T_j = A_j \cap T_j$. Also, $M = A_j + T_j$ since $T_j \geq \mathrm{Rad}\, M$. Since $\{M_1, \ldots, M_{j+1}\}$ is meet-independent in M, $M = M_{j+1} + A_j$. Hence $U_j + A_j = M$ since $M_{j+1} \leq U_j$.

Now $\dfrac{M}{T_j} = \dfrac{A_j + T_j}{T_j} \cong \dfrac{A_j}{A_j \cap T_j}$. The existence of a (canonical) surjection from the semisimple module $\dfrac{M}{\mathrm{Rad}\, M}$ to $\dfrac{M}{T_j}$ shows that $\dfrac{A_j}{A_j \cap T_j}$ is semisimple. Since $M = A_j + U_j$, we have a canonical surjection from $\dfrac{A_j}{A_j \cap T_j}$ to $\dfrac{M}{U_j}$. Therefore $\dfrac{M}{U_j}$ is semisimple, hence semiprimitive, and $U_j \geq \mathrm{Rad}\, M$.

Since at most n of the submodules M_i can fail to contain $\mathrm{Rad}\, M$, there exists a positive integer k such that $M_i \geq \mathrm{Rad}\, M$ for all $i \geq k$. Suppose that $\mathrm{Rad}\left(\dfrac{M}{X}\right) \neq 0$ for some $M_i \leq X < M$. Then there exists $X < Y \leq M$ such that $\dfrac{Y}{X} \hookrightarrow \dfrac{M}{X}$. Since $X \geq \mathrm{Rad}\, M$, $\dfrac{M}{X}$ is semisimple. Hence $Y = X$. Thus, for each $i \geq k$, $\dfrac{M}{M_i}$ is completely semiprimitive.

4. Modules with Chain Conditions

The results of the previous section allow us to derive decomposition results for modules with chain conditions on small submodules.

Suppose that a module M has DCC (respectively ACC) on small submodules. Then each small submodule of M is artinian (respectively noetherian). Moreover, every hollow submodule H of M is artinian (respectively noetherian) since every proper submodule of H, being small in H, is small in M.

Lemma 4.1. *If M is D1 and noetherian then M has finite dual Goldie dimension.*

Proof. Suppose M has an infinite meet-independent family $\{A_1, A_2, \ldots\}$ of proper submodules. Let I_1 be an interior of A_1 in M. For each $j = 2, 3, \ldots$, let I_j be an interior of $\bigcap_{i=1}^{j} A_i$ in M such that $I_j \leq I_{j-1}$. Select modules $B_1, B_2, \ldots$ such that $M = I_1 \oplus B_1$ and $I_j = I_{j+1} \oplus B_{j+1}$ for $j = 1, 2, \ldots$. Now

$B_1 \leq B_1 \oplus B_2 \leq \ldots$, so, since M is noetherian, there exists a positive integer n such that $B_k = 0$ for all $k \geq n$. This contradicts the meet-independence of the family $\{I_1, I_2, \ldots\}$.

Lemma 4.2. *If M is artinian then M has finite dual Goldie dimension.*

Proof. Suppose not. Let $\{A_1, A_2, \ldots\}$ be an infinite meet-independent family of proper submodules of M. Since M is artinian, there exists a positive integer n such that $\bigcap_{i=1}^{n} A_i = \bigcap_{i=1}^{n+1} A_i$. Then $M = A_{n+1} + \bigcap_{i=1}^{n} A_i = A_{n+1} + \bigcap_{i=1}^{n+1} A_i = A_{n+1}$, a contradiction.

Lemma 4.3. *If M is supplemented and satisfies the descending chain condition (DCC) on small submodules, then so also does $\dfrac{M}{A}$.*

Proof. Suppose $\dfrac{M}{A} \geq \dfrac{B_1}{A} \geq \dfrac{B_2}{A} \geq \ldots$ where each $\dfrac{B_i}{A} \stackrel{\circ}{\hookrightarrow} \dfrac{M}{A}$. Let X be a supplement of A in M. Then $\dfrac{M}{A} = \dfrac{A+X}{A} \cong \dfrac{X}{A \cap X}$. Since $\dfrac{B_i}{A} \stackrel{\circ}{\hookrightarrow} \dfrac{M}{A}$, $\dfrac{B_i}{A} \cong \dfrac{C_i}{A \cap X} \stackrel{\circ}{\hookrightarrow} \dfrac{X}{A \cap X}$ for some C_i.

Note that $C_i \stackrel{\circ}{\hookrightarrow} M$. For suppose that $C_i + D = M$. Then $\dfrac{C_i + (D + X \cap A)}{X \cap A} = \dfrac{M}{X \cap A}$. Hence $D + X \cap A = M$ and $D = M$. We have $C_1 \geq C_2 \geq \ldots$. Since M has DCC on small submodules, there exists n such that $C_k = C_{k+1}$ for all $k \geq n$. Then $\dfrac{B_k}{A} = \dfrac{B_{k+1}}{A}$ for all $k \geq n$.

Proposition 4.4. *If M has ACC on small submodules then $\operatorname{Rad} M$ is noetherian.*

Proof. Suppose, by way of contradiction, that $A_1 < A_2 < \cdots < \operatorname{Rad} M$. Let $a_1 \in A_1$ and $a_i \in A_i - A_{i-1}$ for $i \geq 2$. Then $a_1 R < a_1 R + a_2 R < a_1 R + a_2 R + a_3 R < \ldots$. Since every cyclic submodule of $\operatorname{Rad} M$ is small in M, each $a_i R \stackrel{\circ}{\hookrightarrow} M$. Thus, for each n, $\sum_{i=1}^{n} a_i R$ is a small submodule of M. Hence M does not have ACC on small submodules.

Lemma 4.5. *Suppose $K \leq M$ and $M = K + \operatorname{Rad} M$. Then $\operatorname{Rad}\left(\dfrac{M}{K}\right) = \dfrac{M}{K}$.*

Proof. Suppose $L \geq K$ is a maximal submodule of M. Then $L \geq \operatorname{Rad} M$ so $L \geq K + \operatorname{Rad} M = M$, a contradiction. Thus $\dfrac{M}{K}$ has no maximal submodules.

Lemma 4.6. *Suppose M is supplemented and has DCC on small submodules. Then if $\operatorname{Rad} M = M$, M is artinian.*

Proof. Suppose $\operatorname{Soc} M + W = M$. Since $\operatorname{Soc} M$ is semisimple, $\operatorname{Soc} M = \operatorname{Soc} W \oplus U$ for some $U \leq \operatorname{Soc} M$. Then $M = U + W$. If $u + w = 0$ then

$w = -u \in U \leq \operatorname{Soc} M$. Thus $w \in U \cap \operatorname{Soc} W$ so $u = w = 0$. Hence $M = U \oplus W$ and $W \geq \operatorname{Rad} W = \operatorname{Rad} M = M$. Therefore $\operatorname{Soc} M \stackrel{\circ}{\hookrightarrow} M$.

We claim that $\operatorname{Soc} M$ is also essential in M. Suppose not. Then there exists $W \neq 0$ such that $W \oplus \operatorname{Soc} M$ is essential in M. Let Z be a supplement of W in M. Since M has DCC on small submodules, $W \cap Z$ is artinian. Hence, unless it is zero, $W \cap Z$ has a simple submodule S. But then $S \leq W \cap \operatorname{Soc} M = 0$, a contradiction. Thus $M = W \oplus Z$ and $M = \operatorname{Rad} M = \operatorname{Rad} W \oplus \operatorname{Rad} Z$. Now $\operatorname{Rad} W = W \neq 0$ so there exist nonzero small submodules of W summing to W. Each of these is nonzero and artinian and so has a simple submodule. But $\operatorname{Soc} W = 0$, a contradiction.

Next we show that M has finite dual Goldie dimension. Suppose $\{A_i\}_{i=1}^{\infty}$ is a meet-independent family of proper submodules of M. If $\bigcap_{i=1}^{\infty} A_i$ is not small in M, let $A_0 < M$ be a supplement of this module in M. Then $\{A_0, A_1, \ldots\}$ is a meet-independent family whose intersection is small in M. Thus without loss of generality we may assume that $\bigcap_{i=1}^{\infty} A_i$ is small in M. Then $\dfrac{M}{\bigcap_{i=1}^{\infty} A_i}$ is equal to its radical. Since M is supplemented and has DCC on small submodules, $\dfrac{M}{\bigcap_{i=1}^{\infty} A_i}$ also has DCC on small submodules. Hence, by the arguments of the preceding paragraph applied to this module, $\operatorname{Soc}\left(\dfrac{M}{\bigcap_{i=1}^{\infty} A_i}\right)$ is finitely generated. Therefore $\dfrac{M}{\bigcap_{i=1}^{\infty} A_i}$ is finitely cogenerated. Express $\bigcap_{i=1}^{\infty} A_i$ as the intersection of finitely many A_i; say $\bigcap_{i=1}^{n} A_i$. Since $\{A_1, A_2, \ldots\}$ is meet-independent, for each $k \geq 1$, we have $A_{n+k} + \bigcap_{i=1}^{n} A_i = M$. Then $M = A_{n+k} + \bigcap_{i=1}^{\infty} A_i = A_{n+k}$ for all $k \geq 1$. This contradiction establishes that M has finite dual Goldie dimension, say n.

Let $\{A_1, \ldots, A_n\}$ be a meet-independent family in M such that $\bigcap_{i=1}^{n} A_i \stackrel{\circ}{\hookrightarrow} M$ and each $\dfrac{M}{A_i}$ is hollow. Since $\dfrac{M}{\bigcap_{i=1}^{n} A_i}$ is isomorphic to a product of finitely many hollow modules, it is artinian. Being small in M, $\bigcap_{i=1}^{n} A_i$ is also artinian. Thus M is artinian.

Theorem 4.7. *Let M be a D1 module. If M has DCC (ACC) on small submodules then $M = A \oplus S$ where A is artinian (resp. noetherian) and S is semisimple.*

Proof. As in the proof of Theorem 3.3, let K be a supplement of $\operatorname{Rad} M$ in M. Since $K \cap \operatorname{Rad} M \stackrel{\circ}{\hookrightarrow} K$ and M has DCC on small submodules, $K \cap \operatorname{Rad} M$ is artinian. Now $\dfrac{\operatorname{Rad} M}{K \cap \operatorname{Rad} M} \cong \dfrac{K + \operatorname{Rad} M}{K} = \dfrac{M}{K}$. By Lemmas 4.5 and 4.6, $\dfrac{M}{K}$ is artinian. Thus $\operatorname{Rad} M$ is artinian and therefore has finite dual dimension. Hence, by Proposition 3.11, M is eventually completely semiprimitive.

We decompose M as $M = K \oplus T$ where T has finite dual dimension.

Further, as in the proof of Theorem 3.3, $K = K' \oplus T'$, where T' is the direct sum of finitely many hollow modules and K' is semisimple. Thus $M = K' \oplus (T' \oplus T)$. We claim that $A = T' \oplus T$ is artinian. Since $T \cong \dfrac{M}{K}$ is artinian, it suffices to show that T' is artinian. This follows from the fact that every hollow submodule of M is artinian since M has DCC on small submodules.

Now suppose that M has ACC rather than DCC on small submodules. By Proposition 4.4, $Rad\,M$ is noetherian. Letting K be a supplement of $Rad\,M$ as above, write $M = K \oplus T$. Now $T \cong \dfrac{Rad\,M}{K \cap Rad\,M}$ is noetherian. Being a summand of M, T is D1. Hence, by Lemma 4.1, T has finite dual dimension. As in the DCC case, we have $K = K' \oplus T'$ where K' is semisimple and T' is a direct sum of finitely many hollow modules. Each of these hollow modules is noetherian, so T' is noetherian. Thus $M = K' \oplus (T \oplus T')$ is a decomposition of M as the direct sum of a semisimple module and a noetherian module.

References

1. V. Camillo and M. Yousif. *CS modules with acc or dcc on essential submodules.* Comm. Algebra, **19** (1991), 655-662.
2. P. Fleury. *A note on dualizing Goldie dimension* Canad. Math. Bull., **17** (1974), 511-517.
3. K.R. Goodearl. *Ring theory: nonsingular rings and modules.* Marcel Dekker, New York (1976).
4. P. Grzeszczuk and A.R. Puczylowski. *On Goldie and dual Goldie dimension.* J. Pure Applied Algebra, **31** (1984), 47-54.
5. J. K. Haack. *The duals of the Camillo-Zelmanowitz formulas for Goldie dimension.* Canad. Math. Bull., **25** (1982), 325-334.
6. F. Kasch. *Modules and rings.* London Math. Soc. Monographs, **17**, Academic Press, New York (1982).
7. T.Y. Lam. *A first course in noncommutative rings.* Springer Verlag, New York (1991).
8. S.H. Mohamed and B.J. Müller. *Continuous and discrete modules.* London Math. Soc. Lecture Note Series, **147**, Cambridge University Press (1990).
9. K. Varadarajan. *Dual goldie dimension.* Comm. Algebra, **7** (1979), 565-610.

DEPARTMENT OF MATHEMATICS, SOUTHERN ILLINOIS UNIVERSITY AT CARBONDALE, CARBONDALE IL 62901-4408

E-mail address: rbrown@math.siu.edu mwright@math.siu.edu

RIGHT CONES IN GROUPS

H.H. BRUNGS AND G. TÖRNER

ABSTRACT. A right cone C in a group G is a submonoid of G that generates G and $aC \subseteq bC$ or $bC \subset aC$ holds for any a, b in C; such a right cone is closely related to the cones of (right) linearly ordered groups on the one hand and valuation rings, in particular right chain domains, on the other. The ideal theory of right cones is described, the rank one right cones are classified, and three problems are raised.

1. Let G be a group. A right cone C of G is a submonoid of G with

i) $G = \langle C \rangle$, i.e. G is generated by C;
ii) $a^{-1}b \notin C$ implies $b^{-1}a \in C$ for a, b in C.

Part ii) of the definition is equivalent with the following:

ii)' If $a, b \in C$, then $aC \subseteq bC$ or $bC \subset aC$, which means that the lattice of right ideals of C is totally ordered by inclusion.

Lemma. a) If C is a right cone in G then $G = \{ab^{-1} | a, b \in C\}$.
  b) A submonoid C of a group G is a right and left cone of G if and only if $C \cup C^{-1} = G$.

Proof. Let $g \in G$ and $g = a_1^{\varepsilon_1} \cdots a_r^{\varepsilon_r}$ for $a_i \in C$, $\varepsilon_i \in \{1, -1\}$ and r minimal. Suppose that there exists a smallest index i with $\varepsilon_i = -1$. If $i = r$, then $r = 1$ or $r = 2$ and we are done. Otherwise, a term $a_i^{-1} a_{i+1} \notin C$ occurs, which implies $(a_i^{-1} a_{i+1})^{-1} = a \in C$, which contradicts the minimality of r and proves a).

If C is a right cone, every element g in G has the form $g = ab^{-1}$ and $g \notin C$ implies $g^{-1} \in C$ if C is also a left cone. □

If $C = P$ is a right and left cone of G, then P defines a right order and a left order for G if $P \cap P^{-1} = \{e\}$ and the two orders agree if

---

The first author is supported in part by NSERC.

$g^{-1}Pg = P$ for all $g \in G$; in that case P is the cone of all non-negative elements in the linearly ordered group G.

A non-empty subset I of a right cone C is called a right ideal of C if $IC \subseteq C$; left ideals and ideals are defined similarly. A proper ideal I of C is called prime if $aCb \subseteq I$, $a, b \in C$, implies a or b in I, and completely prime if this conclusion follows from $ab \in I$. With $U(C) = C \cap C^{-1}$ we denote the subgroup of units of C, and with $J = J(C) = C \backslash U(C)$ the maximal proper right ideal of C which is also an ideal.

An integral domain R is called a right chain domain if $aR \subseteq bR$ or $bR \subset aR$ hold for a, b in R; such a ring has a skew field F of quotients. A subring R of a skew field F is a right chain domain if and only if $R \backslash \{0\}$ is a right cone of $F \backslash \{0\}$. We say that the right cone C of G is associated with the right chain domain R in F if $G \subseteq F$, $G \cap R = C$, and every element $0 \neq r \in R$ can be written as $r = au$ for $a \in C$, $u \in U(R)$. Further, if $ua = a'u'$ for $a, a' \in C$, $u, u' \in U(R)$, then $CaC = Ca'C$. This last condition ensures that not only the right ideals of C correspond to the right ideals of R, but that in this correspondence ideals and prime ideals correspond to each other.

We say that the right cone C has rank n if C has exactly n completely prime ideals and say that two completely prime ideals $P_1 \supset P_2$ are a prime segment of C if no further completely prime ideal lies between P_1 and P_2; we also call $P_1 \supset \emptyset$ a prime segment if P_1 is a minimal completely prime ideal of C.

Let $P_1 \supset P_2$ be a prime segment where $P_2 = \emptyset$ or P_2 is completely prime and let K be the union of ideals of C properly contained in P_1.

If $K = P_2$, we say that the prime segment is simple. If $P_1 \supset K \supset P_2$ and $P_1 = P_1^2$, then $Q = K$ is prime but not completely prime, since for ideals I_1, I_2 of C that properly contain Q we have $P_1 \subseteq I_1, I_2$ and $P_1 = P_1^2 \subseteq I_1 I_2$; in this case the prime segment is called exceptional. We have described cases b) and c) in the following result.

Theorem. *For a prime segment $P_1 \supset P_2$ of a right cone C in a group G one of the following alternatives occurs:*

  a) *The prime segment is right invariant; i.e. $P_1 a \subseteq aP_1$ for all $a \in P_1 \backslash P_2$.*
  b) *The prime segment is simple.*
  c) *The prime segment is exceptional.*

To complete the proof we use the following two facts that hold for right cones. A proper ideal I of C is completely prime if $a^2 \in I$ implies $a \in I$, and the intersection $\cap I^n$ of all powers of an ideal $I \neq C$ is completely prime if this intersection is not empty.

Considering the cases discussed, before stating the theorem we must prove a) if either $K = P_1$ and P_1 is then the union of ideals I properly contained in P_1 and containing P_2 or $P_1 \supset K \supset P_2$ and $P_1 \neq P_1^2$. It follows that $\cap I^n = P_2$ for the ideals I in the first case and $\cap P_1^n = P_2$ in the second. For any p in $P_1 \backslash P_2$ there exists therefore in either case an ideal I with $p \in I$ and $\cap I^n = P_2$. Under these assumptions we must prove that for p in P_1 and a in $P_1 \backslash P_2$ there exists p' in P_1 with $pa = ap'$. If such a p' does not exist, then $p \notin P_2$ and either $pa = as$ for $s \in C \backslash P_1$, a case considered below, or $paj = a$ for $j \in J(C)$. Let I be an ideal with $p \in I$ and $\cap I^n = P_2$ and from $a = paj = p^n a j^n$ for all $n \geq 1$ we obtain the contradiction $a \in P_2$. It remains to consider the case $pa = as$ for $s \in C$, $s \notin P_1$. Then $p \in I$ with $\cap I^n = P_2 \subset a^2 C$ and $p^n a = a^2 b$ for some $b \in C$ and some n follows. However, $a^2 b = p^n a = as^n$ implies $ab = s^n$, a contradiction since $s \notin P_1$. □

2. We consider rank one right cones $C \supset J$ where the maximal ideal J is the only completely prime ideal of C. It can be shown that in the case where the prime segment $J \supset \emptyset$ is right invariant, the right cone C is right invariant, i.e. $Ca \subseteq aC$ for all a in C or equivalently that all right ideals of C are ideals. In this case, $U(C)a \subseteq aU(C)$ and $\overline{C} = C/U(C) = \{aU(C) | a \in C\}$ with $aU(C)bU(C) = abU(C)$ as operation exists and by Hölder's Theroem ([F66]) $\overline{C}$ is isomorphic to a subsemigroup of $(\mathbb{R}, +)$, the real numbers with addition as operation.

Since the right ideals of C correspond to right ideals of $\overline{C}$ and all right ideals of C are ideals, the ideals of C can be described as follows:

Either J is a principal right ideal and the powers of J are the only ideals of C, or J is not finitely generated as a right ideal; the semigroup of principal right ideals then corresponds to a dense subsemigroup of $(\mathbb{R}^{\geq 0}, +)$ and there is an additional ideal in C corresponding to any non-negative real number.

Examples of rank one right cones $C \supset J$ with $J \supset \emptyset$ simple will be considered in the next section.

If C has rank one and is exceptional, there exist the ideals $C \supset J \supset Q$ and $J = J^2$ and $\cap Q^n = \emptyset$ since $\cap I^n$ is completely prime for an ideal $I \neq C$ of C if this intersection is not $\emptyset$. It follows from the Theorem that there are no further ideals between J and Q. We will show below that Q is a principal right ideal if any power of Q is a principal right ideal. In this case there exists the additional ideal $Q^{n-1}J$ between Q^{n-1} and Q^n for every $n \geq 2$.

It can be proved (see [D94], [BD94]) that in the case where $C \cup C^{-1} = G$ the only other possibility for an ideal between Q^{n-1} and Q^n is an upper neighbor for Q^n for $n = mk$ and a fixed $k \geq 2$.

PROBLEM A. Let $C \supset J \supset Q$ be an exceptional rank one cone. Are the only additional ideals of C, besides C, J and the powers of Q, lower neighbours of Q^n for $n = 1, 2, \ldots$ in case Q is right principal and possibly upper neighbours of Q^{mk} for all $m \geq 1$ and some fixed $k \geq 2$?

For Q^r, $r \geq 1$, we define $\widetilde{Q}^r = \bigcap_{t \in J \backslash Q} tQ^r$ and obtain the next result:

Proposition 1. *Let $C \supset J \supset Q$ be an exceptional rank one right cone. Then:*

a) $\widetilde{Q}^{n-1}$ *is an ideal with* $Q^{n-1} \supseteq JQ^{n-1} \supseteq \widetilde{Q}^{n-1} \supseteq Q^n$ *with* $n \geq 2$.
b) *There are no further ideals between* JQ^{n-1} *and* $\widetilde{Q}^{n-1}$.

Proof. a) Since $J \supseteq tC \supseteq Q$, the containments given in a) follow. To prove that $\widetilde{Q}^{n-1}$ is an ideal, let $r \in C$, $a \in \widetilde{Q}^{n-1}$. Then $ar \in \widetilde{Q}^{n-1}$. If $t \in J \backslash Q$ and $r = tc$ for some $c \in C$, then $ra = tca \in tQ^{n-1}$. If, on the other hand, $t = rj$ for some $j \in J$, then j is not in Q and $a = jq$ for some q in Q^{n-1} and again $ra = rjq = tq$ is in tQ^{n-1}; hence, $\widetilde{Q}^{n-1}$ is an ideal.

To prove b) assume that $JQ^{n-1} \supset I \supset \widetilde{Q}^{n-1}$ for an ideal I of C. Consider $A = \{r \in C | rQ^{n-1} \subseteq I\}$ and $A \supseteq Q$ follows. Since $I \supset \widetilde{Q}^{n-1}$, there exists an element $t \in J \backslash Q$ with $I \supset tQ^{n-1}$ and hence $t \in A$ and $A \supset Q$. However, $JQ^{n-1} \not\subseteq I$ and $J \supset A \supset Q$ follows, a contradiction that proves b). □

In order to discuss the ideals between Q^{n-1} and JQ^{n-1}, we can restrict ourselves to consider the ideals between $Q^{n-1}J$ and JQ^{n-1}, since $Q^{n-1}J$ is either equal to Q^{n-1} or is the lower neighbour of Q^{n-1} in the lattice of right ideals of C. This last case occurs if $Q^{n-1} = rC$ for some r in C. We have the following results:

Proposition 2. *Let $C \supset J \supset Q$ be an exceptional rank one right cone.*

a) *If $Q^{n-1} = rC$ is a principal right ideal for some $n \geq 2$, then $Q = qC$ for some q in C and $Q^n = q^nC$ for all $n \geq 1$.*
b) $JQ^{n-1} = Q^{n-1}$ *implies* $Q^{n-1}J = Q^{n-1}$.

Proof. a) We have $Q^{n-1} = rC$ and $r = q_1 \ldots q_{n-1}$ for some elements q_i in Q. If Q is not a principal right ideal, then there exists $q' \in Q$ with $q'C \supset q_{n-1}C$. Hence, $q_1 \ldots q_{n-2}q'$ is contained in Q^{n-1} but not in rC, a contradiction which proves that $Q = qC$; that $Q^n = q^nC$ follows since Q is an ideal.

b) If $JQ^{n-1} = Q^{n-1}$ and $Q^{n-1}J \neq Q^{n-1}$, then $Q^{n-1} = rC$ and $Q^{n-1}J = rJ$ for some r in C. Then $JQ^{n-1} = JrC = Q^{n-1} = rC$.

It follows that $B = \{c \in C | cr \in rJ\}$ is an ideal of C that contains Q and is properly contained in J; hence, $B = Q$.

Since Q is not completely prime, there exists an element x in C with x^2 in Q, but x not in Q. Since $B = Q$, we have $xr = ru$ for a unit u in C and $x^2r = ru^2$ follows. However, x^2 in B implies $x^2r = rj$ for j in J, a contradiction that proves b). $\square$

With the results from the last two propositions we arrive at the following situation:

$$Q^{n-1} \supseteq Q^{n-1}J \supseteq JQ^{n-1} \supseteq \widetilde{Q}^{n-1} \supseteq Q^n \quad \text{for} \quad n \geq 2,$$

and there are no further ideals between Q^{n-1} and $Q^{n-1}J$ and between JQ^{n-1} and $\widetilde{Q}^{n-1}$.

3. In this section some examples will be considered.

Example A. The group $SL(2, \mathbb{R})$ contains the subgroups $L = \{u = \begin{pmatrix} a & b \\ 0 & a^{-1} \end{pmatrix} | b, 0 < a \in \mathbb{R}\}$ and $M = \{r(\varphi) = \begin{pmatrix} \cos\varphi & -\sin\varphi \\ \sin\varphi & \cos\varphi \end{pmatrix} | \varphi \in \mathbb{R}\}$ and every element g in $SL(2, \mathbb{R})$ can be written uniquely in the form $g = r(\varphi)u$ for $r(\varphi) \in M$, $0 \leq \varphi < 2\pi$, $u \in L$. We consider the covering group $G = \{x^t u | t \in \mathbb{R}, u \in U\}$ of $SL(2, \mathbb{R})$ with $x^{t_1}u_1 = x^{t_2}u_2$ if and only if $t_1 = t_2$ in $\mathbb{R}$ and $u_1 = u_2$ in U; the operation in G is defined by $x^{t_1}u_1 x^{t_2}u_2 = x^{t_1}u_1 x^{2\pi k + \varphi}u_2 = x^{t_1 + 2\pi k + \psi}u_1' u_2$ where $t_2 = 2\pi k + \varphi$, $0 \leq \varphi, \psi < 2\pi$, and $u_1 r(\varphi) = r(\psi)u_1'$ in $SL(2, \mathbb{R})$.

This group contains the exceptional rank one cone $C = \{x^t u | 0 \leq t \in \mathbb{R}, u \in U\}$ with $Q = x^\pi C$ the prime ideal that is not completely prime (see [D94], [BD94] for more details).

Example B. We discuss right cones in groups of affine mappings of ordered vectorspaces over ordered fields.

In the group $G = \{(a,v) | v, 0 < a \in \mathbb{Q}\}$ with $(a,v)(b,w) = (ab, aw + v)$ we consider $P = \{(a,v) \in G | a\sqrt{2} + v \geq \sqrt{2}\}$. Then P is a rank one cone with $J(P) \supset \emptyset$ simple and $(a,v)P \supseteq (b,w)P$ if and only if $a\sqrt{2} + v \leq b\sqrt{2} + w$ in $\mathbb{R}$. That for example $(10, 10)P$ is not an ideal follows from $(\frac{1}{10}, 3)(10, 10) = (1, 4)$ and $(1, 4)P \supset (10, 10)P$; (see [S66]).

More generally, in [BSch95] affine groups of ordered vector spaces V over an ordered field K were considered. Let η be an element in an ordered K-vector space to which the order of V has been extended with $\eta \notin V$.

Let $G = \{(a,v)|\, 0 < a \in K, v \in V\}$ and it follows that $P_\eta = \{(a,v) \in G|\, a\eta + v \geq \eta\}$ is a cone of G. Then η defines a Dedekind cut (U,O) of V with $U = \{v \in V|\, v < \eta\}$ and $O = \{v \in V|\, v > \eta\}$.

The cone P_2 has simple prime segments only (see [BSch95] Th. 4.8)) if and only if the following two conditions hold:

i) For every $0 < r \in V$ there exists $u \in U$ with $u + r \in O$.
ii) U has no largest and O has no smallest element.

The cone P_η is of rank one with one simple prime segment if in addition all positive elements of V are in the same K-archimedean class, i.e. for $0 < v_1, v_2 \in K$ there exist $a_1, a_2 \in K$ with $v_2 < a_1 v_1$, $v_1 < a_2 v_2$.

The groups G occurring in Example B are semidirect products of torsion free abelian groups and the group ring $\mathbb{Q}G$ is therefore an Ore domain and its skew field F of quotients contains a right chain domain R associated with P_η.

That there exists a chain domain associated with the cone in Example A is much more difficult to prove, see [D94].

Problem B. Does there exist for every right cone C of G a right chain domain R associated with C?

4. Let C be a right invariant right cone of G. We observed in Section 2 that then $\overline{C} = C/U(C)$ exists and it follows that $\overline{C}$ is a linearly ordered semigroup with identitye $\overline{e}$ and $\overline{a} < \overline{b}$ if and only if $\overline{b} = \overline{a}\,\overline{d}$ for $\overline{d} \neq \overline{e}$. Such a semigroup is called a right holoid.

Problem C. Is every right holoid H of the form $H = \overline{C}$ for a right invariant right cone C of a group G?

The final example (see [C85] Sec. 8.6 and also [J69]) shows that the right holoid $H_n = \{\alpha|\, \alpha < \omega^n\}$ of ordinal numbers less than a power of ω, the order type of $\mathbb{N}$, is of the form $\overline{C}$ with C a right cone in a free group G generated by $\{x_i|\, i = 1,\ldots,n\}$.

Let U be the subgroup of G generated by elements of the form

$$u_{\beta\alpha_1\ldots\alpha_r} = x_{\alpha_r}^{-1}\ldots x_{\alpha_1}^{-1} x_\beta x_{\alpha_1}\ldots x_{\alpha_r} \quad \text{with} \quad \beta < \alpha_1 \geq \alpha_2 \geq \cdots \geq \alpha_r$$

and let C be the subsemigroup of G generated by the x_i's and U. Since $ux_i = x_i u'$ for $u, u' \in U$ and $x_i x_j = x_j u$ for $i < j$ and some u in U, the principal right ideals of C are of the form $x_{i_1} x_{i_2} \ldots x_{i_r} C$ for $i_1 \geq i_2 \geq \cdots \geq i_r$ and ordered lexicographically.

In [BT95] Problem C is solved for a larger class of right holoids that contains the H_n's.

References

[BD94] Brungs, H.H., Dubrovin, N.I., *Classification of chain rings*, preprint.
[BSch95] Brungs, H.H., Schröder, M., *Prime segments of skew fields*, Can. J. Math. **47** (1995), 1148-1176.
[BT95] Brungs, H.H., Törner, G., *Right chain domains with prescribed value holoids*, J. Algebra **176** (1995), 346-355.
[C85] Cohn, P.M., *Free Ideal Rings and Their Relations*, Academic Press, London, 1985.
[D94] Dubrovin, N.I., *The rational closure of group rings of left orderable groups*, Schriftenreihe des Fachbereichs Mathematik, Vol. **254**, Universität Duisburg, 96pp, 1994.
[F66] Fuchs, L., *Teilweise geordnete algebraische Strukturen*, Vandenhoeck and Ruprecht, Göttingen, 1966.
[J69] Jategaonkar, A.V., *A counter-example in ring theory and homological algebra*, J. Algebra **12** (1969), 418-440.
[S66] Smirnov, D.M., *Right-ordered groups*, Algebra i Logika **5**:6 (1966), 41-59.

Hans-Heinrich Brungs
Dept. of Mathematical Sciences
University of Alberta
Edmonton, Alberta
Canada   T6G 2G1
hbrungs@vega.math.ualberta.ca

Günter Törner
Fachbereich Mathematik
Gerhard-Mercator-Universität
47048 Duisburg
Germany
toerner@math.uni-duisburg.de

DEPARTMENT OF MATHEMATICAL SCIENCES, UNIVERSITY OF ALBERTA, EDMONTON, ALBERTA, CANADA   T6G 2G1

ON EXTENSIONS OF REGULAR RINGS OF FINITE INDEX BY CENTRAL ELEMENTS

W. D. BURGESS AND R.M. RAPHAEL

Dedicated to Bruno Müller on the occasion of his retirement

ABSTRACT. A Von Neumann regular ring R of finite index of nilpotency which is not biregular can have quite complex structure. Even though R can be embedded in a biregular ring of the same index, it need not be "close" to one in structure. It is, however, closely related to a unique smallest overring, $R^\sharp$, which is "almost biregular", i.e., one where supports of elements in the Pierce sheaf are open. It is formed by adjoining certain central idempotents from $Q(R)$. Extensions of R by central elements, particularly idempotents, are examined. Many examples and counterexamples are presented.

Introduction. The class of (von Neumann) regular rings of finite index of nilpotency has been extensively studied. Two chapters in the book by Goodearl on regular rings ([14]) are devoted to them and to closely related rings. Splittings of these rings into direct products of rings of homogeneous index have been examined by various authors (see the extensive bibliography of [14]). Sheaf representations are studied in Burgess and Stephenson, [8], and Carson, [10]. A very thorough study of K_0 of these rings is in Goodearl, [15]. Among regular rings, those of finite index are relatively easy to study; nevertheless, they reveal considerable complexity and many questions about their structure remain unanswered.

Throughout R will stand for a regular ring of finite index.

The easiest of these rings are those which are also biregular (i.e., each principal ideal is generated by a central idempotent). These have a useful description as a sheaf of simple artinian rings (see Pierce, [19] and Dauns and Hofmann, [12], but some of the key ideas go back to Arens and Kaplansky, [1]). The group K_0 of a biregular R also has a nice sheaf representation (Burgess and Goursaud, [5]), and among regular rings of finite index, the biregular ones are precisely those whose K_0 groups are lattice-ordered ([15, Theorem 4.4]).

1991 *Mathematics Subject Classification.* Primary: 16E50, 16E60.
Key words and phrases. Regular ring, finite index, Pierce sheaf, extension.
Les auteurs tiennent à remercier les programmes d'échanges Ontario-Québec du Ministère de l'enseignement supérieur et de la science du Québec et le Ministère de l'éducation de l'Ontario de leur concours. The authors' research was partially supported by Grants A7539 and A7752, respectively, of the NSERC.

A ring R always has a "large" biregular part (over a dense open subset in the base of the Pierce sheaf). The behaviour over the complement is more complex; indeed, the stalks there need not even be artinian. Section 3 has many examples of this.

The complete ring of quotients $Q(R)$ of R is always biregular of the same index as R. However, the elements of $Q(R)$ are not closely related to those of R. The division rings occurring in its stalks may be very much larger than those in the simple images of R. On the other hand the adjunction of central idempotents from $Q(R)$ does not radically change R. If the new ring is S, then R and S have the same simple images and, moreover, the Pierce stalks of S are homomorphic images of those of R. We see below that it is not possible to arrive at a biregular ring by such an adjunction, unless R is already biregular. Nevertheless, the process of adjoining sufficiently many central idempotents from $Q(R)$ does lead to a type of ring, called *almost biregular*, which is much more amenable to study than the original R. Rings in this class are defined by the property (shared by biregular rings) that supports of elements are open. They represent a natural generalization of biregular rings and each R is "close" to an almost biregular one.

Some useful properties of almost biregular rings are derived here. The process of adjoining central idempotents (which leads to an almost biregular ring) is looked at in some detail in Section 1, after some general observations about R. It is shown, in particular, that there is always a unique smallest almost biregular ring, called $R^\sharp$, $R \subseteq R^\sharp \subseteq Q(R)$, formed by adjoining the central covers, from $Q(R)$, of the elements of R.

Section 2 is devoted to the examining what happens when other sorts of elements from $Q(R)$ are adjoined to our ring R. The situation, here, is quite different and even regularity can be lost. Some positive results may be obtained when the elements to be adjoined are from the centre of $Q(R)$ and are integral over the centre of R.

Section 3 presents a scheme for the construction of regular rings of finite index. It is then used to get examples which illustrate the results and show the obstacles to stronger theorems.

Notational conventions. The index of nilpotency of a ring R is denoted $\mathbf{i}(R)$. The centre of R is $Z(R)$ and the set (boolean algebra) of central idempotents is $\mathbf{B}(R)$. Throughout, all references to sheaves and stalks will refer to the Pierce sheaf $(X, \mathcal{R})$, $X = \mathbf{SpecB}(R)$ (see Pierce, [19]). The stalks are the rings $R_x = R/Rx$, for each $x \in X$ (we write r_x for $r + Rx$) and the basic open sets of $\mathcal{R}$ are of the form $\{r_x \mid x \in N\}$, for some $r \in R$ and clopen (closed *and* open) set $N \subseteq X$. For $r \in R$, $\mathrm{supp}(r) = \{x \in X \mid r_x \neq 0\}$. The closure of a subset V of a topological space is denoted $\overline{V}$. A general reference for boolean algebras and Stone duality is Koppelberg, [17].

If S is a ring extension of a ring R and C is a central subset of S so that S is generated, as a ring, by R and C, we write $S = R \diamond C$. If T

is a semisimple artinian ring, its *total index* is the sum of the indices of nilpotence of its simple components.

Section 1 – General observations about regular rings of finite index and their extensions by central idempotents. Suppose R is regular of index n; then [14, Proposition 6.6] says that every non-zero ideal I contains a central idempotent $0 \neq e$ such that Re is biregular. Hence, putting $X = \mathbf{Spec B}(R)$, the set $\{x \in X \mid \text{there is } e \in \mathbf{B}(R) \text{ with } x \in \text{supp}(e) \text{ and } Re \text{ is biregular}\} = \mathbf{bi}(R)$ is dense open in X. Call $\mathbf{bi}(R)$ the *biregular part* of X. Denote by $\mathfrak{b}(R)$ the ideal of R generated by the central idempotents with supports in $\mathbf{bi}(R)$. If $\mathbf{bi}(R) \neq X$ then $\mathfrak{b}(R)$ is a biregular ring without 1. Notice that R_x simple does *not* imply that $x \in \mathbf{bi}(R)$. (See Example 3.2.) The complete left and right rings of quotients coincide ([14, Theorem 6.21]) and are denoted $Q(R)$. It is biregular of index n ([14, Corollary 7.4]). Moreover, $\mathbf{B}(Q(R))$ is the completion of $\mathbf{B}(R)$ ([10, Proposition 3.1]).

Lemma 1.1. *Let R be a regular ring of finite index. Then the Pierce sheaf $(X, \mathcal{R})$ has the property that for $0 \neq r \in R$, $\text{supp}(r)$ has non-empty interior, U_r, and $\text{supp}(r) = \overline{U_r}$. Moreover, $\text{supp}(r) = \overline{U_r \cap \mathbf{bi}(R)}$ and $\text{supp}(r) \cap \mathbf{bi}(R)$ is open.*

Proof. It has already been noticed that supports of non-zero elements have non-empty interiors (by [14, Proposition 7.16]). If $x \in \text{supp}(r)$, let N be a clopen neighbourhood of x with corresponding $e \in \mathbf{B}(R)$ (i.e., $\text{supp}(e) = N$). Then $re \neq 0$ and there is an open $U \subseteq N$ with $U \subseteq \text{supp}(re) \subseteq \text{supp}(r)$ and then $U \subseteq U_r$. However, $\text{supp}(r)$ is closed, so $\text{supp}(r) = \overline{U_r}$. The last remark follows since $\mathbf{bi}(R)$ is dense in X; and if $r \in R$ has $r_u \neq 0$ for some $u \in \mathbf{bi}(R)$, then r is non-zero on a neighbourhood of u. □

We note the following fact; it is a special case of [19, Proposition 1.7].

Lemma 1.2. *Let $R \subseteq S$ be an inclusion of rings with $\mathbf{B}(R) \subseteq \mathbf{B}(S)$. If $x \in \mathbf{Spec B}(R)$ then $Sx = \bigcap_{y \supseteq x} Sy$, where $y \in \mathbf{Spec B}(S)$.*

Proof. Apply Proposition 1.7 of [19] to the S-module $\bar{S} = S/Sx$, noting that for $y \in \mathbf{Spec B}(S)$ with $y \not\supseteq x$, $\bar{S}y = \bar{S}$. □

When R is regular of finite index and $\mathbf{B}(R) = \mathbf{B}(Q(R))$ then supports of elements are clopen since, in this case, $r \in R$ has the same support as it has as an element of the biregular ring $Q(R)$. We next see that the latter property implies that R has stalks artinian and more. (See also [10, Theorem B].)

Propsition 1.3. *If R is regular of index n and the supports of elements of R are clopen then the stalks of R are artinian of total index $\leq n$.*

Proof. Pick any stalk R_x and a finite orthogonal set of idempotents in R_x. These can be lifted to an orthogonal set of idempotents in R which are all non-zero on a neighbourhood N of x. Since N meets $\mathbf{bi}(R)$, the set can have no more than n elements. □

Regular rings of finite index satisfying the condition of (1.3) are called *almost biregular* and are studied in more detail elsewhere ([6]). One characterization from [6] is that R is almost biregular if and only if for each $0 \neq r \in R$ there is $e \in \mathbf{B}(R)$ such that $er = r$ and $_RRrR$ is essential in $_RRe$.

The conclusion of (1.3) can be obtained under weaker hypotheses. It suffices to assume that for any stalk R_x and $r_1, \ldots, r_k \in R$, $k \geq 1$, which are all non-zero in R_x then $\bigcap_{i=1}^{k} \mathrm{supp}(r_i) \cap \mathbf{bi}(R) \neq \emptyset$. We call such rings called *s-rich* (for "support rich"). The example after Theorem 1.5 is s-rich but not almost biregular. The property of having all stalks artinian is much weaker than "s-rich", as examples in Section 3 (e.g., Example 3.5) show.

We have just seen that by adding enough central idempotents from $Q(R)$ we can always get to an almost biregular ring. A variant follows: It is illustrated by Example 3.4.

Propsition 1.4. *Let S be a ring of finite index with artinian stalks. Suppose $R \subseteq S$ is a regular subring such that $\mathbf{B}(R) \subseteq \mathbf{B}(S)$ and, further, for each $x \in \mathbf{Spec}\,\mathbf{B}(R)$ the boolean algebra $\mathbf{B}(S)/x\mathbf{B}(S)$ is finite. Then R has artinian stalks.*

Proof. For each $x \in \mathbf{Spec}\,\mathbf{B}(R)$, there are, by hypothesis, only finitely many $y \in \mathbf{Spec}\,\mathbf{B}(S)$ lying over x, say $y_1, \ldots, y_k$. By Lemma 1.2, $Rx = Sy_1 \cap \cdots \cap Sy_k \cap R$, which shows that R_x embeds in $\prod_{i=1}^{k} S_{y_i}$. Since the stalks of S are artinian, R_x, being regular, is artinian. □

Notice that, in general, if $R \subseteq S$ are rings with $\mathbf{B}(R) \subseteq \mathbf{B}(S)$ and $S = R \diamond \mathbf{B}(S)$, then the induced homomorphisms at the level of stalks are all surjections (and so, in particular, S is also regular of finite index). This is because each element of S can be written in the form $\sum_{i=1}^{k} r_i e_i$, where the $r_i \in R$ and $\{e_1, \ldots, e_k\}$ is an (orthogonal) set from $\mathbf{B}(S)$. In a stalk of S, each of the e_i goes to 0 or 1. Hence, in particular, if R is biregular so is S, and, if R has artinian stalks, so does S. Moreover, by Lemma 1.2, for $x \in \mathbf{Spec}\,\mathbf{B}(R)$, $Rx = \bigcap_{y \in V} Sy \cap R$, where V is the set of $y \in \mathbf{Spec}\,\mathbf{B}(S)$ lying over x. Thus R_x embeds (as a subdirect product) in $\prod_{y \in V} S_y$. This shows that if $\max_{y \in V} \mathbf{i}(S_y) = k$ then $\mathbf{i}(R_x) = k$.

It is shown by Carson (proof of [11, Proposition 8]), that if R is regular of finite index and $R \subseteq S$ with $S = R \diamond \mathbf{B}(S)$, then R and S have the same simple images. It follows that R and S share many elementary properties (see [11, Proposition 8], [14, Lemma 6.9] and [9, Theorem 3.5]), even though the extension is not elementary ([11, Proposition 1]). As already mentioned, S is biregular if R is. The converse is also true, and more detail can be given at the level of individual stalks.

Theorem 1.5. *Suppose R is regular of finite index and $S \supseteq R$ with $\mathbf{B}(R) \subseteq \mathbf{B}(S)$, and $S = R \diamond \mathbf{B}(S)$. Let the induced surjection be $\theta : \mathbf{SpecB}(S) \to \mathbf{SpecB}(R)$. Then,*

(i) *for $x \in \mathbf{SpecB}(R)$, R_x is simple if and only if for all $y \in \theta^{-1}(x)$, S_y is simple (in which case $S_y \cong R_x$);*
(ii) *$\theta^{-1}(\mathbf{bi}(R)) \subseteq \mathbf{bi}(S)$; and*
(iii) *if $R \subseteq S \subseteq Q(R)$, then $\theta^{-1}(\mathbf{bi}(R))$ is dense in $\mathbf{SpecB}(S)$.*

Proof. (i) Suppose, for some $x \in X = \mathbf{SpecB}(R)$, that R_x is not simple but that for all $y \in \theta^{-1}(x)$, S_y is simple. Pick $f^2 = f \in R$ such that $f_x \in \mathbf{B}(R_x)$ is different from 0 or 1.

For each $y \in Y, u \in X$ with $\theta(y) = u$, there is an induced surjection $R_u \to S_y$ (when R_u is simple, this is an isomorphism). In particular, if $y \in \theta^{-1}(x)$, f_y is either 0 or 1. (No notational distinction is made between $f \in R$ and its image in S.)

Since f is not central over any neighbourhood of x, there are, for each neighbourhood N of x, some $u \in N$ and $r \in R$ (depending on u) so that $(fr - rf)_u \neq 0$. But, according to Lemma 1.1 there is an open set $W \subset \mathbf{bi}(R)$ with $u \in \overline{W}$ and $W \subset \mathrm{supp}(fr - rf)$. Hence, there is some $w \in N \cap W$ on which $fr - rf$ is non-zero. This means that there exists a net $\{u_\lambda\}_\Lambda$ in $\mathbf{bi}(R)$ converging to x, so that for each $\lambda \in \Lambda$, f_{u_λ} is not central in R_{u_λ}.

For each $\lambda \in \Lambda$, pick $w_\lambda \in \theta^{-1}(u_\lambda)$. Hence $\{w_\lambda\}_\Lambda$ is a net in Y. Since Y is compact, there is a subnet $\{w_\lambda\}_{\Lambda'}$ which converges to some $z \in Y$. By continuity, $\theta(z) = x$. Since each $u_\lambda \in \mathbf{bi}(R)$, $R_{u_\lambda} \cong S_{w_\lambda}$, for all $\lambda \in \Lambda'$. In particular, $f_{w_\lambda} \neq 0, 1$ for $\lambda \in \Lambda'$. However, by assumption, f_z is 0 or 1. This would mean that f is constantly 0 or 1 on a neighbourhood of z, a contradiction in light of the convergence of the net.

(ii) This is clear. In part (iii), the idempotents from $\mathbf{B}(S)$ come from $\mathbf{B}(Q(R))$, and so each $e \in \mathbf{B}(S)$ is the supremum of the elements $f \in \mathbf{B}(R)$ with $fe = f$. Then for each $0 \neq e \in \mathbf{B}(S)$ there is $0 \neq f \in \mathbf{B}(R)$ with $fe = f$. But then, there is $0 \neq g \in \mathbf{B}(R) \cap \mathfrak{b}(R)$ with $gf = g$. Hence $\theta^{-1}(\mathrm{supp}_X(g)) \subseteq \mathrm{supp}_Y(e)$. □

In the above theorem, the inclusion $\theta^{-1}(\mathbf{bi}(R)) \subseteq \mathbf{bi}(S)$ can be strict. Consider the ring R of sequences from $\mathbf{M}_2(F), F$ a field, which are eventually of the form $(\mathrm{diag}(a, b), \mathrm{diag}(a, a), \mathrm{diag}(a, b), \dots)$, $a, b \in F$ fixed, with the scalar matrices in the even indexed places. If we adjoin $e \in \prod_\mathbf{N} \mathbf{M}_2(F)$, which is 1 in the even places and 0 in the odd ones, then the new ring S has two non-isolated points α_1, α_2 in $\mathbf{SpecB}(S)$ with stalks $F \times F$ and F, respectively. Then $\mathbf{bi}(S) = \mathbf{N} \cap \{\alpha_2\}$, while $\theta^{-1}(\mathbf{bi}(R)) = \mathbf{N}$. Notice that S is almost biregular. In general, in the situation of Theorem 1.5, the biregular ideal of S contains the ideal corresponding to $\theta^{-1}(\mathbf{bi}(R))$, which is the ideal of S generated by the biregular ideal of R. As we have just seen, the inclusion can be strict (but see Proposition 1.8).

By Theorem 1.5, it is not possible to obtain a biregular ring from one which is not by the addition central idempotents. However, we also know that an almost biregular ring may be so obtained. The idempotents to be adjoined form, in general, a set smaller than $\mathbf{B}(Q(R)) \setminus \mathbf{B}(R)$, indeed, sometimes of smaller cardinality (as the example in the previous paragraph shows – 1 vs the power of the continuum).

Recall ([14, page 130]) that each $q \in Q(R)$ has a *central cover*, i.e., $e \in \mathbf{B}(Q(R))$ such that $eq = q$ and e is the unique smallest element of $\mathbf{B}(Q(R))$ with this property. It is denoted $\mathfrak{cc}(q)$.

Lemma 1.6. *Let R be regular of finite index. If for $r \in R$, $\mathrm{supp}(r)$ is clopen with $e \in \mathbf{B}(R)$ having the same support as r, then e is the central cover of r in $Q(R)$. Moreover, if $g \in \mathbf{B}(Q(R))$ then $\mathfrak{cc}(gr) = g\,\mathfrak{cc}(r)$.*

Proof. Clearly $re = r$. Suppose $f \in \mathbf{B}(Q(R))$ with $fr = r$; and $f < e$. Then $e - f \in \mathbf{B}(Q(R))$ and, by density, there is $0 \ne h \in \mathbf{B}(R)$ with $h \le e - f$. Then $(e - h)r = er = r$, which is impossible. For the second part, if $h \in \mathbf{B}(Q(R))$ and $hgr = gr$ then $(hg+(1-g))r = r$ so $(hg+(1-g))\mathfrak{cc}(r) = \mathfrak{cc}(r)$. Hence, $hg\,\mathfrak{cc}(r) = g\,\mathfrak{cc}(r)$. □

This lemma shows that for any R, an almost biregular ring S, $R \subseteq S \subseteq Q(R)$ with $S = R \diamond \mathbf{B}(S)$, must contain the central covers of all the elements of R. This suggests a recipe for constructing a "best approximation" of R by an almost biregular ring.

Theorem 1.7. *Let R be regular of finite index. Define $R^\sharp$ to be the ring generated by R and the central covers from $Q(R)$ of all the elements of R. Then $R^\sharp$ is the unique smallest almost biregular ring among the regular rings S, $R \subseteq S \subseteq Q(R)$. Moreover: (i) If $R^\sharp \subseteq T \subseteq Q(R)$ and $T = R^\sharp \diamond \mathbf{B}(T)$, then T is almost biregular. (ii) Let A be the sub-boolean algebra of $\mathbf{B}(Q(R))$ generated by the central covers of elements of R. Then $\mathbf{B}(R^\sharp) = A$.*

Proof. By the lemma, it will suffice to show that $R^\sharp$ is almost biregular. Since $R^\sharp = R \diamond \mathbf{B}(R^\sharp)$, each element $s \in R^\sharp$ can be expressed (as usual) in the form $s = \sum_{i=1}^m r_i e_i, r_i \in R, e_i \in \mathbf{B}(R^\sharp), \{e_1, \ldots, e_m\}$ orthogonal. Hence, it will suffice to show that supports of elements of R over $Y = \mathbf{Spec}\mathbf{B}(R^\sharp)$ are clopen.

Fix $r \in R$ and its central cover $\mathfrak{cc}(r)$ (which is in $\mathbf{B}(R^\sharp)$). Then $r\mathfrak{cc}(r) = r$ means $\mathrm{supp}_Y(r) \subseteq \mathrm{supp}_Y(\mathfrak{cc}(r))$. If these are not equal, we apply the argument of the lemma to get a contradiction.

(i) This follows from the second part of Lemma 1.6.

(ii) Suppose, with the usual notation, $f = \sum r_i e_i \in \mathbf{B}(R^\sharp)$, where the $r_i \in R$ and the $e_i \in A$. Then, each $r_i e_i \in \mathbf{B}(R^\sharp)$. However, $r_i e_i = \mathfrak{cc}(r_i e_i) = e_i \mathfrak{cc}(r_i) \in A$. □

We have already seen that this process can be more efficient than adding all of $\mathbf{B}(Q(R))$. Much more needs to be done to understand fully the relationship between R and $R^\sharp$. However, both $\mathfrak{b}(R)$ and $\mathbf{bi}(R)$ are unchanged in passing from R to $R^\sharp$ (even though $\mathbf{bi}(R^\sharp)$ may be strictly larger than the copy of $\mathbf{bi}(R)$ in $\mathbf{SpecB}(R^\sharp)$). Also the stalks of $R^\sharp$ do not change as more elements of $\mathbf{B}(Q(R))$ are added (although their multiplicity might increase). Central idempotents may be adjoined at will to a *biregular* ring without changing the stalks or the biregularity ([3, Proposition 2.2]). However, this is not true for "almost biregular" when the new central idempotents do not come from $\mathbf{B}(Q(R))$, as Example 3.3 shows. (On the other hand, "s-rich" – see after (1.3) – is preserved under the adjunction of *any* central idempotents.)

Proposition 1.8. *Let R be regular of finite index and $R^\sharp$ as in Theorem 1.7. (i) Let I be an ideal of R, $I \subseteq \mathfrak{b}(R)$. Then I is an ideal of $R^\sharp$. (ii) For each $u \in \mathbf{bi}(R)$ there is exactly one $y \in \mathbf{SpecB}(R^\sharp)$ lying over u. (iii) Let $T = R^\sharp \diamond A$, where A is a boolean algebra with $\mathbf{B}(R^\sharp) \subseteq A \subseteq \mathbf{B}(Q(R))$. If $y \in \mathbf{SpecB}(T)$ lies over $x \in \mathbf{SpecB}(R^\sharp)$ then $T_y = R^\sharp{}_x$.*

Proof.. (i) The fact that I is an ideal in $R^\sharp$ follows from the fact that for $e \in \mathbf{B}(R) \cap \mathfrak{b}(R), e\,\mathfrak{cc}(r) = \mathfrak{cc}(er)$ and that $er \in \mathfrak{b}(R)$ has a central cover in R.

(ii) Take $u \in \mathbf{bi}(R)$ and $e \in \mathbf{B}(R) \cap \mathfrak{b}(R)$ with $e \notin u$. Then for $y \in \mathbf{SpecB}(R^\sharp)$, y lying over u, $ye \subseteq u$. This shows that if y' lies over u then $y \subseteq y'$.

(iii) We have that T_y is a homomorphic image of $R^\sharp{}_x$. Suppose that for some $f \in R^\sharp, f_x \neq 0$. Let $g = \mathfrak{cc}(f) \in \mathbf{B}(R^\sharp) \subseteq \mathbf{B}(T)$. Then $g \notin x$ and, hence, $g \notin y$. Thus f_x is not sent to 0 in T_y. $\square$

Section 2. Adjoining central regular subrings. Results in Section 1 show that adjoining central idempotents to a regular ring of finite index does not make stalks worse, and can make them more tractable. However, adjoining other sorts of elements of $Q(R)$ can destroy the property that stalks are artinian. In Example 3.1(2), the ring R has a non-artinian stalk; there is a *biregular* ring S with $S \subset R \subset Q(S) = Q(R)$. Even adding a *central* regular subring of $Q(R)$ which extends $Z(R)$ can lead one from a biregular ring R to one which is not even regular: See Example 3.6. Some observations about adjoining central elements from $Q(R)$ can, however, be made.

Proposition 2.1. *Let R be a regular ring of finite index whose stalks are artinian. Suppose that, for each $x \in X = \mathbf{SpecB}(R)$, the field $Z(R)_x$ is perfect. Let C be a subring, $Z(R) \subseteq C \subseteq Z(Q(R))$ with C integral over $Z(R)$. Then the subring $R \diamond C$, generated by R and C, is regular.*

Proof. We have that $Z(R) \subseteq C \subseteq Z(R \diamond C)$ and so $\mathbf{B}(R) \subseteq \mathbf{B}(C) \subseteq \mathbf{B}(R \diamond C)$. Let $R' = R \diamond \mathbf{B}(R \diamond C)$ and $C' = C \diamond \mathbf{B}(R \diamond C)$. Pick $z \in \mathbf{SpecB}(R \diamond C)$, and let $x = z \cap \mathbf{B}(R)$, $y = z \cap \mathbf{B}(C)$. There are natural surjective homomorphisms $R_x \to R'_z$ and $C_y \to C'_z$. Now $R' \diamond C' = R \diamond C$ and we wish to show that $(R' \diamond C')_z$ is regular.

First, C' is regular by [20, Lemma 1.9]. Put $F_x = Z(R)_x$. The stalk $(R \diamond C)_z$ is a homomorphic image of $R'_z \otimes_{F_x} C'_z = T_z$, and so it suffices to show that T_z is regular. Any fixed element of T_z is in a subring of the form $R'_x \otimes_{F_x} D$, where D is a *finite* extension field of F_x. But, by [2, Corollaire 3, page 93], $R'_x \otimes D$ is semiprimitive, and then [16, footnote, page 116] shows that it is artinian. Hence each element is in a regular subring of T_z, and so T_z is regular. It follows that $R \diamond C$ is regular. □

An extension $R \diamond C$ of R by a central subring C of $Q(R)$ may be regular even though C is not integral over $Z(R)$ – see Example 3.7, where the regular ring $R \diamond \mathbf{B}(Q(R))$ has centre C which is not integral over $Z(R)$.

Proposition 2.2. *Suppose R is regular of finite index whose stalks are artinian and let C be a regular ring with $Z(R) \subseteq C \subseteq Z(Q(R))$. (i) Assume that for each $x \in X = \mathbf{Spec}\,\mathbf{B}(R)$, $Z(R)_x$ is the centre of each simple component of R_x, and each is finite dimensional over $Z(R)_x$. Then the subring $R \diamond C$ of $Q(R)$, generated by R and C, is regular with stalks artinian. (ii) If, in addition to the hypotheses of (i), R is almost biregular, then $R \diamond C$ is almost biregular. (iii) When R is Azumaya over its centre then $R \diamond C$ is also Azumaya with centre C.*

Proof. (i) We begin as in the previous proof. Set $R' = R \diamond \mathbf{B}(R \diamond C)$ and $C' = C \diamond \mathbf{B}(R \diamond C)$. Then $R \diamond C = R' \diamond C'$. Pick $z \in \mathbf{Spec}\,\mathbf{B}(R \diamond C)$ and set $x = z \cap \mathbf{B}(R)$. We have that $Z(R) \subseteq Z(R')$ and R'_z is a homomorphic image of R_x. Hence $F_x = Z(R)_x$ is the centre of each component of R'_z and each such component is finite dimensional over F_x.

Finally, $(R' \diamond C')_z$ is a homomorphic image of $R'_z \otimes_{F_x} C'_z$, which is artinian and semisimple by [16, Theorem 1, page 114; footnote, page 116].

(ii) We can use the same notation as in (i) because R' remains almost biregular (Theorem 1.7). Again, $(R' \diamond C')_z$ is a homomorphic image of $R'_z \otimes_{F_x} C'_z$, and the latter is a direct product of at most k simple artinian rings where k is the number of simple factors of R'_z. Moreover, $R'_z \subseteq (R' \diamond C')_z$ because R and R' have the "same" stalks by Proposition 1.8(iii) and $R' \diamond C'$ has the same central idempotents as R'. Since C' is central, $(R' \diamond C')_z$ will have at least as many simple components as R'_z. Hence, $(R' \diamond C')_z \cong R'_z \otimes_{F_x} C'_z$.

The simple components of $(R' \diamond C')_z$ have matrix units which, as we shall see, may be taken to be refinements of those in R'_z, and these will all have clopen support. A primitive idempotent $u \in R'_z$ splits into a sum of *isomorphic* orthogonal primitive idempotents in $(R' \diamond C')_z$, say $u = u_1 + \cdots + u_p$, thus there is a clopen neighbourhood N of z on which liftings of the u_i remain orthogonal isomorphic idempotents and on which a lifting of u is non-zero. Hence, the liftings of the u_i are *all* non-zero on N.

Using this fact, each element of $(R' \diamond C')_z$ can be expressed as a combination of elements of $(R' \diamond C')_z$ (the various matrix units) which lift to elements with clopen support, and with "coefficients" which are zero or elements invertible with respect to the central idempotents of R'_z. These latter have support on a neighbourhood of z. This shows that $R \diamond C$ is almost biregular.

(iii) By [21, Theorem 6.1] a ring is an Azumaya algebra over a commutative regular ring if and only if it is a biregular ring that is finitely generated over its centre. In that case, each R_x is a finitely generated simple algebra with centre $Z(R)_x$. Thus, as we have seen, $R \diamond C$ is regular with stalks artinian. Moreover, by [13, Lemma 5.1 and Lemma 3.5], $R \otimes_{Z(R)} C$ is an Azumaya algebra with centre C. In the natural homomorphism $R \otimes_{Z(R)} C \to R \diamond C$, the kernel is generated by an ideal of C, and, hence, must be trivial. □

Hypotheses such as those of Proposition 2.2(ii) cannot be dispensed with. The adjunction of a central integral element from $Z(Q(R))$ can change a biregular ring into one which is not even almost biregular. See Example 3.8.

Section 3 – Examples. The examples presented in this section serve to illustrate some of the results of the earlier sections and to show some of the complexity of regular rings of finite index. Even so, all the examples presented here have features which make them relatively simple.

To what extent can a regular ring of finite index have stalks which are *not* artinian? By a special case of [8, Proposition 2], if S_n, $n \in \mathbf{N}$ is an indecomposable ring and T any ring equipped with a homomorphism $\varphi : T \to \Pi/\bigoplus = \prod_{n \in \mathbf{N}} S_n / \bigoplus_{n \in \mathbf{N}} S_n$ such that $\mathbf{B}(T) \cap \varphi^{-1}(\pi(\mathbf{B}(\Pi)) = \{0, 1\}$, then these data give rise to a ring R with $\mathbf{SpecB}(R) = \mathbf{N} \cup \{\infty\}$, where $R_n = S_n$ for all $n \in \mathbf{N}$ and $R_\infty = T$. If we now take the S_n to be simple artinian rings of bounded index and T a regular ring of finite index then in order to use the construction $\ker \varphi$ must be 0; otherwise it would contain a non-zero central idempotent, contradicting the condition on φ. Hence φ must be an injection. (A generalization of [8, Proposition 2] is found in [7, Theorem 3.9].)

An example of a regular ring of finite index with a stalk which is not artinian is [10, Example, page 525]. The following is a recipe for constructing regular rings of finite index with one stalk which can be made quite complicated. Many variants are possible.

Example 3.1. *Let F be a field and fix $m \in \mathbf{N}$. Let T be any regular subring of $\prod_\mathbf{N} \mathbf{M}_m(F)$ which contains $\bigoplus_\mathbf{N} \mathbf{M}_m(F)$. Define $\varphi : T \to \prod_\mathbf{N} \mathbf{M}_{2m}(F) = \Pi$ as follows. Choose a set of disjoint infinite subsequences of $\mathbf{N}$, $\mathcal{S}_1, \mathcal{S}_2, \ldots$ so that $\bigcup_{i \in \mathbf{N}} \mathcal{S}_i = \mathbf{N}$. Put $\varphi(t)_n = \begin{pmatrix} t_i & 0 \\ 0 & t_{i+1} \end{pmatrix}$ if $n \in \mathcal{S}_i$. Then φ followed by the natural epimorphism from the product to the product divided by the direct sum, gives a ring R, as in [8, Proposition 2], on*

$\mathbf{N} \cup \{\infty\}$ *(the one-point compactification of the discrete $\mathbf{N}$), with stalks $R_n = \mathbf{M}_{2m}(F)$ and $R_\infty = T$. In this case $R/\mathfrak{b}(R) = T$.*

Proof. Let π be the epimorphism from the direct product Π to the product modulo the direct sum, $\Pi/\bigoplus$. Any component t_i of $t \in T$ will appear in infinitely many components of $\varphi(t)$ in the upper left corner, and, except for t_1, in infinitely many different ones, in the lower right corner. If $\pi\varphi(t) \in \pi(\mathbf{B}(\Pi))$ then if $t_1 = 1, t_2 = 1$ as well, and so on, for all the components of t. Similarly if $t_1 = 0$. □

We now list some special cases, variants and remarks based on this construction.

1. If T is not biregular (e.g., the ring of sequences from $\mathbf{M}_2(F)$ which are eventually constant and diagonal), then R has a stalk which is not biregular. Moreover, $R/\mathfrak{b}(R) = R_\infty$ is not biregular.
2. Note that in these examples if φ is restricted to the field F (as the diagonal in T), we get a biregular ring S. Then $S \subseteq R$, $\mathbf{B}(S) = \mathbf{B}(R)$ and $Q(S) = Q(R) = \prod_{\mathbf{N}} \mathbf{M}_{2m}(F)$.
3. A variant of the above is to take $\varphi : T \times F \to \prod_{\mathbf{N}} \mathbf{M}_{m+1}(F)$ defined by $(\varphi(t, \alpha))_n = \begin{pmatrix} t_i & 0 \\ 0 & \alpha \end{pmatrix}$, if $n \in S_i$. This builds a ring R of index $m+1$ from one of index m. The stalk R_∞ here $T \times F$.

Example 3.2. *(in which R has a simple stalk which is not over a point in $\mathrm{bi}(R)$).* Let F be a field and T be the ring of sequences from $\mathbf{M}_2(F)$ which are eventually constant and diagonal. Put $R = (\bigoplus_{\mathbf{N}} T) + F$. Here $\mathrm{Spec}\mathbf{B}(R) = \bigcup_{\mathbf{N}}(\mathbf{N} \cup \{\infty_n\}) \cup \{\infty\}$. For $m_n \in \bigcup_{\mathbf{N}} \mathbf{N}, R_{m_n} = \mathbf{M}_2(F)$, $R_{\infty_n} = F \times F$ and $R_\infty = F$. However, any neighbourhood of ∞ meets some of the ∞_n.

Example 3.3. There is an almost biregular ring R which has an extension by central idempotents that is not almost biregular. (cf., Theorem 1.7(i).)

Proof. Let F be a field and R the ring of sequences from $\mathbf{M}_2(F)$ which are eventually constant and of the form $\mathrm{diag}(a, b)$. Let $T = R \times \prod_{\mathbf{N}} F$ and $\alpha : R \to T$ defined by $\alpha(r) = (r, b, a, a, a, \dots)$, where $r \in R$ is eventually constant of the form $\mathrm{diag}(a, b)$. Let the sequences which make up T be considered as functions on $A \cup \{u\} \cup B$, where A corresponds to the components of elements of R, u to the component "b" and B to the remaining components.

Let $E = \{e \in \mathbf{B}(T) \mid \mathrm{supp}_A(e) \text{ is cofinite, } e_u = 0 \text{ and } \mathrm{supp}_B(e)$ is cofinite$\}$. This will be abbreviated by saying that e has support (cofinite, 0, cofinite). We put $S = \alpha(R)[E]$. A central idempotent of S must have components which are each either 0 or 1. It is readily seen that $g \in \mathbf{B}(S)$, with $g_u = 1$, has support (finite, 1, finite) or (cofinite, 1, cofinite) – with an obvious extension of the terminology.

Now let $r \in R$ be constantly diag$(0,1)$. Then the only elements e of $\mathbf{B}(S)$ with $e\alpha(r) = \alpha(r)$ are those with support $(A, 1, \text{cofinite})$. There is no smallest such. □

The ring S, above, is, however, s-rich.

Example 3.4. *Let F be a field and for each $n \in \mathbf{N}$, let R_n be the ring of eventually constant sequences from F. Put $R = (\bigoplus_{n \in \mathbf{N}} R_n) + F$, that is, the ring of sequences $(s_n)_{n \in \mathbf{N}}$, $s_n \in R_n$ which are eventually a constant scalar sequence. Adjoin the central idempotents e_n from $\mathbf{B}(\prod_{\mathbf{N}} \times \mathbf{N} F) = \mathbf{B}(Q(R))$, where $(e_n)_{ij}$ is 0 if $i \neq n$ or j is odd, and is 1 if $i = n$ and j is even. Then $S = R[e_1, e_2 \ldots]$ is an infinite extension but for each $x \in \mathbf{SpecB}(R)$, there are only finitely many $y \in \mathbf{SpecB}(S)$ lying over x.*

Proof. We see that

$$\mathbf{SpecB}(R) = \left(\bigcup_{n \in \mathbf{N}} \mathbf{N}_n \cup \{\infty_n\} \right) \cup \{\infty\}$$

and

$$\mathbf{SpecB}(S) = \left(\bigcup_{n \in \mathbf{N}} \mathbf{N}_n \cup \{\alpha_{n1}, \alpha_{n2}\} \right) \cup \{\infty\},$$

where $\mathbf{N}_n$ is a copy of $\mathbf{N}$. Note that α_{n1}, α_{n2} lie over ∞_n. (See Proposition 1.4.) □

Example 3.5. *(of a regular ring of index 2 where the stalks are artinian but the number of simple components in them is unbounded.) Let F be a field and $\Pi = \prod_{\mathbf{N}} \mathbf{M}_2(F)$ and $\bigoplus = \bigoplus_{\mathbf{N}} \mathbf{M}_2(F)$. For each $n > 1$ let $\phi_n : F^n \to \Pi$ be defined by*

$$(a_1, \ldots, a_n) \mapsto \left(\begin{pmatrix} a_1 & 0 \\ 0 & a_2 \end{pmatrix}, \ldots, \begin{pmatrix} a_1 & 0 \\ 0 & a_n \end{pmatrix}, \ldots \right),$$

where the pattern is repeated. The only central idempotents in the image, modulo the direct sum, are 0 and 1. Hence we get a ring S_n over $\mathbf{N} \cup \{\infty_n\}$ with simple stalks except at ∞, where the stalk is F^n. To complete the example, let $R = (\bigoplus_{n>1} S_n) + F$. Then each F^n appears as a stalk; nevertheless, all the stalks are artinian.

Example 3.6. *There is an example of a regular biregular ring R of finite index (even abelian regular) so that the ring obtained by adjoining the centre of $Q(R)$ is not regular.*

Proof. Consider a division ring D with centre $\mathbf{C}$, the field of complex numbers, $D \neq \mathbf{C}$ (for example the division ring of quotients of the first complex

Weyl algebra). Let X be an extremally disconnected infinite boolean space and R the ring of sections of the simple sheaf of D over X ([19, Definition 11.2]). Then $Z(R)$ is the ring of sections of the simple sheaf of **C** over X, which is not self-injective ([4, Theorem 1]). This means that $C = Z(Q(R))$ is strictly larger than $Z(R)$ and for some $x \in X$, C_x is a transcendental extension of **C**. However, $(R \diamond C)_x$ is a homomorphic image of $R_x \otimes_\mathbf{C} C_x = D \otimes_\mathbf{C} C_x$, which is simple ([16, Theorem 1, page 114]) but not regular ([18, Corollary 1.4]). Hence $R \diamond C$ is biregular but not regular. □

Example 3.7. *(i) There exists a regular ring of finite index R and a boolean algebra of idempotents A, with $\mathbf{B}(R) \subseteq A \subseteq \mathbf{B}(Q(R))$, so that $\mathbf{B}(R \diamond A) \neq A$. (ii) There is an example of a regular ring of finite index R so that $Z(R \diamond \mathbf{B}(Q(R))) \neq Z(R) \diamond \mathbf{B}(Q(R))$, and, moreover, the extension is not integral.*

Proof. (i) Fix a field F. Take the ring

$$R = \bigoplus_{n \in \mathbf{N}} \mathbf{M}_2(F) +$$

$$\{(\mathrm{diag}(a,b), \mathrm{diag}(a,a), \mathrm{diag}(a,b), \mathrm{diag}(b,b), \ldots) \mid a, b \in F\}.$$

Let $e = (0, 1, 0, 1, 0, 1, \ldots)$ and let $A = \mathbf{B}(R)[e]$. Then $\mathbf{B}(R \diamond A)$ also contains $f = (0, 0, 0, 1, 0, 0, 0, 1, 0, 0, 0, 1, \ldots) \notin \mathbf{B}(R)[e]$.

(ii) The type of ring in the second part is illustrated by the basic construction of Example 3.1 using $T = \prod_\mathbf{N} F$ where $n = 1$. This procedure gives a ring R where $Z(R)$ is the ring of scalar sequences which are eventually constant. Hence $Z(R) \diamond \mathbf{B}(Q(R))$ is the ring of scalar sequences with finitely many values. On the other hand, $Z(R \diamond \mathbf{B}(Q(R)))$ contains scalar sequences which are constant on each $\mathcal{S}_i$, and, hence, may have infinitely many values. Since this latter ring has homomorphic images which are non-standard models of F (if F is infinite), it follows that the embedding $Z(R) \diamond \mathbf{B}(Q(R)) \subset Z(R \diamond \mathbf{B}(Q(R)))$ is not integral. □

Example 3.8. *There is a biregular regular ring of finite index R and $c \in Z(Q(R))$ so that c is integral over $Z(R)$ but the regular ring $R[c]$ is not almost biregular.*

Proof. Let R be the ring of sequences of 2×2 real matrices which are eventually "constant" of the following form: $q_1 + q_2 A$ in the even-numbered places and $q_1 + q_2 \sqrt{2}$ in the odd numbered places, where q_1 and q_2 are rational (all scalars are considered to be scalar matrices) and $A = \begin{pmatrix} 0 & 2 \\ 1 & 0 \end{pmatrix}$. It is important that $A^2 = 2$. This ring is biregular with $\mathbf{Spec}\mathbf{B}(R) = \mathbf{N} \cup \{\infty\}$; $R_n = \mathbf{M}_2(\mathbf{R})$ for $n \in \mathbf{N}$ and $R_\infty = \mathbf{Q}[\sqrt{2}]$, **Q** the field of rational numbers.

We now adjoin $c \in Z(Q(R))$ which is constantly $\sqrt{2}$. A typical element of $S = R[c]$ is a sequence of elements of $\mathbf{M}_2(\mathbf{R})$ which is eventually "constant" of the form
$$\begin{pmatrix} q_1 + q_1'\sqrt{2} & 2q_2 + 2q_2'\sqrt{2} \\ q_2 + q_2'\sqrt{2} & q_1 + q_1'\sqrt{2} \end{pmatrix}$$
in the even places and $q_1 + 2q_2' + (q_2 + q_1')\sqrt{2}$ in the odd, $q_1, q_2, q_1', q_2' \in \mathbf{Q}$. These even-numbered components will be zero only if $q_1 = q_2 = q_1' = q_2' = 0$ while the odd-numbered ones are zero if $q_1' = q_2$ and $q_2' = -(1/2)q_1$.

Now suppose that we have an element $s \in S$ which is eventually 0 in the odd places but non-zero in the even. Then the even places cannot be 1. Hence, the only central idempotent e of S which has $es = s$ is 1 almost everywhere. This is not the central cover of s in $Q(R)$. Thus, S is not almost biregular. (The only non-simple stalk of S is $\mathbf{Q}[\sqrt{2}] \times \mathbf{Q}[\sqrt{2}]$ at "∞".) $\square$

REFERENCES

[1.] R.F. Arens and I. Kaplansky, *Topological representation of algebras*, Trans. Amer. Soc. **63** (1948), 457-481.

[2.] N. Bourbaki, *Algèbre, Chapitre 8*. Hermann, Paris, 1958.

[3.] W.D. Burgess, *Minimal rings, central idempotents and the Pierce sheaf*, Contemp. Math. **171** (1994), 51-67.

[4.] W.D. Burgess, K.A. Byrd and R. Raphael, *Self-injective simple Pierce sheaves*, Arch. Math. **42** (1984), 354-361.

[5.] W.D. Burgess and J.-M. Goursaud, K_0 *d'un anneau dont les localisés centraux sont simples artiniens*, Canad. Math. Bull. **25** (1982), 344-347.

[6.] W.D. Burgess and R. Raphael, *On almost biregular rings of finite index*, preprint, 1996.

[7.] W.D. Burgess and R. Raphael, *Ideal extensions of rings – some topological aspects*, Comm. Algebra **23** (1995), 3815-3830.

[8.] W.D. Burgess and W. Stephenson, *Pierce sheaves of non-commutative rings*, Comm. Algebra **4** (1976), 51-75.

[9.] W.D. Burgess and W. Stephenson, *An analogue of the Pierce sheaf for non-commutative rings*, Comm. Algebra **6** (1978), 863-886.

[10.] A.B. Carson, *Representations of regular rings of finite index*, J. Algebra **39** (1976), 512-526.

[11.] A.B. Carson, *Representations of regular rings of finite index - II*, Comm. Algebra **21** (1993), 4173-4177.

[12.] J. Dauns and K.H. Hofmann, *The representation of biregular rings by sheaves*, Math. Zeit. **91** (1966), 103-123.

[13.] F. DeMeyer and E. Ingraham, *Separable algebras over commutative rings*, Lecture Notes in Mathematics, vol. 181, Springer-Verlag, 1971.

[14.] K.R. Goodearl, *von Neumann Regular Rings*. Pitman, London, 1979; Second Ed., Krieger, Melbourne, FL, 1991.

[15.] K.R. Goodearl, K_0 of regular rings with bounded index of nilpotence, Contemp. Math. **171** (1994), 173-199.
[16.] N. Jacobson, *Structure of Rings*. Amer. Math. Soc. Colloquium Publications vol. 37, Providence, RI, 1956.
[17.] S. Koppelberg, *General Theory of Boolean Algebra* (J.D. Monk, ed.), Handbook of Boolean Algebra, Part I, North-Holland, Amsterdam, 1989.
[18.] P. Menal, *On tensor products of algebras being von Neumann regular or self-injective*, Comm. Algebra **9** (1981), 691-697.
[19.] R.S. Pierce, *Modules over commutative regular rings*, Memoirs Amer. Math. Soc. **70**, 1967.
[20.] R. Raphael, *Algebraic extensions of commutative regular rings*, Canad. J. Math. **22** (1970), 1133-1155.
[21.] G. Szeto, *A characterization of Azumaya algebras*, J. Pure Appl. Algebra **9** (1976), 69-71.

DEPARTMENT OF MATHEMATICS AND STATISTICS, UNIVERSITY OF OTTAWA, OTTAWA, CANADA K1N 6N5
E-mail address: wdbsg@uottawa.ca

DEPARTMENT OF MATHEMATICS, CONCORDIA UNIVERSITY, MONTREAL, CANADA H4B 1R6
E-mail address: raphael@alcor.concordia.ca

INTERSECTIONS OF MODULES

JOHN DAUNS

ABSTRACT. Necessary and sufficient condition are found on a ring R or on a module M so that every submodule $K < M$ can be represented as an irredundant intersection $K = \cap \{K_i \mid i \in I\}$ of submodules $K < K_i < M$, where either the K_i or their quotients M/K_i are required to satisfy some prescribed property. The phenomenon of irredundancy of such intersections is investigated.

INTRODUCTION

For modules over a commutative ring, various Noether-Lasker type theorems have been considered ([F], [Ki], [Ma], [McCoS], [McC2], [N] and [O]). Some attempts have been made to extend this to two sided ideals of a noncommutative ring ([B], [C] and [Mu]), and very recently to modules over a noncommutative ring ([D11]).

It is known that R is a V-ring if and only if every right ideal $K < R$ is an intersection $K = \cap\{K_i \mid i \in I\}$ of maximal right ideals $K \subseteq K_i < R, i \in I$ ([MV; p. 186, Thm 2.1]). Here as a special case of the type of theory developed here, among other things we will characterize that proper subclass of V-rings R for which every right ideal is an *irredundant* intersection of maximal right ideals. A method is developed whereby conditions or criteria are found on the ring R in order that every submodule $K < M$ of any module M be an irredundant intersection $K = \bigcap_{i \in I} K_i$ of submodules $K_i < M$ such that all M/K_i belong to a prespecified class $\mathcal{S}$ of R-modules, where $\mathcal{S}$ is closed under isomorphic copies and nonzero submodules (Theorem 2.2 and Main Corollary 2.3). If quotients of modules in $\mathcal{S}$ contain submodules in $\mathcal{S}$, then the irredundant intersection property is equivalent to the existence of a smooth chain $0 = R_0 < R_1 < \cdots < R_\alpha < \cdots < R_\tau = R$, where $R_{\alpha+1}/R_\alpha \in \mathcal{S}$, $\alpha \leq \tau$ (Corollary 2.4). The latter is satisfied, in particular, if $\mathcal{S}$ are the simple modules, or quotient finite dimensional modules. As shown in 2.10 here, such an R need not itself be q.f.d., but only locally q.f.d. in the above sense. It would be interesting to see how the various remarkable characterizations of q.f.d. rings ([AJL], [Ca], [Ku], [L] and [S]) extend to the locally q.f.d. ones.

If we are interested in V-rings, then $\mathcal{S}$ are the simple R-modules. A nonzero module V is *atomic* if for any $0 \neq x, 0 \neq y \in V$, for some index set Γ, there is an embedding $xR \hookrightarrow E(\bigoplus_\Gamma yR)$, where 'E' denotes injective hulls. Of possible future interest are the following special choices of $\mathcal{S}$:

(i) atomic modules,

(ii) atomic modules which contain no uniform submodules, i.e., *continuous* atomic modules,

---

1991 Mathematics Subject Classification. 16D10.

(iii) uniform modules,

(iv) uniform modules which do or alternatively, do not contain simple modules, and

(v) the set of simple modules in the already mentioned case of V-rings; or

(vi) quotient finite dimensional modules.

However, none of the previous choices are generally applicable, except to somewhat special rings. And therefore it is shown how it is always possible to represent a submodule $K < M$ of any module M as an irredundant intersection $K = \bigcap\{K_i \mid i \in I\}$. Here $\Gamma = \{\alpha(i) \mid i \in I\}$ is a pairwise orthogonal set of saturated classes of R-modules (with $\alpha(i) \cap \alpha(j) = \{(0)\}$ for $i \neq j$), and with $M/K_i \in \alpha(i)$. It is known that many such sets $\Gamma = \{\alpha(i) \mid i \in I\}$ always exist in great abundance ([D10], [D9]). This approach is quite different from the previous one because now "$\mathcal{S}$" is no longer predominantly used. A theorem of Osofsky ([O, p. 761, Thm A]) is a special case of the present more general approach (Proposition 3.3). Here saturated classes of modules provide the rather general, but also unifying concept. They have shown themselves to be useful in many other contexts. Page and Zhou ([PZ]) call them 'natural classes'. For any saturated class Δ, every module M contains a Δ-maximal submodule (Definition 1.3). They are used extensively not only here, but also by Zhou ([Z1], [Z2]), who calls them 'type submodules'. For torsion free modules M, they were indispensible in [D6, p. 108, 3.5 and 3.7] and [D7, p. 331, 2.2].

In many cases above, it will be relatively easy to conclude that $K = \bigcap\{K_i \mid i \in I\}$, where all the $K_i < M$ must be of some restricted kind. More effort, and additional restrictive hypotheses will be necessary to show that the intersection is irredundant. Irredundant intersections are of interest because sometimes they give rise to essential subdirect product representations $M/K \hookrightarrow \prod_{i \in I} M/K_i$ as in [D5] and [D12].

1. SATURATED CLASSES

In this section some properties of saturated classes of modules are developed. Only those facts are considered which are needed later.

SYMBOLS AND TERMINOLOGY. Modules M are right unital over an arbitrary ring R. Annihilators, ideals and submodules are denoted by '$\perp$', $\triangleleft$, $<$ or $\leq$, and large or essential submodules by '$<<$'. The notation '$A <\!\!\not< B$' means that $A < B$ are modules, and that A is not large in B. The right R-singular sumbodule of M is $ZM = Z(M) = \{m \in M \mid m^\perp << R\} \leq M$, while $ZM << Z_2 M = Z_2(M) \leq M$ is the second singular submodule. A module M is *torsion free* (notation: t.f.) if $ZM = 0$, and *torsion* if $M = Z_2 M$. Right R-injective hulls are denoted by both '$\widehat{}$' or 'E', e.g.,

$\widehat{M} = EM = E(M)$. We use 'E' if M is given by a complicated formula, and the same for using or not using '( )' after E, Z or Z_2. The category of right R-modules is denoted by $\mathcal{M}od_R$.

We summarize some relevant notions from [D9, p. 47, Def. 3.1 and 3.2]. For more details, see [D13, 2.9-2.15] and/or [D11, 2.1, 2.2, 2.6-2.9].

Definition 1.1. By a class Δ of modules will always be meant a nonempty class $\Delta \neq \emptyset$ closed under isomorphic copies. The *complementary class* of Δ is $c\Delta = c(\Delta) = \{W_R \in \mathcal{M}\ od_R \mid \forall\, 0 \neq V \leq W, V \notin \Delta\}$. Note that $(0) \in c\Delta$. A class of modules is a *saturated class* if it is closed under (i) submodules, (ii) direct sums and (iii) injective hulls. For any ring R, $\Sigma(R)$ denotes the set of all saturated classes of right R-modules. (See [D11, Def. 2.3 and Lemma 2.7].) For any class $\mathcal{S}$ of R-modules, $\langle \mathcal{S} \rangle$ denotes the saturated class of right R-modules that $\mathcal{S}$ generates. For $M \in \mathcal{M}od_R$, abbreviate $\langle \{M\} \rangle = \langle M \rangle$.

Definition 1.2. A module V is *atomic* if $V \neq 0$, and if for any $0 \neq v \in V$, there is some index set Γ such that there exists an embedding $V \hookrightarrow E(\bigoplus_\Gamma vR)$.

Three special saturated classes are defined as follows. A module M is (i) *saturated simple*, (ii) *discrete* or (iii) *molecular* if M contains an essential sum of, respectively, (i) simple, (ii) uniform or (iii) atomic modules. The complementary saturated class of the discrete ones are the *continuous* modules, and of the molecular ones – the *bottomless* modules.

Definitions and Notations 1.3. For any saturated class Δ, its complimentary class $c\Delta$ is also saturated. By Zorn's lemma, any module M contains complement submodules $M_{(\Delta)} \leq M, M_{(c\Delta)} \leq M$ which are characterized or defined by simply being maximal with respect to $M_{(\Delta)} \in \Delta$ and $M_{(c\Delta)} \in c\Delta$. In view of 1.1 (iii), they necessarily are complements. In general, they are not unique and the notation '$M_{(\Delta)}$' will denote any one such. Any submodule of M of the form $M_{(\Delta)}$ is called a Δ-*maximal* submodule. For any right complement submodule $C \leq M$ such that $M_{(\Delta)} \oplus C << M$, automatically $C = M_{(c\Delta)} \in c\Delta$ is a $c\Delta$-maximal submodule. Since $c(c\Delta) = \Delta$, the above also holds for $c\Delta$ in place of Δ.

Define $\Delta_M = \{A \leq M \mid A \in \Delta\}$, and $max\Delta_M$ to be the set of maximal elements of Δ_M, i.e., $max\Delta_M = \{B \leq M \mid B \in \Delta, \forall\, B \stackrel{C}{\neq} C \leq M, C \notin \Delta\}$. Similarly for $c\Delta_M = (c\Delta)_M = c(\Delta)_M$ and $max\ c\Delta_M$.

Lemma 1.4. Assume that $\mathcal{S}$ is any nonempty class of nonzero R-modules closed under isomorphic copies and nonzero submodules. For an R-module M, the following are equivalent.

(i) $M \in \langle \mathcal{S} \rangle$.

(ii) $\exists$ cyclics $x_\gamma R \in \mathcal{S}$ and $\exists\, M \hookrightarrow E(\bigoplus_\gamma x_\gamma R)$

(iii) $\exists\, v_\alpha \in M, v_\alpha R \in \mathcal{S}$, and $\bigoplus_\alpha v_\alpha R << M$.

Proof. (iii) $\Longrightarrow$ (ii) $\Longrightarrow$ (i) are straightforward. (i) $\leftrightarrow$ (iii). Let Y denote the class of all modules M satisfying (iii). Since by definition, $\langle S \rangle$ is the intersection of all saturated classes containing S, and since $S \subseteq Y \subseteq \langle S \rangle$, it suffices to show that $Y = \langle S \rangle$. In order to show that Y is closed under submodules, it suffices to show that for any $M < E(\bigoplus_\gamma x_\gamma R), x_\gamma R \in S$, M satisfies (iii).

For $V, W \leq M \leq E(\bigoplus_\gamma x_\gamma R)$, define $V \stackrel{\preceq}{=} W$ if $W = V \oplus U \leq M$, where both V and U are direct sums of cyclics belonging to S. Application of Zorn's lemma to the inductive partial order '$\stackrel{\preceq}{=}$' shows that there exists a $V = \bigoplus_\alpha v_\alpha R \leq M$ with $v_\alpha \in M$ and $v_\alpha R \in S$, and with V a maximal element in the partial order '$\stackrel{\preceq}{=}$'. If $V <\not\preceq M$, then $V \oplus \xi R \leq M$ for some $0 \neq \xi \in M$. But for any $0 \neq \xi \in M \leq E(\bigoplus_\gamma x_\gamma R)$, there exists an index γ and $a, b \in R$ with $0 \neq \xi a \in M, 0 \neq x_\gamma b \in x_\gamma R \subset E(\oplus x_\gamma R)$, with $(\xi a)^\perp = (x_\gamma b)^\perp$, and $x_\gamma bR \cong \xi aR \in S$. (See one of [D13, 1.3] or [D11, 1.3].) Thus $V \prec V \oplus x_\gamma R \leq M$ is a contradiction. Hence $M \leq \widehat{V} = E(\bigoplus_\alpha v_\alpha R), v_\alpha R \in S$ shows that $M \in Y$, and Y is closed under submodules. That Y is closed under injective hulls and direct sums is clear. Thus (i) $\Longleftrightarrow$ (ii) $\Longleftrightarrow$ (iii).

Lemma 1.4 also holds under the hypothesis that $(0) \in S$ and S is closed under all submodules. The next definition and observation will show how many examples of sets of the form 'Γ' used later here can always be obtained for any classes or concrete examples of rings R.

Definition and Facts 1.5. It is known that $\Sigma(R)$ is a complete Boolean lattice with $0 = \{(0)\} < 1 = \mathcal{M}od_R \in \Sigma(R), \alpha < \beta \in \Sigma(R) \Longleftrightarrow \alpha \subset \beta$, $\alpha \wedge \beta = \alpha \cap \beta, \alpha \wedge c\alpha = 0, \alpha \vee c\alpha = 1$, while for any finite or infinite $S \subseteq \Sigma(R), \vee S = sup\, S = \langle S \rangle \in \Sigma(R)$. (See [D13, 2.15, Theorem I] or [D11, 2.10].)

Two R-modules M and N are said to be *orthogonal* if M and N do not have any nonzero isomorphic submodules, denoted by $M \perp N$. Abbreviate $\langle \{M\} \rangle = \langle M \rangle$. Then $M \perp N$ if and only if $\langle M \rangle \perp \langle N \rangle$ in $\Sigma(R)$, i.e., $\langle M \rangle \wedge \langle N \rangle = 0$. (In [D2] and later in other places, the author used "$\|$" for the present "$\perp$".)

A pairwise disjoint set of saturated classes with supremum one is a subset $\emptyset \neq \Upsilon = \{\alpha, \beta, \cdots\} \subset \Sigma(R)$ such that for any $\alpha \neq \beta \in \Upsilon, \alpha \wedge \beta = 0$ and $\vee \Upsilon = sup\, \Upsilon = 1 = \mathcal{M}od_R$.

A completely equivalent reformulation of the latter is that for every $M \in \mathcal{M}od_R$, and for any choice or selection of α-maximal submodules $\{M_{(\alpha)} \mid \alpha \in \Upsilon\}$, we have $\Sigma\{M_{(\alpha)} \mid \gamma \in \Upsilon\} = \bigoplus\{M_{(\alpha)} \mid \alpha \in \Upsilon\} << M$.

For $K < M$, we now discard all $\alpha \in \Upsilon$ with $(M/K)_{(\alpha)} = 0$, and then keep only those modules in α which are submodules of injective hulls of certain direct sums of cyclics of M. Define $\Gamma = \{\gamma \in \Sigma(R) \mid \exists \alpha \in \Upsilon, \gamma = \alpha \cap \langle M/K \rangle$ and $(M/K)_{(\alpha)} = (M/K)_{(\gamma)} \neq 0\}$. The latter is more of a conceptual and notational convenience than anything else, and the reader may equally well prefer to work with the isomorphic set $\{\alpha \in \Upsilon \mid$

$(M/K)_\alpha \neq 0\} \cong \Gamma$, under $\alpha \longrightarrow \alpha \cap \langle M/K \rangle$. (In lattice language, $\langle M/K \rangle = sup\{\alpha \cap \langle M/K \rangle \mid \alpha \in \Upsilon\} = \vee \Gamma = sup\, \Gamma \in \Sigma(R)$.) There is also the option of working with the same fixed Υ for all M, and all $K < M$, provided that it is understood that in direct sums zero direct summands are deleted, or in intersections submodules equal to all of M are removed.

The next lemma can be obtained from [D13, Lemma 2.8 (2), (i)] (see also [D11, 2.7]), and its proof is omitted.

Lemma 1.6. For any saturated class Δ, for any module M, and for any complement submodule $N \leq M$, the following are all equivalent.

(i) $N \in max\Delta_M$;

(ii) $N \in \Delta$ and $\exists D \leq M, N \oplus D << M, D \in c(\Delta)$;

(iii) $N \in \Delta$, and $\forall D \leq M, N \cap D = 0 \Longrightarrow D \in c(\Delta)$;

(iv) $N \in \Delta, M/N \in c(\Delta)$.

Later on, the next lemma will tell us whether a certain submodule in an intersection of submodules is unique, or whether it can be replaced by some other appropriate submodule without changing the whole intersection. Absolute values denote the cardinality of a set.

Lemma 1.7. For a module M and a saturated class Δ, the following conditions are equivalent.

(i) $\mid max(c\Delta)_M \mid \geq 2$

(ii) $\exists\, 0 \neq K \leq M, K \in c\Delta$, and $0 \neq \varphi : K \longrightarrow W$, where $W \leq M$, and $W \in \Delta$

(iii) $\forall\, 0 \neq N \in max(c\Delta)_M, \exists\, 0 \neq K \leq N$, and $0 \neq \varphi : K \longrightarrow W, M \geq W \in \Delta$.

For any class $\mathcal{S}$ of R-modules closed under isomorphic copies and nonzero submodules such that $\mathcal{S} \subseteq \Delta = \langle \mathcal{S} \rangle$, (i), (ii) and (iii) are equivalent to (iv), and equivalent to (v).

(iv) $\exists\, 0 \neq L \leq M,\ L \in c\Delta$, and $\exists\, 0 \neq \psi : L \longrightarrow V$, where $V \leq M, V \in \mathcal{S}$.

(v) $\forall\, 0 \neq N \in max\,(c\Delta)_M, \exists\, 0 \neq L \leq N$, and $\exists\, 0 \neq \psi : L \longrightarrow V, M \geq V \in \mathcal{S}$.

Proof. Trivially, (iv) $\Longrightarrow$ (ii). (ii) $\Longrightarrow$ (iv). By 1.4 (iii), there exists $\bigoplus_\alpha w_\alpha R << W$ with all $w_\alpha R \in \mathcal{S}$. Since $0 \neq \varphi K \subseteq W, L = \varphi^{-1}[\varphi K \cap \bigoplus_\alpha w_\alpha R] \neq 0$, and for an appropriate projection π_α of the direct sum of cyclics onto the summand $w_\alpha R, 0 \neq \pi_\alpha \varphi : L \longrightarrow w_\alpha R \in \mathcal{S}$ as required.

(ii) $\Longrightarrow$ (i). For any module M and elements $a, b \in M$ with $a^\perp \subseteq b^\perp$ and $aR \cap bR = 0$, we always have $a^\perp = (a-b)^\perp$. Now take $a \in K$ with

$0 \neq b = \varphi a \in W$. Then $aR \in c\Delta, \varphi aR \in \Delta$ and hence $aR + \varphi aR = aR \oplus \varphi aR$. Thus, $aR \cong (a - \varphi a)R \in c\Delta$, and by Zorn's lemma, there exist $M_1, M_2 \in max(c\Delta)_M$ with $aR \subseteq M_1 \leq M$ and $(a - \varphi a)R \subseteq M_2 \leq M$. Since $\varphi aR \subseteq W \in \Delta$, and $\varphi aR \subseteq M_1 + M_2$, necessarily $M_1 + M_2 \notin (c\Delta)_M$, and in particular, $M_1 \neq M_2$. Thus $| max(c\Delta)_M | \geq 2$.

(i) $\implies$ (ii). Let $M_1 \neq M_2 \in max(c\Delta)_M$. Since $M_2 < M$ has no proper essential extension in M, $M_2 \not<_e M_1 + M_2$, and there exists $0 \neq W \leq M_1 + M_2$ with $W \oplus M_2 \leq M_1 + M_2$. If $\pi: M_1 \longrightarrow (M_1 + M_2)/M_2$ is the quotient map modulo M_2, then $W \cong (W \oplus M_2)/M_2 \leq (M_1 + M_2)/M_2 \leq M/M_2 \in \Delta$ by Lemma 1.6(iv), and hence $W \in \Delta_M \subset \Delta$. Alternatively, by 1.6(iii), necessarily $W \in \Delta_M \subset \Delta$. Define $K \leq M_1$ by $K = \pi^{-1}[(W \oplus M_2)/M_2]$. Since $K \leq M_1 \in (c\Delta)_M$, also $K \in (c\Delta)_M$. Let φ be the restriction of π to K followed by the isomorphism $0 \neq \varphi: K \xrightarrow{\pi} (W \oplus M_2)/M_2 \cong W$.

So far it has been shown that (i) $\iff$ (ii) $\iff$ (iii) $\iff$ (iv). Since for any $0 \neq K \in (c\Delta)_M, \exists K \leq N \in max(c\Delta)_M$, and since for any $N_1, N_2 \in max(c\Delta)_M, \widehat{N}_1 \cong \widehat{N}_2$, it can be shown that (ii) $\iff$ (iii) and (iv) $\iff$ (v), and the details of the proof are omitted.

With $\mathcal{S}$ the uniform modules, 1.7(iv) gives the already known result [O, p. 761, Theorem A(c)]. If M is t.f., then for any $\Delta \in \Sigma(R)$, $| max(c\Delta)_M | = | max\Delta_M | = 1$.

2. IRREDUNDANCY

In order to prove a statement to be valid for all $K < M$ it is not allowable to assume (supposedly without loss of generality) that $K = 0$ and that $M = M/K$ because the zero submodule $(0) < M/K$ may possess properties which $0 \neq K < M$ does not have. One example of this is that $(0) < M/K$ is a complement, whereas the inverse image of a complement submodule of M/K need to be a complement submodule in M. (E.g., See 2.2(b) $\implies$ (c) below.) The converse also applies; M/K may have properties which M does not.

Irredundancy Criterion 2.1. Let $K < M$ and $K \overset{\subseteq}{\neq} K_i < M, i \in I$ be given such that $K = \bigcap \{K_i \mid i \in I\}$. For each $i \in I$, define $A_i = \bigcap \{K_\mu \mid i \neq \mu \in I\}$. Let the canonical subdirect product image of subsets of M/K be denoted by "$\sim$" or "$(\ )^\sim$". Let $\pi_i: P \equiv \prod \{M/K_i \mid i \in I\} \longrightarrow M/K_i$ be the projections and view $M/K_i < P$. Thus $A_i/K \cong (A_i/K)^\sim \leq (M/K)^\sim < P$, and $(A_i/K)^\sim \leq M/K_i$. Then the following four conditions (i), (ii), (iii) and (iv) are equivalent.

(i) $K = \cap \{K_i \mid i \in I\}$ is irredundant.

(ii) $\exists\, 0 \neq V_i \leq M/K, \widetilde{V}_i \subseteq M/K_i$ all $i \in I$.

(iii) $0 \neq (A_i/K)^\sim = (M/K)^\sim \cap M/K_i$.

(iv) $\forall i \in I, ker[(1 - \pi_i) \mid (M/K)^\sim] \neq 0$, i.e., $((M/K)^\sim) \cap (M/K_i) \neq 0$.

Proof. First note that the equality in (iii) always holds. (i) $\Longrightarrow$ (ii). By hypothesis now all $A_i/K \neq 0$, and simply take $V_i = A_i/K$. (ii) $\Longrightarrow$ (iii). The hypothesis $0 \neq \widetilde{V_i} \subseteq M/K_i$ automatically entails that $0 \neq V_i \subseteq A_i/K$ and hence $\widetilde{V_i} \subseteq (A_i/K)^\sim = (M/K)^\sim \cap M/K_i \overset{\subseteq}{\leq} M/K_i$. (iii) $\Longrightarrow$ (i). The hypothesis $0 \neq (A_i/K)^\sim \cong A_i/K$ trivially means that for all $i \in I, K \neq A_i = \cap \{K_\mu \mid i \neq \mu \in I\}$, in other words, that the intersection in (i) is irredundant.

(i) $\Longrightarrow$ (iv). The kernel of the restriction $(1-\pi_i) \mid (M/K)^\sim$ of $(1-\pi_i)$ to $(M/K)^\sim$ is precisely $(A_i/K)^\sim$ which is nonzero for all $i \in I$ if and only if no K_i can be omitted from the intersection $K = \cap \{K_i \mid i \in I\}$ without enlarging it.

Having established the above irredundancy criterion for a given fixed submodule $K < M$, we now require that it be satisfied for all submodules $K < M$ of still a given single module M.

Theorem 2.2. Let $\mathcal{S}$ be any nonvoid class of nonzero right R-modules closed under isomorphic copies, and nonzero submodules only. For a given R-module M, the following five conditions (a), (b), (c), (d) and (e) are all equivalent.

(a) Every nonzero quotient module of M contains a submodule of $\mathcal{S}$.

(b) Every nonzero quotient module of M contains an essential direct sum of modules of $\mathcal{S}$.

(c) $\forall K < M, \exists K \subseteq K_i < M$, where $\exists 0 \neq V_i << M/K_i$ with $V_i \in \mathcal{S}$, $i \in I$, such that $K = \cap\{K_i \mid i \in I\}$ is irredundant.

(d) $\exists 0 \neq V_i \leq W_i << U_i$, with $V_i \leq W_i \in \mathcal{S}$, $i \in I$, and $\exists$ an embedding of M/K into $\bigoplus_{i \in I} V_i \leq M/K \leq \prod_{i \in I} U_i$.

(e) $\forall K < M, \exists$ a smooth ordinal indexed chain of submodules $K = M_0 \subset M_1 \subset \cdots \subset M_\alpha \subset \cdots \subset M_\tau = M$ with $M_{\alpha+1}/M_\alpha \in \mathcal{S}$ for all $\alpha < \tau$ for some ordinal τ.

Proof. Since (a) $\Longleftrightarrow$ (e), and (d) $\Longrightarrow$ (a) are trivial, it suffices to prove that (a) $\Longrightarrow$ (b) $\Longrightarrow$ (c) $\Longrightarrow$ (d). (a) $\Longrightarrow$ (b). Given $K < M$, by any one of several possible Zorn's lemma arguments there exists a maximal direct sum $\bigoplus_{i \in I} V_i \leq M/K$ of submodules $0 \neq V_i \in \mathcal{S}$. Take any D with $(\bigoplus_{i \in I} V_i) \oplus D << M/K$. If $D \neq 0$, let $\bigoplus_{i \in I} V_i << N/K < M/K$ be any complement submodule of M/K. Then $D \cong (N/K \oplus D)/(N/K) << (M/K)/(N/K) \cong M/N$. Thus by hypothesis (a), M/N and hence also its essential submodule D contains a nonzero submodule of $\mathcal{S}$, in view of the fact that $\mathcal{S}$ is closed under nonzero submodules. This contradicts the maximality of $\bigoplus_{i \in I} V_i \leq M/K$. Thus $D = 0$.

(b) $\Longrightarrow$ (c). For $K < M$, by (b), take $\bigoplus_{i \in I} V_i << M/K, 0 \neq V_i \in \mathcal{S}$. Define $K_i/K < M/K$ to be any complement submodule of M/K with

$\oplus \{V_\mu \mid i \neq \mu \in I\} \ll K_i/K$. (Note that $K_i < M$ need not be a complement.) Hence $V_i \oplus (K_i/K) \ll M/K$. By several applications of the modular law for any finite subset $\{1, \cdots, n\} \subset I$ of indices, we have $(V_1 \oplus \cdots \oplus V_n) \oplus \bigcap_{i=1}^n (K_i/K) \ll M/K$. From the latter it follows that $(\bigoplus_{i \in I} V_i) \oplus \bigcap_{i \in I} (K_i/K) \leq M/K$. First, since the first direct sum is already large in M/K, the intersection $\bigcap_{i \in I} K_i$ equals $\bigcap_{i \in I} K_i = K$. And secondly, it is irredundant because $0 \neq V_i \subseteq \bigcap \{K_v/K \mid i \neq v \in I\}$ by definition of the K_v. Since essential extensions modulo complement submodules such as $K_i/K < M/K$ remain essential, and since $V_i \in \mathcal{S}$, for any $i \in I$ also $V_i \cong [V_i \oplus (K_i/K)]/(K_i/K) \ll (M/K)/(K_i/K) \cong M/K_i$.

(c) $\implies$ (d). By (c), let $U_i = M/K_i$ By (c), let $0 \neq W_i \ll M/K_i$ with $W_i \in \mathcal{S}$. By use of the irredundancy hypothesis (c), define $0 \neq V_i \equiv [(\bigcap_{i \neq \mu \in I} K_\mu)/K]^\sim \cap (M/K_i) \cap W_i < (M/K)^\sim$, where as before in 2.1, $\sim$ denotes the canonical subdirect product representation of M/K or subsets thereof. Since $\Sigma_{i \in I} M/K_i = \bigoplus_{i \in I} M/K_i$ and $0 \neq V_i \leq M/K_i$, also $\Sigma_{i \in I} V_i = \bigoplus_{i \in I} V_i \leq M/K \leq \prod_{i \in I} U_i$.

The next corollary, in which also M is allowed to vary, is the third stage in our bootstrap argument.

Main Corollary 2.3. For $\mathcal{S}$ as above, the following two conditions (i) and (ii) are equivalent.

(i) $M = R_R$ satisfies 2.2(a)–(e) above.

(ii) $\forall M \in \mathcal{M}od_R$, M satsifies 2.2(a)–(e).

Proof. (i) $\implies$ (ii). Condition 2.2(a) will be verified for any $K < M$. For any $0 \neq m + K \in M/K$, by (i) there exists a $B \leq R$ with $0 \neq B/m^{-1}K \leq R/m^{-1}K \cong (mR+K)/K$ with $B/m^{-1}K \in \mathcal{S}$. But then $B/m^{-1}K \cong (mB+K)/K \leq M/K$. Trivially, (ii) $\implies$ (i).

In order to prove that either M or R_R satisfies 2.2(e), for classes $\mathcal{S}$ of modules which are preserved under homomorphic images, it suffices to verify condition 2.2(e) for the single submodule $K = 0$.

Corollary 2.4. For a class $\mathcal{S}$ of nonzero right R-modules closed under isomorphic copies and nonzero submodules let (f) denote the following property of a module M.

(f) $\exists$ a smooth ordinal indexed chain $0 = M_0 < \cdots < M_\alpha < \cdots < M_\tau = M$ with $M_{\alpha+1}/M_\alpha \in \mathcal{S}$ for all $\alpha < \tau$ for some ordinal τ, such that every nonzero homomorphic image of every $M_{\alpha+1}/M_\alpha$ contains a nonzero submodule of $\mathcal{S}$. Then

(i) $(f) \implies M$ satisfies 2.2(a)–(e).

(ii) If $M = R_R$ satisfies (f), then every R-module M satisfies 2.2(a)-(e). In particular, if

(iii) $\exists\, 0 = R_0 < \cdots < R_\alpha < \cdots < R_\tau = R$, $R_{\alpha+1}/R_\alpha \in \mathcal{S}$, and every nonzero homomorphic image of every $R_{\alpha+1}/R_\alpha$, for $\alpha < \tau$ contains a module of $\mathcal{S}$, then every R-module satisfies 2.2(a)-(e).

Proof. (ii) and (iii). If (i) holds, then (ii) and (iii) follow from Corollary 2.3. (i) Condition 2.2(a) will now be verifies for an arbitrary $K < M$ assuming only (f). Let $\alpha \leq \tau$ be smallest such that $M_\alpha \backslash K \neq \emptyset$. Then α is not a limit ordinal, $M_{\alpha-1} \backslash K = \emptyset$, $M_{\alpha-1} \subseteq K$, and hence the composite $M_\alpha/M_{\alpha-1} \longrightarrow (M_\alpha + K)/M_{\alpha-1} \longrightarrow (M_\alpha + K)/K$ is surjective and nonzero. By our additional hypothesis, $(M_\alpha + K)/K$ contains a nonzero submodule of $\mathcal{S}$.

It is known that R is a V-ring if and only if every right ideal of R is an intersection of maximal right ideals of R ([MV, p. 186, Theorem 2.1(4)]). If below in 2.5, in addition it is assumed at the outset that R is a V-ring, then 2.5 gives a characterization of the proper subclass of the class of all V-rings in which every right ideal is not only an intersection of maximal right ideals, but is an irredundant intersection. Modules possessing such properties as in 2.5 below have been considered previously by several authors including [S, p. 228, Prop. 2.1], who called them min-modules, and [DS, p. 28, Proposition 1] who called them semi-artinian. Also, below in 2.5(iii) the chain could be replaced by a more general chain with completely reducible quotients by use of essentially [D1, p. 237, Theorem 2], the biggest such being the ascending socle series.

Corollary 2.5. For a ring R, the three conditions (i), (ii) and (iii) are equivalent.

(i) $\forall\, K < R$, R/K contains a simple module.

(ii) Every $K < R$ is an irredundant intersection $K = \bigcap\{K_i \mid i \in I\}$ of right ideals $K_i < R$ such that there exists a simple module $V_i << R/K_i$, $i \in I$.

(iii) $\exists$ a smooth ascending ordinal indexed right socle series $0 = S_0 < S_1 < \cdots < S_\alpha < \cdots < S_\tau = R$, $S_{\alpha+1}/S_\alpha$ is simple, $\alpha < \tau$ for some ordinal τ.

(iv) Furthermore, if R satisfies the above 2.5(i), (ii) and (iii), then every R-module satisfies 2.2(a)-(e).

We now give an example of a V-ring R which does *not* satisfy any of 2.5(i), (ii) and (iii).

Example 2.6. Let F be a universal differential field with a nonzero derivation δ, and $R = F[x; \delta]$ the twisted polynomial ring in which $cx = xc + c^\delta$ for $c, c^\delta = \delta c \in F$. (See [Co, p. 77, Theorem 1.4 and p. 76].) Up to isomorphism, R has only one simple module isomorphic to $R/(x-c)R$ for any $c \in F$, and R_R itself contains no simple modules. If every intersection of the form $\bigcap\{(x - c(i))R \mid i \in I\}$ were irredundant then by successively removing terms in the intersection we could keep on enlarging it, thus obtaining an infinite ascending chain of right ideals. This contradicts the fact that R is Noetherian.

It would be of interest to reestablish the above via a direct computation by giving an explicit description of $c_i \in F$ such that $0 = \bigcap_{i=1}^{\infty}(x - c_i)R = \bigcap_{i=n}^{\infty}(x - c_i)R$ for all $n = 1, 2, \cdots$. (See [D, p. 333, AIV-5.1 and p. 370, AIV-5.9].) Note that R_R is a uniform module which contains no simple submodules.

The next logical step is to move from the simple modules to let $\mathcal{S}$ be the uniform modules, and obtain as a special case of Corollary 2.3 above a result proved by L. Fuchs.

Corollary 2.7 (L. Fuchs). For a ring R the following are all equivalent.

(i) Every submodule of any R-module M is an irredundant intersection of irreducible submodules of M.

(ii) For any $0 \leq K < R$, R/K contains a uniform submodule.

Many rings including all those possessing Krull dimension have the property that every nonzero quotient of a uniform module contains a uniform submodule. This class satisfies the conditions 2.7(i), (ii) as well as 2.8(i), (ii) and (iii).

Example 2.8. Let $\mathcal{S}$ be the class of nonzero quotient finite dimensional (abbreviation: q.f.d.) modules. Then for a ring R, the following are all equivalent.

(i) $\forall\, M \in \mathcal{M}od_R$, M satifies 2.2(a), (b), (c), (d) and (e).

(ii) $\exists\, 0 = R_0 < \cdots < R_\alpha < \cdots < R_\tau = R$, $R_{\alpha+1}/R_\alpha$ is q.f.d., $\alpha < \tau$.

(iii) $\forall\, K < R$, $\exists$ an irredundant $K = \bigcap\{K_i \mid i \in I\}$, $K \subsetneqq K_i < R$, R/K_i is q.f.d., $i \in I$.

It would be interesting to obtain examples and various internal characterizations of the class of rings R satisfying (iii) both with and without the irredundance requirement. Such characterizations should take into account various known internal characterizations of q.f.d. rings found in [AJL], [Ca], [Ku], [L], [Ma] and [S]. The related and similar question of finding conditions under which a complement submodule $K < M$ is an irredundant intersection of maximal complement submodules of M was raised and answered in [D2, p. 38, 2.13(a)].

Above in 2.8(iii), every K_i is a finite and hence irredundant intersection of irreducible right ideals. Hence, if R satisfies 2.8(iii), then every $K < R$ is an intersection of irreducible right ideals, but possibly redundant.

Application 2.9. For any cardinal $\aleph_0 \leq \aleph$, a module M is $\aleph^<$ - q.f.d. if for any submodule $0 \leq K < M$, every direct sum of submodules of M/K has strictly less than $\aleph$ nonzero terms. Note than when the Goldie dimension $Gd\ M$ is not an inaccessible cardinal, then this simply says that $Gd\ M/K < \aleph$, see [DF]. Thus $\aleph_0^<$ -q.f.d. is the same as q.f.d. Define a submodule $K < M$ to be a co-$\aleph^<$ - q.f.d. if M/K is $\aleph^<$ -q.f.d. It is an immediate consequence of 2.3 that if R_R is $\aleph^<$ - q.f.d. then for any module M, and any submodule $0 \leq K < M$, K is an irredundant intersection of co-$\aleph^<$ - q.f.d. submodules. See [D10, p. 2880, 1.4; p. 2882, 2.2; p. 2885, 2.9, Theorem II] for some characterizations of such rings.

Construction 2.10. Let R_i be algebras over a commutative ring R_∞. For $1 = e_i \in R_i$ and $e = (e_i)_{i=1,2,\cdots} \in \prod_1^\infty R_i$, let R be the ring $R = \bigoplus_1^\infty R_i + R_\infty \cdot e < \prod_1^\infty R_i$. Let now $\mathcal{S}$ be either the class of simple, uniform or atomic modules, and then for any ring such as R or R_i, $\mathcal{S}_R$ the same class of R-modules. Form the following R-modules:

$$0 = R_0 < R_1 \cdots < \bigoplus_1^n R_i < \cdots < R\ ;\ \bigoplus_1^n R_i / \bigoplus_1^{n-1} R_i \cong R_n$$

and

$$R/\bigoplus_1^\infty R_i \cong R_\infty\ .$$

Now, in addition, assume that the rings R_i have the further property that for any right ideals $K_i \subset R_i$ the quotient R_i/K_i contains an element of $\mathcal{S}_{R_i}$ for $i = 1, 2, \cdots, \infty$, which also then autommatically will have to belong to $\mathcal{S}_R$. Each R_i/K_i becomes an R-module in view of the above ascending series for R_R. Since now R satisfies 2.4(f), we can conclude that every R-module M satisfies 2.2(a)-(e), and in particular that for any R-module M, every submodule $K < M$ is an *irredundant* intersection of maximal, irreducible or co-atomic submodules $K \stackrel{\subseteq}{\neq} K_i < M$, with $K = \bigcap \{K_i \mid i \in I\}$ and $M/K_i \in \mathcal{S}_R$. Note that if we were to take all the $R_i, i = 1, 2, \cdots, \infty$ to be domains, then the above R satisfies 2.8(i), (ii) and (iii) but that R itself is not q.f.d.

3. INTERSECTIONS

The choice of the index set Γ needed to represent any submodule $K < M$ as $K = \bigcap_{\gamma \in \Gamma} K_\gamma$ is constrained by two requirements. First, for $\alpha \neq \beta \in \Gamma, M/K_\alpha$ and M/K_β must have no nonzero isomorphic submodules. Secondly, in some sense Γ must be big enough, or complete in order that for every $K < M, M/K$ embeds subdirectly in $M/K \hookrightarrow \prod_{\gamma \in \Gamma} M/K_\gamma$. As in 1.5, Γ will be derived from any pairwise disjoint set of saturated classes

Υ of R-modules with supremum one by downsizing Υ to M/K. It has been shown that such sets (and even classes) Υ exist in abundance for every ring R ([D6, 7, 9, 11 and 13].)

First, a constructive procedure is given whereby any submodule $K < M$ can be represented as $K = \bigcap_{\gamma \in \Gamma} K_\gamma$. Then, secondly, the subsequent theorem gives some of the properties of the latter.

Construction 3.1. For $K < M$, let Υ and Γ be as in 1.5. Let $K < W_\gamma \leq M, \gamma \in \Gamma$ be any choice of submodules of M such that $W_\gamma/K = (M/K)_{(\gamma)} \leq M/K$ is a γ-maximal submodule (as in 1.3 and 1.6). By the results of [D13, 3.1, Theorem II] or [D11, 3.1(i)] applied to M/K (in place of M in [D13] or [D11]) it follows that $\Sigma\{W_\gamma/K \mid \gamma \in \Gamma\} = \oplus\{W_\gamma/K \mid \gamma \in \Gamma\} << M/K$. For every $\gamma \in \Gamma$, let $K_\gamma/K \leq M/K$ be any complement submodule of M/K such that $\oplus\{W_\mu/K \mid \gamma \neq \mu \in \Gamma\} << K_\gamma/K$, while $K_\gamma/K \oplus W_\gamma/K << M/K$ where now both direct summands are complements. (Note that we may not let $K = 0$ without loss of generality, because in general $K_\gamma, W_\gamma \leq M$ need not be complements.)

Theorem 3.2. In the notation of 3.1, for any Υ and all $K < M$, there exists a particular choice of submodules $W_\gamma, K_\gamma \leq M, \gamma \in \Gamma$ such that the following hold.

(i) $K = \bigcap\{K_\gamma \mid \gamma \in \Gamma\}$ is irredundant.

(ii) $M/K_\gamma \in \gamma; K_\gamma/K < M/K$ is $c(\gamma)$-maximal; $0 \neq W_\gamma/K < M/K$ is γ-maximal; $\forall \alpha \neq \beta \in \Gamma, M/K_\alpha$ and M/K_β do not have nonzero isomorphic submodules.

(iii) $W_\gamma = \bigcap\{K_\mu \mid \gamma \neq \mu \in \Gamma\}, K \overset{\subsetneq}{\neq} W_\gamma, \gamma \in \Gamma$. If the canonical subdirect product images of subsets of M/K are denoted by "$\sim$" or "$(\ )^\sim$" (see 1.2), then

(iv) $M/K \cong (M/K)^\sim$ and $\bigoplus_{\gamma \in \Gamma}(W_\gamma/K)^\sim << (M/K)^\sim \leq \prod_{\gamma \in \Gamma} M/K_\gamma$.

(v) $W_\gamma/K \cong (W_\gamma/K)^\sim = (W_\gamma + K_\gamma)/K_\gamma = (M/K)^\sim \cap M/K_\gamma << M/K_\gamma \leq \prod_{\gamma \in \Gamma} M/K_\gamma$.

(vi) $\bigoplus_{\gamma \in \Gamma}[(M/K)^\sim \cap M/K_\gamma] << \bigoplus_{\gamma \in \Gamma} M/K_\gamma$.

Proof. First of all, it follows from [D13, 3.1, Theorem II(1)] or [D11, 3.1(i)] that $\Sigma\{W_\gamma/K \mid \gamma \in \Gamma\} = \oplus\{W_\gamma/K \mid \gamma \in \Gamma\} << M/K$. (ii) Since by 3.1 $W_\gamma/K \oplus K_\gamma/K << M/K$ where both are complements, it follows from 1.6 that $K_\gamma/K < M/K$ is $c(\gamma)$-maximal since the first direct summand by construction is γ-maximal. Since Γ is pairwise orthogonal and $M/K_\alpha \in \alpha, M/K_\beta \in \beta, \alpha \wedge \beta = \langle M/K_\alpha \rangle \wedge \langle M/K_\beta \rangle = 0$ and $M/K_\alpha \perp M/K_\beta$.

(i) and (iv). Repeated use of the modular law shows that for any finite number n of distinct indices $\gamma(1), \cdots, \gamma(n) \in \Gamma, [(K_{\gamma(1)} \cap \cdots$

$\cap K_{\gamma(n)})/K] \oplus [W_{\gamma(1)}/K \oplus \cdots \oplus W_{\gamma(n)}/K] << M/K$ (see [D9, p. 57, 6.3]). The latter implies that $\cap \{K_\gamma \mid \gamma \in \Gamma\} = K$, and that $M/K \cong (M/K)^\sim$ with $\bigoplus_{\gamma \in \Gamma}(W_\gamma/K)^\sim << (M/K)^\sim \leq \prod_{\gamma \in \Gamma} M/K_\gamma$. By 2.1(ii), the intersection is irredundant.

(iii) Later a second proof of (iii) will be given. Define $A_\gamma = \cap \{K_\mu \mid \gamma \neq \mu \in \Gamma\}$ for $\gamma \in \Gamma$. Since by construction 3.1, $W_\gamma \subseteq K_\mu$ for $\gamma \neq \mu$, $W_\gamma \subseteq A_\gamma$. Since $W_\gamma/K \leq M/K$ is a γ-maximal submodule, it is a complement in M/K, and hence $W_\gamma/K <\!\!\not< A_\gamma/K \leq M/K$. Consequently $W_\gamma/K \oplus L/K \leq A_\gamma/K$ for some $0 \neq L/K \in c(\gamma)$ by 1.6. Since $L/K \cong (L/K)^\sim$, and by definition of A_γ, $(A_\gamma/K)^\sim \subseteq M/K_\gamma < \prod_{\gamma \in \Gamma} M/K_\gamma$, also $(L/K)^\sim \subseteq M/K_\gamma \in \gamma$. Thus $(L/K)^\sim \in \gamma \cap c(\gamma) = \{(0)\}$, a contradiction.

The next observation will be needed also in (v). Since $K_\gamma/K \leq M/K$ is a complement, it follows that modulo this complement the large submodule $W_\gamma/K \oplus K_\gamma/K << M/K$ will have a large image $W_\gamma/K = W_\gamma/(W_\gamma \cap K_\gamma) \cong (W_\gamma + K_\gamma)/K_\gamma << M/K_\gamma$.

(iii) Alternate proof of (iii). For $A_\gamma = \cap\{K_\mu \mid \gamma \neq \mu \in \Gamma\}$, always $(A_\gamma/K)^\sim \subseteq M/K_\gamma \in \gamma$. Thus $(W_\gamma/K)^\sim = (W_\gamma + K_\gamma)/K_\gamma \subseteq (A_\gamma + K_\gamma)/K_\gamma = (A_\gamma/K)^\sim$. Note that $A_\gamma/(A_\gamma \cap K_\gamma) = A_\gamma/K$. Thus $(W\gamma/K)^\sim << M/K_\gamma$ and $(W_\gamma/K)^\sim \subseteq (A_\gamma/K)^\sim \leq M/K_\gamma \in \gamma$. But $(W_\gamma/K)^\sim \leq (M/K)^\sim$ is a γ-maximal submodule, while the above shows that also $(A_\gamma/K)^\sim \in \gamma$. Hence $(W_\gamma/K)^\sim = (A_\gamma/K)^\sim$ or $W_\gamma = A_\gamma$.

(v) and (vi) As in the proof of 2.2, always $(M/K)^\sim \cap M/K_\gamma = (A_\gamma/K_\gamma)^\sim$. In view of $A_\gamma = W_\gamma$, we get that

$$W_\gamma/K \cong (W_\gamma/K)^\sim = (W_\gamma + K_\gamma)/K_\gamma$$

$$= (M/K)^\sim \cap M/K_\gamma << M/K_\gamma \leq \prod_{\gamma \in \Gamma} M/K_\gamma \ .$$

Upon summing the latter over all $\gamma \in \Gamma$, we get the last part.

A key ingredient in 3.2 is irredundancy, spelled out in great deail in 3.2(ii)-(vi), and obtained from 2.1 and 2.2. This irredundancy though is achieved at a cost – very special kinds of $c(\gamma)$-maximal submodules $K_\gamma/K < M/K$ had to be used, instead of working with completely arbitrary $c(\gamma)$-maximal submodules $L_\gamma/K < M/K$.

Proposition 3.3. For $\Upsilon, K < M$ and Γ as in 1.5 (or in 3.1), let $K < L_\gamma < M$ be any submodules such that $L_\gamma/K < M/K$ are $c(\gamma)$-maximal, $\gamma \in \Gamma$. Then

(i) $\cap\{L_\gamma \mid \gamma \in \Gamma\} = K$;

(ii) $M/L_\gamma \in \gamma \in \Gamma$;

(iii) $\exists\, K_\gamma \neq L_\gamma, K_\gamma/K, L_\gamma/K \in max(c(\gamma))_{M/K} \iff \forall N/K \in$

$max(c(\gamma))_{M/K}$, $\exists\, K \subsetneq L \leq N$ and $0 \neq \psi : L/K \longrightarrow W$, $M/K \geq W \in \gamma$.

Proof. (ii) and (iii). These follow from 1.6 and 1.7. For simplicity, let $K = 0$; the same proof will work for M/K in place of M. (i) If $I = \cap_{\gamma \in \Gamma} L_\gamma \neq 0$, take $\bigoplus_{\gamma \in \Gamma} I_{(\gamma)} << I$. Select $\xi \neq 0$ with n minimal such that $0 \neq \xi = \xi_i + \cdots + \xi_n \in \bigoplus_{\gamma \in \Gamma} I_{(\gamma)}$, $0 \neq \xi_i \in I_{(\gamma(i))}$ for distinct $\gamma(1), \cdots, \gamma(n) \in \Gamma$. Then $\xi^\perp = \xi_1^\perp = \cdots = \xi_n^\perp$. But then for $i \neq j$

$$I_{(\gamma(i))} \supset \xi_i R \cong \xi_j R \subset I_{(\gamma(j))}$$

will contradict the fact that $\gamma(i) \wedge \gamma(j) = 0$, unless $n = 1$. Thus $0 \neq \xi R = \xi_1 R \subseteq I_{(\gamma(1))}$. But $0 \neq \xi R \subseteq \cap_{\gamma \in \Gamma} L_\gamma \subseteq L_{\gamma(1)}$, and $L_{\gamma(1)} \cap I_{(\gamma(1))} = 0$ by [D13, 2.8(2)(ii)] or [D11, 2.7(i)] or 1.6 gives a contradiction. Thus $\cap_{\gamma \in \Gamma} L_\gamma = 0$, and (i) follows.

An immediate corollary of the last proposition is a theorem proved in [O, p. 761, Thm A] in which irredundancy is not concluded because it is not needed later for finite intersections in [O, p. 764, Thm B]. More general, but weaker, analogues of the later are considered in [D14] for inifinte intersections, where irredundancy is decidedly needed, but hard to get.

In view of the fact that in any module for any saturated class γ, any two γ-maximal submodules have isomorphic injective hulls, the hypothesis in the next corollary is independent of the particular choice of γ and $c(\gamma)$-maximal submodules.

Corollary 3.4. For $\Upsilon, K < M$, and Γ as in 1.5, let $K < L_\gamma < M, \gamma \in \Gamma$ be any submodules such that $L_\gamma/K < M/K$ are $c(\gamma)$-maximal submodules. Then $K = \cap \{L_\gamma \mid \gamma \in \Gamma\}$ is irredundant provided any one of the following hypotheses hold:

(i) $\oplus \{M/L_\gamma \mid \gamma \in \Gamma\} << \prod \{M/L_\gamma \mid \gamma \in \Gamma\}$;

(ii) $\forall \alpha \neq \beta \in \Gamma, Hom_R(E[(M/K)_{(\alpha)}], E[(M/K)_{(\beta)}]) = 0$.

Proof. For any $\gamma \in \Gamma$, let $W_\gamma/K < M/K$ be any γ-maximal submodules. Hence $L_\gamma/K \oplus W_\gamma/K << M/K$ and $W_\gamma/K << (M/K)/(L_\gamma/K) \cong M/L_\gamma$. Thus $E[(M/K)_{(\gamma)}] \cong E(W_\gamma/K) \cong E(M/L_\gamma)$.

If for some $\alpha \in \Gamma, K = \cap \{L_\mu \mid \alpha \neq \mu \in \Gamma\}$, there is a subdirect product embedding $W_\gamma/K < M/K \hookrightarrow \prod \{M/L_\mu \mid \alpha \neq \mu \in \Gamma\}$. Hence for some $\alpha \neq \beta \in \Gamma, W_\alpha/K$ has a nonzero projection into M/L_β which contradicts (ii). If (i) holds, the projection argument ([D13, 1.3] or [D11, 1.3]) shows that for some $\alpha \neq \beta \in \Gamma, W_\alpha/K$ and M/L_β have nonzero isomorphic submodules. This contradicts the fact that $\alpha \wedge \beta = 0$.

Application 3.5. For $\Upsilon, K < M, K_\gamma \leq M, \gamma \in \Gamma$, as in 3.2, assume in addition that M/K is torsion free, which is equivalent to $K < M$ being a complement with $Z_2 M \subseteq K$. Then in 3.3, $K = \bigcap \{L_\gamma \mid \gamma \in \Gamma\}$ is irredundant by 3.2 because $L_\gamma = K_\gamma, \gamma \in \Gamma$ are unique. This follows from the fact that $(M/K)_{(\gamma)} = K_\gamma/K \leq M/K$ are unique. In this case, also 3.4(ii) holds.

References

[AJL] A.H. Al-Huzali, S.K. Jain and S.R. López-Permouth, *Rings whose cyclics have finite Goldie dimension*, Jour. Alg. **153** (1992), 37-40.

[B] W.E. Barnes, *Primal ideals and isolated components in noncommutative rings*, Trans. Amer. Math. Soc. **82** (1956), 1-16.

[Ca] V.P. Camillo, *Modules whose quotients have finite Goldie dimension*, **68** (1977), 337-338.

[Co] J. Cozzens, *Homological properties of the rings of differential polynomials*, Bull. Amer. Math. Soc. **76** (1970), 75-79.

[Cu] C.W. Curtis, *On additive ideal theory in general rings*, Amer. Jour. Math. **76** (1952), 687-700.

[D1] J. Dauns, *Chains of modules and completely reducible quotients*, Pacific Jour. Math. **17** (1966), 235-242.

[D2] J. Dauns, *Uniform modules and complements*, Houston Jour. Math. **6** (1980), 31-40.

[D3] J. Dauns, *A Concrete Approach to Division Rings*, Heldermann Verlag, Berlin, 1982.

[D4] J. Dauns, *Sums of uniform modules*, in: *Advances in Non-Commutative Ring Theory*, Springer Verlag, New York, LNM **848**, 1981, pp. 235-254.

[D5] J. Dauns, *Semiprime modules and rings*, in: *Non-Commutative Ring Theory*, Proceedings, edited: S.K. Jain and S.R. López-Permouth, Springer Verlag, New York, LNM **1448**, 1990, pp. 41-62.

[D6] J. Dauns, *Torsion free types*, Fundamenta Mathematicae **139** (1991), 99-117.

[D7] J. Dauns, *Classes of modules*, Forum Math. **3** (1991), 327-338.

[D8] J. Dauns, *Modules and Rings*, Cambridge University Press, Cambridge and New York, 1994.

[D9] J. Dauns, *Direct sums and subdirect products*, in: *Methods in Module Theory*, Proceedings, edited: G. Abrams, J. Haefner and K.M. Rangaswamy, Mercel Dekker, New York, 1992, pp. 39-65.

[D10] J. Dauns, *Generalized finiteness conditions*, Comm. Alg. **22** (1994), 2877-2895.

[D11] J. Dauns, *Unsaturated classes of modules*, in: *Abelian Groups and Modules, Proceedings of the International Conference at Colorado Springs*, edited: D. Arnold and K.M. Rangaswamy, Marcel Dekker, New York, 1996.

[D12] J. Dauns, *Semiprime torsion free rings*, Math. Jour. Okayama University **37** (1995), 75-90.

[D13] J. Dauns, *Module types*, Rocky Mountain Jour. Math., to appear.

[14] J. Dauns, *Primal modules*.

[DF] J. Dauns and L. Fuchs, *Infinite Goldie dimensions*, Jour. Algebra **115** (1988), 297-302.

[DS] N.V. Dung and P.F. Smith, *On semi-artinian V-modules*, Jour. Pure Applied Alg. **82** (1992), 27-37.

[F] L. Fuchs, *On primal ideals*, Proc. Amer. Math. Soc. **1** (1950), 1-8.

[Ki] D. Kirby, *Coprimary decomposition of Artinian modules*, Jour. London Math. Soc. (2), **6** (1973), 571-576.

[Ku] R.P. Kurshan, *Rings whose cyclic modules have finitely generated socle*, Jour. Algebra **15** (1970), 376-385.

[L] S.R. López-Permouth, *Rings characterized by their weakly-injective modules*, Glasgow Math. Jour. **34** (1992), 349-353.

[Ma] H. Marubayashi, *A note on locally inform rings and modules*, Proc. Japan Acad. **47** (1971), 11-14.

[McCS] R.L. McCasland and P.F. Smith, *Prime submodules of Noetherian modules*, Rocky Mountain Jour. Math. **23** (1993), 1041-1062.

[McC1] N.H. McCoy, *Prime ideals in general rings*, Amer. Jour. Math. **71** (1949), 823-833.

[McC2] N.H. McCoy, *Rings and Ideals*, Carus Math. Monographs **8**, Math. Assoc. of America, Washington, DC, 1948.

[MV] G.O. Michler and O.E. Villamayor, *On rings whose simple modules are injective*, Jour. Algebra **25** (1973), 185-201.

[N] D.G. Northcott, *Lessons on Rings, Modules and Multiplicities*, Cambridge University Press, Cambridge, 1968.

[O] B.L. Osofsky, *Noether Lasker primary ecomposition revisited*, Amer. Math. Monthly **101(9)** (1994), 759-770.

[PZ] S. Page and Y. Zhou, *On direct sums of injective modules and chain conditions*, Canad. J. Math. **46** (1994), 634-647.

[S] R.C. Shock, *Dual generalizations of the Artinian and Noetherian conditions*, Pacific Jour. Math. **54** (1974), 228-235.

[Z1] Y. Zhou, *Decomposing modules into direct sums of submodules with types*.

[Z2] Y. Zhou, *Nonsingular rings with finite type dimension*.

CURRENT ADDRESS: Department of Mathematics, Tulane University, New Orleans, LA 70018.

MINIMAL COGENERATORS OVER OSOFSKY AND CAMILLO RINGS

CARL FAITH

Department of Mathematics
Rutgers University

*This paper is dedicated to Dr.Mitchell Stokes
who started the whole thing*

ABSTRACT. The direct sum C of the injective hulls $E(V_i)$ of the set $\{V_i\}_{i \in I}$ of non-isomorphic simple right R-modules is a minimal right co-generator for R. While the injective hull $E(C)$ is the unique (up to isomorphism) minimal injective right cogenerator, Osofsky [0] showed C is not necessarily unique even for commutative R, but that it is when R is either right Noetherian, semilocal, or C is quasi-injective. In this paper, we call a ring R a right *Osofsky* ring when C is the unique minimal right cogenerator, and show that rings studied by Camillo [C1] with the property that $\text{Hom}_R(E(V_i), E(V_j)) = 0$ for $i \neq j$, are right Osofsky. We call these right Camillo rings, and show that commutative SISI rings of Vámos [V], and locally perfect commutative rings, in fact, any 0-dimensional ring, among others, are Camillo, hence Osofsky rings.

0. INTRODUCTION

A **right Osofsky** ring is a ring which has an up to isomorphism unique minimal cogenerator. Any semilocal or right Noetherian ring is right Osofsky [O]. A ring R is **right Camillo** if $\hom_R(E(V), E(W)) = 0$ for the injective hulls $E(V)$ and $E(W)$ of any two non-isomorphic simple right modules V and W. By Theorem 5.4 any right Camillo ring is right Osofsky, and they include:

(1) Any right V-ring (=every simple right module is injective), e.g. von Neumann regular (= VNR) rings with Artinian primitive factor rings (See Corollary 1.6, Theorem 5.6 and Theorem 5.8);

1991 *Mathematics Subject Classification.* Primary 16D10, 16D50, 16E50 Secondary 16L30, 13C05, 13C99.

Dr. Stokes conjectured the existence of non-Osofsky rings and this was verified by Osofsky in [0].

Typeset by $\mathcal{A}\mathcal{M}\mathcal{S}$-TEX

(2) Any commutative ring with subdirectly irreducible factor rings all local rings;
(3) Commutative SISI rings (= all subdirectly irreducible factor rings are self-injective) (5.9);
(4) Any commutative ring whose modules all have maximal submodules, equivalently locally perfect (6.1).

(2) is Camillo's Theorem [C1], and in Theorem 5.1* we characterize arbitrary non-commutative right Camillo rings, by a condition similar to (2), namely, every simple epic image of a cyclic subdirectly irreducible module is a copy of its socle.

We also consider module theoretic properties: Let C denote the direct sum $\oplus_{i \in I} E(V_i)$, where $\{V_i\}_{i \in I}$ is a representative set of simple right modules, $C_1 = \oplus_{i \in I_1} E(V_i)$, where $E(V_i) \neq V_i \forall i \in I$, and $C_0 = \oplus_{i \in I_0} V_i$, where $I_0 = I - I_1$. Then R is right Osofsky whenever C_1 is injective, and this occurs, e.g. whenever R is right Noetherian, or $|I_1| < \infty$, or $E(C_1)$ is completely decomposable (see Theorem 4.3ff).

A necessary condition for C_1 to be injective has been given by Camillo [C2], namely, there exists a cofinite subset B of I_1, such that

$$R/\operatorname{ann}_R(\oplus_{b \in B} E(V_b))$$

satisfies the *acc* on right annihilators (Corollary 4.6). By a theorem of Zimmerman-Huisgen [Z], this condition is also sufficient when R is VNR, in which case I_1 is finite (Corollary 4.7B).

1. Quasi-injective Cogenerators are Injective

As pointed out by Osofsky in her introduction to [0], p.2071, R is right Osofsky whenever R has a quasi-injective minimal cogenerator M, in which $M \approx C$ is injective (Theorem 1.5). Also, if C splits into a direct sum of an injective and a semisimple module, then R is again right Osofsky (Theorem 4.3B).

Below $A \underset{\text{ess}}{\subset} B$ denotes that A is an essential submodule of B. We need the next results in the following sections.

Theorem 1.1. (Utumi-Osofsky). *If F is a right quasi-injective module over R, and $\Gamma = \operatorname{End} F_R$, then Γ has Jacobson radical*

(a) $J(\Gamma) = \{\gamma \in \Gamma \mid \ker \gamma \underset{\text{ess}}{\subset} F\}$

and

(b) $\overline{\Gamma} = \Gamma/J(\Gamma)$ *is VNR.*

Moreover,

(c) $\overline{\Gamma}$ *is right self-injective.*

Note: (a) and (b) of this theorem is due to Utumi and (c) is due to Utumi in case $F = R$, and in general to Osofsky. See [F2], p.76, Theorem 19.27.

The following corollary is a theorem of Matlis-Utumi:

Theorem 1.2. (Matlis-Utumi). *If E is an indecomposable quasi-injective right R-module, then $\Gamma = \operatorname{End} E_R$ is a local ring with Jacobson radical*

$$J(\Gamma) = \{\gamma \in \Gamma \mid \ker \gamma \neq 0\}$$

Example 1.3. Let $\Lambda = \operatorname{End} C_R$, and let $\lambda \in \Lambda$ be a monic. Since $E = E(C)$ is injective, there is an extension of λ to $\lambda' \in \Gamma = \operatorname{End} E_R$, and if λ'' is another extension of λ, then $\ker(\lambda' - \lambda'') \supset C$, hence $\lambda' - \lambda'' \in J(\Gamma)$. By the Kasch-Sandomierski theorem, the socle of a module is the intersection of all essential submodules, hence

(1) $$J(\Gamma) = \operatorname{ann}_\Gamma S,$$

where $S = \oplus_{i \in I} V_i$ is the direct sum of the simple R-modules exactly one V_i from each isomorphy class. Moreover, $C = \oplus_{i \in I} E(V_i), E = E(C) = E(S)$, and

(2) $$\bar{\Gamma} = \Gamma / J(\Gamma) \approx \operatorname{End} S_R.$$

(2) follows from (1) and the fact that every $\alpha \in \operatorname{End} S_R$ has an extension to an element of $\Gamma = \operatorname{End} E_R$. Furthermore, since $\hom_R(V_i, V_j) = 0$ for all $i \neq j \in I$, then

$$\bar{\Gamma} \approx \Pi_{i \in I} D_i$$

is a product of sfields $D_i = \operatorname{End}(V_i)_R, \forall i \in I$, hence right and left self-injective. Thus, $\bar{\Gamma}$ is strongly regular, that is, an Abelian VNR; in particular $\bar{\Gamma}$ is Dedekind Finite. (See [F2], p.85, Props. 19.39 and 19.40: $\bar{\alpha}\bar{\beta} = 1 \Rightarrow \bar{\beta}\bar{\alpha} = 1$ in $\bar{\Gamma}$ since, $\bar{\Gamma}$ has, e.g. no nilpotents elements.)

It follows from this, for each monic $\lambda \in \Lambda = \operatorname{End} C_R$, that λ' is monic on E, hence has a left inverse, say $\mu \in \Gamma$. Since $\bar{\mu}\bar{\lambda}' = 1$ in $\bar{\Gamma}$ and $\bar{\Gamma}$ is Dedekind Finite, then $\bar{\lambda}'\bar{\mu} = 1$, whence $\lambda'\mu = 1$, i.e. each monic λ of C extends to an automorphism λ' of E, that is, a unit of Γ.

Let $\rho : \Gamma \to \Lambda$ be the ring homomorphism given by restriction, and suppose $\phi \in \Lambda$ and $\phi\lambda = 0$. Then $\ker \phi' \supseteq \lambda C$, hence $\phi' \in J(\Gamma)$, hence $\phi \in \rho(J(\Gamma))$.

Thus, while the monic λ extends to a unit λ', (1) may, and in examples of non-Osofsky rings does, have left annihilator $^\perp\lambda \neq 0$ in Λ. Moreover $^\perp\lambda \subseteq \rho(J(\Gamma))$.

Proposition 1.4.A. *If R is not right Osofsky, then there exists a monic $\lambda \in \Lambda$ that is not a unit in Λ, but λ extends to a unit λ' in*

$$\Gamma = End E_R$$

where $E = E(C)$, and there is a chain

(1) $\qquad (\lambda')^{-n}C \supset \cdots \supset (\lambda')^{-1}C \supset C \supset \lambda C \supset \cdots \supset \lambda^n C \supset \cdots$

Moreover, if $C \supset M \supset \lambda C$ and M is a submodule, then

$$(\lambda')^{-n}C \supset (\lambda')^{-n}M \supset (\lambda')^{-n+1}C \supset \cdots \supset C \supset M \supset \lambda C \supset$$
$$\cdots \supset \lambda^n C \supset \lambda^n M \supset \cdots .$$

Furthermore,

$$\bigcup_{n \geq 0} (\lambda')^{-n} C = \bigcup_{n \geq 0} (\lambda')^{-n} M$$

and

$$\bigcap_{n \geq 0} \lambda^n C = \bigcap_{n \geq 0} \lambda^n M.$$

Proof. This follows from Example 1.3.

The next corollary follows from Proposition 1.4A.

Corollary 1.4B. *If $C/\text{soc}C$ is Noetherian equivalently, if C has acc on essential submodules, then R is right Osofsky. (cf. Corollary 4.5.)*

Remark on 1.4B. By the Kasch-Sandomierski theorem, every essential submodule of C contains $\text{Soc}C$. But since $\text{soc}C$ is essential in C, then a submodule M is essential iff $M \supseteq \text{soc}C$. This explains the "equivalently" in 1.4B.

Theorem 1.5. *The right universal injective module $E = E(C)$ of R is the unique UTI minimal injective right cogenerator. Moreover, any quasi-injective minimal right cogenerator M is isomorphic to C.*

Proof. By a theorem of Bumby and Osofsky (see, e.g. [F1], p.171, Prop.3.60), any two quasi-injective modules E_1 and E_2 are isomorphic iff each can be embedded into each other. This suffices for the first assertion. If M is a minimal cogenerator then we have embeddings $\mu : M \hookrightarrow C$ and $\lambda : C \hookrightarrow M$ and these are extendable to elements μ' and λ' in $\Gamma = \text{End} E$. We may assume

$$E \supseteq C \supseteq M \supset \lambda C.$$

In Example 1.3, we saw that λ' is a unit in Γ. By a theorem of Johnson and Wong (see, [F2], p.63, Prop.19.2 and Cor. 19.3.), any quasi-injective module is fully invariant in its injective hull, so if M is quasi-injective, then $\lambda' M \subseteq M$ and $(\lambda')^{-1} M \subseteq M$, hence $\lambda' M = M$. Then $M \supset \lambda C$ implies $(\lambda')^{-1} M = M \supseteq C$, whence $C = M$, as required. $\square$

Corollary 1.6 (Osofsky [0]). *If R has a quasi-injective minimal cogenerator, then R is right Osofsky. In particular, any right V-ring is right Osofsky.*

2. When Monics are Automorphisms

We let $\Lambda = \operatorname{End}(C_R)$, and call Λ the **right universal ring** of R.

Lemma 2.1. *If every monic endomorphism $\lambda \in \Lambda$ is a unit, then R is an Osofsky ring.*

Proof. If M is another minimal cogenerator, then $M \hookrightarrow C$. But C is a minimal cogenerator, hence there is a nonzero monic endomorphism $\lambda \in \Lambda$ with $\lambda(C) \subseteq M$. By the hypothesis, λ is an automorphism of C, hence $M = \lambda C = C$ as asserted.

Theorem 2.2. *If R has a right minimal cogenerator M such that every monic endomorphism of M is an automorphism, then R is right Osofsky.*

Proof. In this case we have $C \hookrightarrow M$, hence, there is a monic $\omega \in \Omega = \operatorname{End} M_R$ such that $\omega M \subseteq C$. But $\omega M = M$, by the unit hypothesis, so $M = C$. Then R is right Osofsky by Lemma 2.1.

3. When Right Regular Elements are Units

An element x of a ring is **right regular** if its right annihilator $x^\perp$ is zero. A right and left regular element is said to be **regular**. We let A^* denote the set of regular elements of a ring A, and $\mathbf{U}(A)$ the units group of A.

Proposition 3.1. *If every right regular element of the right universal ring Λ of R is a unit, then R is right Osofsky.*

Proof. Let $\lambda \in \Lambda$ have $\ker \lambda = 0$. If $\lambda\mu = 0$ for $\mu \in \Lambda$, then $\lambda(\mu E) = 0$, so the fact that $\ker \lambda = 0$ implies $\mu E = 0$, whence $\mu = 0$. Thus, λ is right regular, hence a unit, so R is right Osofsky by Lemma 2.1.

Proposition 3.2. *If R has a right minimal cogenerator C_0 such that every right regular element of $\Lambda_0 = \operatorname{End}(C_0)_R$ is a unit, then R is right Osofsky.*

Proof. Same proof as in 3.1 (cf. 2.1) with C_0 replacing C: For, if M is any minimal cogenerator, then we may assume $C_0 \supset M$, hence $M \supseteq \lambda(C_0)$ for some $\lambda \in \Lambda_0$ with $\ker \lambda = 0$. But λ is right regular hence a unit of Λ so $\lambda C_0 = C_0 = M$.

BASS AND HAMSHER MODULES

A module M is a **Bass** (resp.**Hamsher**) module, if every nonzero quotient (resp. " sub") module has a maximal submodule. It is known ([F7]) that R is a right max ring iff R has a Hamsher right injective cogenerator. In this case every nonzero right R-module is Bass and Hamsher.

Theorem 3.3. *If C is right Bass, then every left regular element of Λ is an epimorphism.*

Proof. Let $\lambda \in \Lambda$ be such that $\lambda C \neq C$. Then, there is a maximal submodule $C' \supseteq \lambda C$. Since, every simple module embeds in E, then there exists $\mu \in \Lambda$ with $\ker \mu = C'$, that is, $\mu\lambda = 0$. This shows that every non-epimorphism λ is a left zero divisor.

Corollary 3.4. *If every right regular element of Λ is left regular, and if C is a Bass module, (e.g., if R is right max), then R is right Osofsky.*

Proof. If $\lambda \in \Lambda$ is monic, then λ is right regular, hence left regular, so $\lambda C = C$ by Theorem 3.3. It follows that λ is a unit of Λ, hence R is right Osofsky by Lemma 2.1.

4. COMPLETELY INDECOMPOSABLE MODULES

A right R-module M is **indecomposable** if 0 and M are the only direct summands, equivalently, $\mathrm{End}M_R$ has only the trivial idempotents 0 and 1. Thus, M is indecomposable whenever $\mathrm{End}M_R$ is a local ring. Any **uniform** module (= submodules $A \neq 0$ and $B \neq 0 \Rightarrow A \cap B \neq 0$) is patently indecomposable, and hence so is any subdirectly irreducible module. In fact an injective module F is indecomposable iff F is uniform and, by the Matlis-Utumi Theorem, iff F has local endomorphism ring. However, indecomposable modules–even indecomposable Artinian modules–need not have local endomorphism rings. However, any Artinian module has semilocal endomorphism ring (Camps [CA]).

A module M is **completely decomposable** iff M is indecomposable, or a direct sum of indecomposable modules.
We need the following

Theorem 4.1. ([F-W]). *Let F be a completely decomposable injective module, say $F = \oplus_{a \in A} F_a$, where $F_a \forall a \in A$ is indecomposable. Then the f.a.e.c.'s on a submodule S:*

(a) S *is injective*
(b) S *is a direct summand*
(c) $S = \oplus_{b \in B} S_b$, *where S_b is indecomposable injective.*

When this is true, then $|B| \leq |A|$, and each $S_b \approx$ some F_a.

Proof. This is stated somewhat differently in [F-W], and in [F2],p.147,Theorem 21.15. □

At this juncture, it is germane to cite a theorem of Camillo which applies to Theorem 4.1.

Theorem 4.2. (Lenzing [L]-Camillo [C2]). *If a direct sum $M = \oplus_{a \in A} M_a$ of right R-modules splits off in their direct product $\Pi_{a \in A} M_a$, then there is a cofinite subset B of A, such that*

$$R/\operatorname{ann}_R(\oplus_{b \in B} M_b)$$

satisfies the acc on annihilator right ideals.

Remark. Lenzing's theorem ([L]) is the special case where $M_a = M_b$ $\forall a, b \in A$ is faithful.

NOTATION 4.3.A. Let C_0 denote the sum of all simple injective submodules if such exist, or 0 if none exist, and let C_1 be either 0, or the direct sum of all non-simple indecomposable injective submodules in C. Then, $C = C_1 \oplus C_0$, hence

$$E = E(C_1) \oplus E(C_0)$$

The next theorem in the case $C_0 = 0$ generalizes Theorem 1.5, and in the case $C = C_0$ generalizes the second statement of Corollary 1.6.

Theorem 4.3.B. *If $E(C_1)$ is completely decomposable, e.g. if $E(C)$ is, then R is right Osofsky, and C_1 is injective. Moreover, R is right Osofsky whenever C_1 is injective.*

Proof. Let λ be a monic of C, and let S be the socle. Since $C = C_0 \oplus C_1$, and $S \supseteq C_0$, then $S = C_0 \oplus S_1$, where $S_1 = C_1 \cap S$. Since S is direct sum of (all) non-isomorphic simple modules, then $\lambda S = S$, in fact $\lambda V = V$ for every simple module V contained in C, so $\lambda C_0 = C_0$.

By Theorem 4.1, C_1 is injective, hence so is λC_1, whence $\lambda C_1 = C_1$, and R is right Osofsky by Lemma 2.1.

The following is a rewording of Theorem 4.3B and its proof.

Corollary 4.4. *If the direct sum C_1 of all the non-simple indecomposable injectives in C is injective (or has completely decomposable injective hull), then R is right Osofsky.*

Proof. C_1 is completely decomposable. □

Corollary 4.4 implies the known fact that any right Noetherian or semilocal ring is right Osofsky.

Corollary 4.5. *If all but a finite number of simple right R-modules are injective, then R is right Osofsky. Moreover, if R is VNR, this condition is necessary for C_1 to be injective.*

Proof. In this case, $C = C_1 \oplus C_0$, where C_1 is a finite direct sum of subdirectly irreducible injectives, whence injective, so Theorem 4.3 applies. The necessity of this condition when R is VNR, follows from [Z], as stated in the Introduction. See Corollary 4.7B below. □

Corollary 4.6. *If C decomposes as in Theorem 4.3, and if $I_1 \subseteq I$ is such that*
$$C_1 = \oplus_{i \in I_1} E(V_i)$$
then there is a cofinite subset B of I_1 such that
$$R/\operatorname{ann}_R(\oplus_{b \in B} E(V_b))$$
satisfies the ascending chain condition on right annihilators.

Proof. This is a corollary of Camillo's Theorem 4.2. □

Theorem 4.7.A. (Zimmerman-Huisgen [Z]). *If R is a VNR ring, then a direct sum $M = \oplus_{a \in A} M_a$ of right R-modules splits off in the direct product $\prod_{a \in A} M_a$ iff there is a cofinite subset B of A such that*
$$\overline{R} = R/\operatorname{ann}_R(\oplus_{b \in B} M_b)$$
is semisimple Artinian. In this case, $M_B = \oplus_{b \in B} M_b$ is an injective semisimple right R-module, and there are just finitely many homogeneous components.

Below, as before, $\{V_i\}_{i \in I}$ denotes a complete isomorphy class of simple right R-modules.

Corollary 4.7.B. *If R is a VNR ring, then a direct sum $\oplus_{a \in A} E(V_a)$, where $A \subseteq I$, is injective iff A is finite.*

Proof. For by Theorem 4.7.A, A has a finite cofinite subset B.

Theorem 4.8. *If R is VNR, then R has a completely decomposable injective cogenerator F iff R is semisimple Artinian. Moreover, $E(C_1)$ is completely decomposable iff C_1 is a finite direct sum of indecomposable modules, that is, iff I_1 is finite.*

Proof. Follows from the fact that any right cogenerator is faithful (over any ring), and hence by Camillo's Theorem 4.2, R has the acc on right annihilators, which in a VNR implies R is semisimple. The converse of the first statement follows from the fact that every module is completely decomposable when R is semisimple. The last statement follows from Corollary 4.7.B.

5. Camillo Rings

A ring R is **right Camillo** provided $\hom_R(E(V), E(W)) = 0$ for non-isomorphic simple right R-modules V and W. Camillo [C1] characterized commutative Camillo rings:

Theorem 5.1. *A commutative ring R is Camillo iff every subdirectly irreducible factor ring of R is a local ring.*

Corollary 5.2. *A commutative Camillo ring R is a subdirect product of subdirectly irreducible local rings.*

Proof. Every ring is a subdirect product of subdirectly irreducible factor rings, a classic result that owes to G. D. Birkhoff.

Remark. The property of being a Camillo ring is not necessarily inherited by the polynomial ring (F[6]).

For non-commutative rings we have a similar characterization in Theorem 5.1* below.

A module M is *local* if M has a unique maximal submodule, equivalently, $M/(\mathrm{rad}\, M)$ is simple, i.e. $\mathrm{rad}\, M$ is the unique maximal submodule. The *top of a module* is top $M = M/\mathrm{rad}\, M$ (possibly top $M = 0$). Dually $\mathrm{Soc}\, M$ is the *bottom of* M, possibly 0.

Since a *subdirectly irreducible* ($= SDI$) module M is one with a unique minimal submodule, equivalently $\mathrm{soc}\, M$ is simple and an essential submodule, it can be seen that local and SDI are dual concepts for cyclic modules, hence SDI is also called *colocal* (e.g. in [C1]).

We say that a module M is *matched* if every simple module in the top appears in the bottom and conversely. Note, if R is a commutative Camillo ring, then by Theorem 5.1 every cyclic colocal module R/I is local, hence matched because (even noncommutative) local rings have unique simple modules.

A slight modification of Camillo's proof of Theorem 5.1 yields:

Theorem 5.1*. *A ring R is right Camillo iff every cyclic colocal module is matched.*

Proof. Assume all cyclic colocal modules are matched, and let

$$f : E(R/H) \to E(R/N)$$

be a nonzero homomorphism, where H and N are maximal right ideals. Also let $x \in f^{-1}(1 + N)$, that is, x maps onto the coset $1 + N$ of R/N in $E(R/N)$. Then the cyclic module xR is colocal with $\mathrm{socle} = R/H$, and an simple epic image $\approx R/N$. Since xR is matched then $R/N \approx R/H$, so R is right Camillo.

Conversely, if R has a cyclic colocal module $M = R/I$ that is not matched, then there is a maximal right ideal $N \supset I$ such that R/N is not isomorphic to the socle of M, say $R/N \not\approx R/M$, where M is a maximal right and $R/M = \operatorname{soc} M$. Then there is a nonzero map $f : E(R/N) \to E(R/N)$ extending the map $R/I \to R/N$ having kernel N/I. Since $R/M \not\approx R/N$, R is not right Camillo. □

Theorem 5.3. *If R is right Camillo, the right universal ring $\Lambda = EndC_R = \prod_{i \in I} \Lambda_i$ is a product of local rings, $\Lambda_i = EndE(V_i)_R$ $\forall i \in I$.*

Proof. This follows from the Matlis-Utumi Theorem 1.2 and the structure of the endomorphism ring of a direct sum of modules.

Theorem 5.4. *If R is right Camillo, then R is right Osofsky.*

Proof. By the Matlis-Utumi Theorem 1.2, $\Lambda_i = \operatorname{End} E(V_i)_R$ of Theorem 5.3 is a local ring with radical

$$J(\Lambda_i) = \{\lambda_i \in \Lambda_i | \ker \lambda_i \neq 0\}$$

hence $\lambda = (\lambda_i) \in \Lambda$ is monic iff $\ker \lambda_i = 0$, that is, $\lambda_i^{-1} \in \Lambda_i$ for all i, in which case λ is a unit with $\lambda^{-1} = (\lambda_i^{-1})$. Then R is right Osofsky by Lemma 2.1.

Example 5.5. Any right V-ring R is right Camillo.

Proof. Right V-rings have the defining property that every simple right R-module is injective. Since $\hom_R(V_1, V_2) = 0$ for any two non-isomorphic simple modules V_1 and V_2, then R is right Camillo. Also, 5.1* applies.

Theorem 5.6. *If R is a VNR ring with all primitive factor rings Artinian, then R is a right V-ring, hence an Osofsky-Camillo ring.*

Proof. This is a result of [F4] (p.165, Proposition 23) where it is shown that if $\bar{R} = R/P$ is simple or semi-simple Artinian, every right $\bar{R}$-module M is a (pull back) injective R-module, when R is VNR. Since $R/\operatorname{ann}_R V$ is simple Artinian for each simple right R-module V, this implies that every such V is injective. Hence, R is a right V-ring.

Theorem 5.7. *Any VNR PI-ring R is a right V-ring, hence an Osofsky-Camillo ring.*

Proof. By a theorem of Kaplansky, any primitive PI-ring is Artinian, so Corollary 5.6 applies. (Note, any PI for R must have one of its coefficients a unit in order to insure that any factor ring is PI.)

Theorem 5.8. *A right self-injective VNR ring of bounded index is a right and left V-ring, hence Osofsky ring.*

Proof. By [G], p.79, Theorem 7.20, R has Artinian primitive factor rings, so Theorem 5.6 applies.

Sisi Rings

A ring R is **right SISI** (after Vámos [V]) provided that every subdirectly irreducible (= SDI) factor ring is right self-injective. These include the following commutative rings:

(1) Noetherian,
(2) von Neuman regular (= VNR) rings, and
(3) Morita rings.

Other commutative examples include:

(4) Any ring R such that R_M is Noetherian for each maximal ideal M (= R is locally Noetherian). Note that (4) contains all the rings in (1) and (2), while (3) are the rings with a Morita duality, and coincide with the rings that are linearly compact in the discrete topology (Ánh [A]).

Theorem 5.9. *Any commutative SISI ring is a Camillo, hence Osofsky ring.*

Proof. By the Matlis-Utumi Theorem 1.2, every subdirectly irreducible factor ring $\bar{R}$ is a local ring, so Camillo's Theorem and Theorem 5.4 apply.

6. Max Rings

Hamsher [H] characterized a right max ring R as ring with that $J = $ rad$R$ is left vanishing (= right T-nilpotent) and R/J is right max. **Left vanishing** means for each sequence $\{j_n\}_{n=1}^{\infty}$ of elements of J, some finite "left" product

$$j_m j_{m-1} \cdots j_2 j_1 = 0$$

Furthermore, Hamsher [H] also characterized commutative rings via left vanishing of J and von Neumann regularity of R/J.

A commutative ring R is 0-*dimensional* if every prime ideal is maximal, equivalent $J = $ rad R is nil and R/J is VNR. By Hamsher's theorem, any max ring is 0-dimensional.

Theorem 6.1A. *Every commutative 0-dimensional ring R is a Camillo, hence Osofsky, ring.*

Proof. Let $\bar{R}$ be any subdirectly irreducible factor ring of R. Every factor ring of a 0-dimensional ring is a 0-dimensional ring, hence the radical $\bar{H}$ of $\bar{R}$ is nil, and $R_2 = \bar{R}/\bar{H}$ is VNR. Since $\bar{H}$ is nil, idempotents of R_2 lift. Since a SDI ring has only trivial idempotents, then so does R_2, that is, R_2 is a field, hence $\bar{R}$ is a local ring. This proves that R is Camillo.

Corollary 6.1B. *Any max ring hence any locally perfect commutative ring is Camillo.*

Proof. See [F8].

Corollary 6.2. *Every commutative max ring is a subdirect product of subdirectly irreducible perfect local rings.*

Proof. By Bass [B], a ring is right perfect iff R is a semilocal ring with left vanishing radical. By the proof of Theorem 6.1, every subdirectly irreducible factor ring $\bar{R}$ is local with vanishing radical, hence perfect. Since any commutative ring R is a subdirect product of its SDI factor rings, the proof is complete.

Example 6.3.A. The converse of Corollary 6.1 and 6.2 fail: (1) Any local ring R is Camillo, but is a max ring iff rad$R$ is vanishing. (2) Consider $R = \prod_{n=1}^{\infty} \mathbb{Z}_{p^n}$, the product of the Artinian local rings $\mathbb{Z}_{p^n} = Z/(p^n)$. Then R is a product of perfect local rings, but R is not a max ring, since, e.g. the radical of R, namely $J = \prod_{n=1}^{\infty} p\mathbb{Z}_{p^n}$, is not even a nil ideal (not to mention vanishing).

Example 6.3.B. Apropos of [F7] we remark that [F4] has an example of a VNR ring R that is a right and left max ring which is a right but not V-ring. (cf. [F-1], [F-2] and [N-P] in this connection: the ring R has finite Loewy length, so these results apply to conclude R is left max.)

Theorem 6.4. *A ring R is right max iff $R(V) = R/\operatorname{ann}_R E(V)$ is right max for all simple right R-modules V. Moreover R is a subdirect product of the rings $\{R(V)\}$.*

Proof. Any factor ring of a max ring is a max ring, so the necessity is obvious. The converse follows from Theorem 1 of [F7] since the condition that $R/\operatorname{ann}_R E(V)$ is right max implies that $E(V)$ is right Hamsher for all simple V. Since $C = \oplus_V E(V)$ is a cogenerator, C is faithful, hence $\bigcap_V \operatorname{ann}_R E(V) = 0$, so R is a subdirect product as stated.

7. OPEN QUESTIONS

1. Characterize right Osofsky rings ideal-theoretically.

2. Are all right max rings right Osofsky?

3. Characterize the VNR rings which are right Osofsky.

    Note: a right ideal I of R is said to be *colocal* (resp. *coirreducible*) if R/I is a co-local (uniform) right module. In [Y], Yamagata proved that a direct sum M of indecomposable injective modules is always up to isomorphism unique iff R satisfies the acc on coirreducible right ideals.[1] D. Herbera[2] has pointed out that the arguments of [Y] show that R is right Osofsky if R satisfies the acc on co-local right ideals.

CORRECTION

In [O] it is claimed that Prop. 3.55 on p.167 of [F1] states that C is unique. However, as the proof shows, injectivity, is assumed! What happened is a series of misprints (which robbed $\hat{C}$ of its hat (= carot)). The Russian translation ought to be corrected as well. (Note that the proof of Theorem 1.5 is much better than the one given in [F1]!)

Page	Line	IS	OUGHT
167	-17	cogenerator	injective cogenerator
167	-17	C	$\hat{C}$
167	-3	$h : C \to C$	$h : \hat{C} \to \hat{C}$
168	1	$H \approx C$	$H \approx \hat{C}$
168	1	C	$\hat{C}$
168	2	$H = C$	$H = \hat{C}$

Acknowledgement: The author wishes to thank the referee for many suggestions for improving the readability of this paper, and Barbara Miller for taking great pains in putting it into AMS tex and the proper format.

REFERENCES

A. P.N.Ánh, *Morita duality for commutative rings*, Comm. Alg. (6) **18** (1990), 1781–1788.

B. H. Bass, *Finitistic dimension and a homological generalization of semiprimary rings*,
Trans.Amer.Math.Soc. **95** (1960), 466–488.

C1. V. Camillo, *Homological independence of injective hulls of simple modules over commutative rings*, Comm. in Alg. **6** (1978), 1459–1469.

C2. _____, *On Zimmermann-Huisgen's Splitting Theorem*, Proc.Amer.Math.Soc. **94** (1985), 206–208.

C-F1. V. Camillo and K. Fuller, *On Loewy length of rings*, Pac.J.Math. **53** (1974), 347–354.

C-F2. _____, *A note on Loewy rings and chain conditions on primitive ideals*, in Lecture Notes in Math., **700** (1979), Springer-verlag Berlin, Heidelberg and New York, pp. 75–85.

CA. R. Camps, *Cancellation of Modules: Artinian Modules and Rings of Continuous Functions*, Ph. D. Thesis, Univ. Autónoma de Barcelona, Bellaterra 08193, SPAIN.

---

[1] And in this case every direct summand of M is a direct sum of indecomposable injectives ([Y]).

[2] During her Fulbright Postdoctoral year at Rutgers in 1993-1994.

F1. C. Faith, *Algebra I: Rings, Modules, and Categories*, Grundl. der Math.Wiss.Bd. 190, Springer-Verlag, Basel,Berlin,Heidelberg, and New York, 1972, Corrected Reprint 1981.

F2. _____, *Algebra II: Ring Theory; same series as [F2], Bd.191*, 1976.

F3. _____, *Rings with ascending chain condition on annihilators*, Nagoya Math. J. **27** (1966), 179–191.

F4. _____, *Modules finite over endomorphism ring*, Lecture Notes in Math. **246** (1972), 145–189.

F5. _____, *Linearly compact injective modules and a theorem of Vámos*, Pub.Mat. Univ.Autónoma de Barcelona **30** (1986), 127–148.

F6. _____, *Polynomial rings over Jacobson-Hilbert rings*, Pub.Mat.Univ.Autónoma de Barcelona **33** (1989), 85–97.

F7. _____, *Rings whose modules have maximal submodules*, Pub. Math. **39** (1995), 201–214.

F8. _____, *Locally perfect commutative rings are those whose modules have maximal submodules*, Comm. in Algebra **23** (1995), 4885–4886.

F-W. C. Faith and E.A. Walker, *Direct sum representations of injective modules*, J. Algebra **5** (1967), 203–221.

G. K. R. Goodearl, *von Neumann Regular Rings*, London,San Francisco,Melbourne, Pitman (1979).

H. R. Hamsher, *Commutative rings over which every module has a maximal submodule*, Proc.Amer.Math.Soc. **18** (1967), 1133–37.

L. H. Lenzing, *Direct sums of projective modules as direct summands of the direct product*, Comm. Algebra **4** (1976), 231–248.

M. E. Matlis, *Injective modules over Noetherian rings*, Pac. J. Math. **8** (1958), 511–528.

N-P. C. Nastasescu and N. Popescu, *Anneaux semi-artiniens*, Bull.de la Soc.Math. de France **96** (1968), 357–368.

O. B. Osofsky, *Minimal cogenerators need not be unique*, Comm.Alg.(7) **19** (1991), 2071–2080.

P. Z. Papp, *On algebraically closed modules*, Pub.Math.Debrecen. (1958), 311–327.

V1. P. Vámos, *Classical rings*, J. Algebra **34** (1975), 114–129.

V2. _____, *A note on the quotients of indecomposable injective modules*, Canad.Bull.Math. **12** (1969), 661–665.

V3. _____, *Test modules and cogenerators*, Proc.Amer.Math.Soc. **56** (1976), 8–10.

Y. K. Yamagata, *A note on a problem of Matlis*, Proc. Japan Acad. **49** (1973), 145–147.

Z. Birge Zimmermann-Huisgen, *The sum-product splitting theorem*, Proc.Amer.Math.Soc **83** (1981), 251–254.

Uniform Modules Over Goldie Prime Serial Rings

FRANCO GUERRIERO

ABSTRACT. We investigate the uniseriality of uniform injective modules over serial rings. Let R be an arbitrary ring and fix a decomposition, of the identity, $1 = e_1 + e_2 + \cdots + e_n$ into orthogonal idempotents. For any uniform injective module V_R, we prove that there exists $e = e_i$ such that, with $A = eRe$, $V_R \cong \hom_A(Re, Ve)$. Moreover, Ve is a uniform injective A-module. We also show that if R is Goldie prime serial, then V is uniserial if and only if Ve is uniserial as an A-module.

1. INTRODUCTION

A module is **uniform** if the intersection of any two nonzero submodules is again nonzero. A module is **uniserial** if its submodules form a single chain and is **serial** if it is the direct sum of uniserial submodules. Sufficient conditions for certain uniform modules to be uniserial have appeared in several papers [4-8]. Warfield has shown that over a left Noetherian serial ring every uniform is uniserial [6]. Wright has also provided some sufficient conditions for certain uniform modules over serial rings to be uniserial [5,7,8]. Some of Wright's work has been generalized by Müller and Singh [4].

These results provide only sufficient conditions. We will give a condition which is both necessary and sufficient for a uniform injective (equivalently, indecomposable injective) module over a Goldie prime serial ring to be uniserial. Specifically, we show that the uniseriality of a uniform injective module, over a Goldie prime serial ring, is equivalent to the uniseriality of a particular uniform injective module over a certain valuation ring. Since the injective hull of a uniform module is again uniform, our result also provides a sufficient condition for an arbitrary uniform module over a Goldie prime serial ring to be uniserial.

All rings are assumed to have an identity and modules are assumed to be unitary. A **ring is serial** if it is serial as a left and right module over itself. For an R-module V and a subset $X \subseteq R$, the **annihilator of X in V** is $\text{ann}_V(X) = \{v \in V \mid vX = 0\}$.

1991 *Mathematics Subject Classification.* Primary 16P60, 16D80, 16L99.

2. Structure of Uniform Injective Modules

In this section we will present a result which, to a certain extent, describes the structure of a uniform injective module over an arbitrary ring. Let R be a ring, $0 \neq e \in R$ an idempotent, $A = eRe$, and E an A-module. We will relate some properties of E and the R-module $\hom_A(Re, E)$.

Proposition 2.1. *If E is injective, then $\hom_A(Re, E)$ is injective.*

Proposition 2.2. *E is uniform if and only if $\hom_A(Re, E)$ is uniform.*

For Proposition 2.1 see [1, 3.51.2, p165]. Proposition 2.2 is well-known.

Proposition 2.3. *Let E_1 and E_2 be A-modules such that, as R-modules, $\hom_A(Re, E_1) \cong \hom_A(Re, E_2)$. Then $E_1 \cong E_2$.*

Proof. For $i = 1, 2$, let $H_i = \hom_A(Re, E_i)$ and let $f = 1 - e$. Note that
$$K_i = \operatorname{ann}_{H_i}(f) = \{\varphi \in H_i \mid \varphi f = 0\} = \{\varphi \in H_i \mid \varphi(fRe) = 0\}.$$
It is readily verified that K_1 is a unitary A-module. Furthermore, as A-modules, $K_1 \cong E_1$ via $\Phi : K_1 \to E_1$ by $\Phi(\varphi) = \varphi(e)$. For example, to show that Φ is surjective let $x \in E_1$. Consider the map $\varphi_x : Re \to E_1$ by $\varphi_x(re) = xere$. Then $\varphi_x(fRe) = xe(fRe) = 0$; hence, $\varphi_x \in K_1$ and $\Phi(\varphi_x) = \varphi_x(e) = xe = x$. Similarly, $E_2 \cong K_2$. Moreover, the R-isomorphism $\hom_A(Re, E_1) \cong \hom_A(Re, E_2)$, induces an A-isomorphism $K_1 \cong K_2$.

Definition. For an R-module V and idempotent $e \in R$, we say that e is **faithful (to V)** if $\operatorname{ann}_V(Re) = \{0\}$.

Let R be a ring and fix an arbitrary decomposition, of the identity, $1 = e_1 + e_2 + \cdots + e_n$ into orthogonal idempotents.

Lemma 2.4. *Let V be a uniform injective R-module. There exists an idempotent $e = e_j$ which is faithful to V. Moreover, for any faithful idempotent e, Ve is a uniform injective A-module.*

Proof. For each i, let $K(e_i) = \operatorname{ann}_V(Re_i)$. Surely each $K(e_i)$ is an R-submodule of V. If $v \in \bigcap_{i=1}^n K(e_i)$, then
$$v = v(e_1 + e_2 + \cdots + e_n) = ve_1 + \cdots + ve_n = 0.$$
That is, $\bigcap_{i=1}^n K(e_i) = 0$. The uniformity of V, implies $K(e_j) = 0$ for some j. Therefore, $e_j = e$ is faithful.

Let e be any faithful idempotent. Consider arbitrary $ve, we \neq 0$ in Ve. Because V_R is uniform, there exist $r, s \in R$ such that $0 \neq ver = wes$. Since $erRe \neq 0$, we can choose $0 \neq x \in R$ such that

$$0 \neq ver(xe) = wes(xe) = ve(erxe) = we(esxe).$$

Thus, Ve is uniform.

Let I be any right ideal of A and $f \in \hom_A(I, Ve)$. Consider the right ideal $IR \leq R$. Define $F : IR \to V$ by

$$F\left(\sum_{i=1}^{k} x_i r_i\right) = \sum_{i=1}^{k} f(x_i) r_i = \sum_{i=1}^{k} f(x_i) er_i \quad \text{for} \quad x_i \in I \quad \text{and} \quad r_i \in R.$$

$\sum_{i=1}^{k}(x_i r_i) = 0$, then, for all $s \in R$, $0 = \left(\sum_{i=1}^{k} x_i r_i\right) se = \sum_{i=1}^{k}(x_i er_i se)$. Because $er_i se \in A$ for all i,

$$0 = f\left(\sum_{i=1}^{k}(x_i er_i se)\right) = \sum_{i=1}^{k} f(x_i) er_i se = \left(\sum_{i=1}^{k} f(x_i) er_i\right) se.$$

Since $s \in R$ was arbitrary, $\left(\sum_{i=1}^{k} f(x_i) r_i\right) Re = 0$. Our assumption on e implies that $\sum_{i=1}^{k} f(x_i) r_i = F\left(\sum_{i=1}^{k} x_i r_i\right) = 0$. This proves that F is well-defined. We may now conclude that $F \in \hom_R(IR, V)$. Since V_R is injective, we may extend F to all of R. Consequently, there exists $v \in V$ such that $F(z) = vz$ for all $z \in IR$.

Define $\varphi : A \to Ve$ by $\varphi(a) = vea$. Then $\varphi \in \hom_A(A, Ve)$ and, for $x \in I \subseteq IR$, $\varphi(x) = vex = vx = F(x) = f(x)$. This proves that Ve is injective.

Theorem 2.5. *Let V be an R-module, the following are equivalent.*
(1) V is uniform and injective.
(2) There exists $e = e_j$ such that, with $A = eRe$, Ve is a uniform injective module and $V_R \cong \hom_A(Re, Ve)_R$.

Proof. (1)$\Rightarrow$(2) The existence of an idempotent $e = e_j$ such that Ve is a uniform injective A-module is guaranteed by Lemma 2.4.

For each $v \in V$, define

$$\varphi_v : Re \to Ve \quad \text{by} \quad \varphi_v(re) = vre.$$

Clearly, $\varphi_v \in \hom_A(Re, Ve)$. It follows that $\Phi : V \to \hom_A(Re, Ve)$ defined $\Phi(v) = \varphi_v$ is an R-module homomorphism. Furthermore,

$$\ker \Phi = \{v \in V \mid \varphi_v(Re) = vRe = 0\} = \text{ann}_V(Re) = \{0\}.$$

That is, Φ is an injection. Since V is injective, $\hom_A(Re, Ve)_R \cong V \oplus Y$ for some module Y.

Because Ve is uniform, $\hom_A(Re, Ve)_R$ is uniform (Proposition 2.2). We conclude that $Y = 0$ and $V_R \cong \hom_A(Re, Ve)_R$.

The converse is a direct consequence of Proposition 2.1 and Proposition 2.2.

The next result shows that any nonzero idempotent of R is faithful to some uniform injective R-module.

Proposition 2.6. *Let e be a nonzero idempotent in R, let $A = eRe$, and let E be any (uniform injective) A-module. Then e is faithful to $\hom_A(Re, E)$.*

Proof. Let $\varphi \in V$ and suppose that $\varphi Re = 0$. Then, for all $r \in R$, $0 = (\varphi re)(e) = \varphi(re)$. This implies that $\varphi = 0$; hence, e is faithful.

Remark. The above results imply that a uniform injective module V_R is isomorphic to $\hom_A(Re, Ve)$ if and only if e is faithful to V. We will show that faithful idempotents need not be unique, not even up to isomorphism.

3. UNIFORM INJECTIVE MODULES OVER GOLDIE PRIME SERIAL RINGS

Proposition 3.1. *Let R be a ring, e an idempotent, $A = eRe$, and E a A-module. If $\hom_A(Re, E)$ is uniserial, then E_A is uniserial.*

Proof. Let $0 \neq u, v \in E$. Define $\varphi_u : Re \to E$ by $\varphi_u(re) = u(ere)$ for all $r \in R$; similarly, define φ_v. Then $0 \neq \varphi_u, \varphi_v \in \hom_A(Re, E)$. By uniseriality there exists $x \in R$, such that

$$\varphi_u x = \varphi_v \quad \text{or} \quad \varphi_v x = \varphi_u.$$

Suppose the latter, then $u = ue = \varphi_u(e) = (\varphi_v x)(e) = \varphi_v(xe) = v(exe)$. This shows that E is uniserial.

Definition. *Let D be a division ring. A subring $A \subset D$ is a **valuation** (of D) if, for all $0 \neq d \in D$, either $d \in A$ or $d^{-1} \in A$.*

It is immediate that if $A \subset D$ is a valuation, then the modules D_A and $_A D$ are both uniserial.

Let R be a Goldie prime serial ring. Then $1 = e_1 + e_2 + \cdots + e_n$ where the e_i are orthogonal indecomposable idempotents. Using the results in [3], we get that $R \cong (X_{ij}) \subseteq M_n(D)$, where $X_{ij} \cong e_i Re_j$, and D is a divison ring; thus, R is a tiled order. This means that the X_{ij} are subsets of D and the operations are the natural ones obtained from the matrix operations. We will always assume that R is a proper tiled order; that is, R is not simple Artinian.

Using [3, Theorem 2, and the preliminaries] we get $0 \neq X_{ij} \subset D$ for all i, j. Furthermore, the $X_{ii} = A_i$ are valuations on the division ring D.

Lemma 3.2. *Let $R = (X_{ij}) \subset M_n(D)$ be a Goldie prime serial ring. Then there exists a ring R_1 such that*
(1) $R \cong R_1 = (Y_{ij}) \subset M_n(D)$;
(2) $Y_{11} = A_1 \subset D$;
(3) *for all* $i = 1, \cdots, n$, $Y_{i1} \subseteq A_1$.

Proof. Let $d_1 = 1$; then $d_1^{-1} X_{11} = A_1$. For each $i = 2, \ldots, n$ we may choose $0 \neq d_i \in D - X_{i1}$ ($X_{i1} \subset D$). By the uniseriality of D_{A_1}, $X_{i1} \subset d_i A_1$. Therefore, for all $i = 1, 2, \ldots, n$ there exists $0 \neq d_i \in D$ such that $d_i^{-1} X_{i1} \subseteq A_1$. Let T be the diagonal matrix with the d_i^{-1} along the main diagonal and zeros elsewhere. Let $R_1 = TRT^{-1}$; then $R_1 = (Y_{ij})$ where $Y_{ij} = d_i^{-1} X_{ij} d_j$. In particular, $Y_{i1} = d_i^{-1} X_{i1} \subseteq A_1$ for all i.

Remark. Lemma 3.2 allows us to assume, without loss of generality, that whenever we consider a Goldie prime serial ring, $X_{i1} \subseteq A_1$ holds for all $i = 1, 2, \ldots, n$.

Let A_R be an R-module and B an abelian group such that $A \cong B$ as abelian groups via a group isomorphism Φ. By defining $b*r = \Phi(\Phi^{-1}(b) r)$ for all $b \in B$ and $r \in R$, B is an R-module and $A_R \cong B_R$. We will refer to this as the **action on B induced by Φ**.

Let $R = (X_{ij}) \subset M_n(D)$ be a Goldie prime serial ring and let $e_1 = e$ be the idempotent having a 1 in the upper left hand corner and zeros elsewhere. Identify eRe with $A_1 = A$, which is a valuation on D; as A-modules $Re \cong A \oplus X_{12} \oplus \cdots \oplus X_{n1}$. For an (injective) A-module, E,

$$\hom_A(Re, E) \cong \hom_A(A, E) \oplus \hom_A(X_{21}, E) \oplus \cdots \oplus \hom_A(X_{n1}, E)$$

via the isomorphism $\Phi : \varphi \longmapsto (\varphi \circ \iota_A, \varphi \circ \iota_{X_{21}}, \ldots, \varphi \circ \iota_{X_{n1}})$, where $\iota_{X_{j1}}$ is the inclusion. This is a group isomorphism. The right side of the above expression becomes an R-module via the action induced by Φ. We will now describe this action explicitly.

For each j, let $\hom_A(X_{j1}, E)$ be denoted by $X_{j1}^{\#}$. Let $x \in X_{ij}$; define $\alpha_x : X_{j1} \to X_{i1}$ by $\alpha_x(y) = xy$ for all $y \in X_{j1}$. Surely $\alpha_x \in \hom_A(X_{j1}, X_{i1})$. $\alpha_i \in X_{i1}^{\#}$, then $\alpha_i \circ \alpha_x \in X_{j1}^{\#}$. Denote the map $\alpha_i \circ \alpha_x$ by $\alpha_i \circ x$. Let $\alpha_1, \alpha_2, \cdots, \alpha_n) \in \bigoplus_{j=1}^{n} X_{j1}^{\#}$ and $\overline{x} = (x_{ij}) \in R$. Define $\alpha_1, \alpha_2, \cdots, \alpha_n) \circ \overline{x} = (\beta_1, \beta_2, \cdots, \beta_n)$ where, for each $j = 1, 2, \cdots, n$,

$$\beta_j = \sum_{i=1}^{n} \alpha_i \circ x_{ij}.$$

By our earlier discussion, $\beta_j \in X_{j1}^{\#}$ for each j. Whence, we get a map

$$\circ : \left(\bigoplus_{j=1}^{n} X_{j1}^{\#} \right) \times R \longrightarrow \bigoplus_{j=1}^{n} X_{j1}^{\#}.$$

Lemma 3.3. *The action induced by Φ is the same map as $\circ$.*

The proof is rather straight forward and will be omitted.

Lemma 3.4. *Let j be given and let $\varphi \in X_{j1}^{\#}$. Then, there exists $u_j \in E$, such that $\varphi(z) = u_j z$ for all $z \in X_{j1}$.*

The proof follows easily using the injectivity of E.

In the following theorem we assume that R is a Goldie prime serial ring with a decomposition $1 = e_1 + e_2 + \cdots + e_n$ into indecomposable orthogonal idempotents. We let $e = e_1$, $A = eRe$, and E an arbitrary injective right A-module. For $\varphi_i \in X_{i1}^{\#}$, the above lemma shows that φ_i is multiplication by an element of E. We shall denote this by $\varphi_i = \alpha_{u_i}$ where $\varphi_i(z) = \alpha_{u_i}(z) = u_i z$.

Theorem 3.5. *Let R be a Goldie prime serial ring. Then $\hom_A(Re, E)$ is uniserial if and only if E_A is uniserial.*

Proof. ($\Rightarrow$) This is proposition 3.1.
($\Leftarrow$) Let i and j be given. Suppose $\varphi = (0, \cdots, \varphi_i, 0, \cdots, 0)$ and $\psi = (0, \cdots, \psi_j, 0, \cdots, 0)$ are in $\bigoplus_{k=1}^{n} X_{k1}^{\#}$. Then $\varphi_i = \alpha_{u_i}$ and $\psi_j = \psi_{v_j}$ for some $u_i, v_j \in E$. Uniseriality implies that $u_i = v_j b$ or $v_j = u_i b$ for some $b \in A$. Suppose the latter:
<u>Case(1)</u>: Suppose $b \in X_{ij}$. Let

$$B = (b_{lk}) \in R \quad \text{where} \quad b_{lk} = \begin{cases} b & \text{if } l = i \text{ and } k = j \\ 0 & \text{otherwise} \end{cases}.$$

Then $\varphi \circ B = (0, \cdots, \alpha_{u_i} \circ b, 0, \cdots 0)$. If $z \in X_{j1}$, then

$$(\alpha_{u_i} \circ b)(z) = \alpha_{u_i}(bz) = u_i(bz) = (u_i b) z = v_j z = \psi_j(z).$$

Therefore, $\varphi \circ B = \psi$.
<u>Case(2)</u>: Suppose $b \notin X_{ij}$. Then $b^{-1} \in X_{ji}$ [3, Theorem 2 and preliminaries]. Let

$$B = (b_{lk}) \in R \quad \text{where} \quad b_{lk} = \begin{cases} b^{-1} & \text{if } l = j \text{ and } k = i \\ 0 & \text{otherwise} \end{cases}.$$

Then $\psi \circ B = (0, \cdots, \alpha_{v_j} \circ b^{-1}, 0, \cdots, 0)$.

We claim that $\alpha_{v_j} \circ b^{-1} = \alpha_{u_i}$. To show this, let $z \in X_{i1} \subseteq A$. Then $z = (bb^{-1}) z = b (b^{-1} z)$. Furthermore, $b^{-1} z \in X_{ji} X_{i1} \subseteq X_{j1} \subset A$. Hence, for all $z \in X_{i1}$,

$$\alpha_{u_i}(z) = u_i z = u_i \left(b \left(b^{-1} z \right) \right) = (u_i b) \left(b^{-1} z \right) = v_j \left(b^{-1} z \right) = \left(\alpha_{v_j} \circ b^{-1} \right) (z).$$

Consequently, $\psi \circ B = \varphi$.

These two cases allow us to conclude that φR and ψR are comparable.

Consider now an arbitrary $\varphi = (\varphi_1, \varphi_2, \cdots, \varphi_n) \in \bigoplus_{j=1}^{n} X_{j1}^{\#}$; let $\widehat{\varphi_i} = (0, \cdots, \varphi_i, 0, \cdots, 0)$ for each $i = 1, 2, \cdots, n$. By the above, there exists i such that $\widehat{\varphi_j} R \subseteq \widehat{\varphi_i} R$ for all j. Thus,

$$\varphi R \subseteq \sum_{j=1}^{n} \widehat{\varphi_j} R \subseteq \widehat{\varphi_i} R = (\varphi e_i) R \subseteq \varphi R.$$

This implies $\varphi R = \widehat{\varphi_i} R$; similarly, for $\psi = (\psi_1, \psi_2, \cdots, \psi_n) \in \bigoplus_{j=1}^{n} X_{j1}^{\#}$, $\psi R = \widehat{\psi_j} R$ for some j. Therefore, ψR and φR are comparable.

Hill has proven the following result [2]:

Theorem [2, Theorem 1.7]. *Let R be a left serial ring and suppose that for each primitive idempotent e, eRe has indecomposable injective left modules uniserial. The following are equivalent.*
  (a) The injective hull of each simple left R-module is uniserial.
  (b) Every indecomposable injective left R-module is uniserial.
  (c) Every finitely generated left R-module is serial. Under the above conditions R is a serial ring.

Our Theorem 3.5 implies that, in case R is Goldie prime serial, the conditions of Hill's Theorem are not only equivalent to each other but actually hold true. It then follows that, under these conditions, every uniform module is uniserial.

Example. This example shows that two nonisomorphic idempotents can be faithful to the same module.

Let A be a valuation (on some division ring D) and let the maximal ideal of A be denoted by $\mathbf{m}$. Suppose that E_A is a uniform injective A-module such that $\text{ann}_E(\mathbf{m}) = \{0\}$. Let $R = \begin{pmatrix} A & A \\ \mathbf{m} & A \end{pmatrix}$ (which is Goldie prime serial), $e = \begin{pmatrix} 1 & 0 \\ 0 & 0 \end{pmatrix}$, and $f = \begin{pmatrix} 0 & 0 \\ 0 & 1 \end{pmatrix}$. Using previous results, $V_1 = \text{hom}_A(Re, E)$ and $V_2 = \text{hom}_A(Rf, E)$ are uniform injective R-modules with e faithful to V_1 and f faithful to V_2. Using the fact that $\text{ann}_E(\mathbf{m}) = \{0\}$, it is easy to show that $V_1 \cong E \oplus E \cong V_2$ as abelian groups. In both cases the action of R on $E \oplus E$ is given by formal matrix multiplication. Therefore, $V_1 \cong V_2$ as R-modules. It follows that e and f are both faithful to the same module.

To complete this example we need only show that there is such a valuation and uniform injective module. Let k be a field, let Q denote the rationals and Q^+ the positive rationals. Let

$$A = k\langle Q^+ \rangle = \left\{ f = \sum a_\alpha x^\alpha \mid \alpha \in Q^+,\ a_\alpha \in k,\ \text{supp}(f) \text{ is well ordered} \right\}.$$

Then A is a valuation in its quotient field, D, the Laurent series ring. The maximal ideal of A is $\mathbf{m} = \{f \in A \mid a_0 = 0\}$.

Let $I = xA$ and use $\overline{A}$ to denote A/I. It follows that $E = E(\overline{A})$ is a uniform injective A-module. It is now easy to show that $\text{ann}_E(\mathbf{m}) = \{0\}$.

Acknowledgments

The contents of this note form a part of the author's Ph.D. thesis. It is with great pleasure that I thank my supervisor Bruno J. Müller for his guidance and support.

References

1. Faith, C., *Algebra: Rings, Modules and Categories I*, Springer Verlag, 1973

2. Hill, D.A., *Injective Modules over Non-Artinian Serial Rings*, J. Austral Math. Soc. (Series A) 44 (1988), 242-251.

3. Müller, B.J., *Goldie-Prime Serial Rings*, Contemporary Mathematics 130 (1992), 301-310.

4. Müller, B.J. and Singh, S., *Uniform Modules over Serial Rings*, J. Algebra 144 (1991), 94-109.

5. Upham, M.H., *Serial Rings with Right Krull Dimension One*, J. Algebra 109 (1987), 319-333.

6. Warfield, R.B., *Serial Rings and Finitely Presented Modules*, J. Algebra 37 (1975), 187-222.

7. Wright, M.H., *Certain Uniform Modules over Serial Rings are Uniserial*, Comm. Algebra 17(2) (1989), 441-469.

8. Wright, M.H., *Uniform Modules over Serial Rings with Krull Dimension*, Comm. Algebra 18(8) (1990), 2541-2557.

Department of Mathematics and Statistics, McMaster University, 1280 Main Street West, Hamilton, Ontario, Canada L8S 4K1

CO– VERSUS CONTRAVARIANT FINITENESS OF CATEGORIES OF REPRESENTATIONS

B. Huisgen–Zimmermann and S. O. Smalø

ABSTRACT. This article supplements recent work of the authors. (1) A criterion for failure of covariant finiteness of a full subcategory of Λ-mod is given, where Λ is a finite dimensional algebra. The criterion is applied to the category $\mathcal{P}^\infty(\Lambda\text{-mod})$ of all finitely generated Λ-modules of finite projective dimension, yielding a negative answer to the question whether $\mathcal{P}^\infty(\Lambda\text{-mod})$ is always covariantly finite in Λ-mod. Part (2) concerns contravariant finiteness of $\mathcal{P}^\infty(\Lambda\text{-mod})$. An example is given where this condition fails, the failure being, however, curable via a sequence of one-point extensions. In particular, this example demonstrates that curing failure of contravariant finiteness of $\mathcal{P}^\infty(\Lambda\text{-mod})$ usually involves a tradeoff with respect to other desirable qualities of the algebra.

1. Introduction and Terminology

Functorial finiteness conditions for certain categories of finitely generated representations of an algebra may have a major impact also on the non-finitely generated representations, as was shown by the authors in [10]. More precisely: Let Λ be an artin algebra, and let $\mathcal{P}^\infty(\Lambda\text{-mod})$ and $\mathcal{P}^\infty(\Lambda\text{-Mod})$ be the full subcategories of the categories Λ-mod of finitely generated left Λ-modules and the full module category Λ-Mod, respectively, consisting of the objects of finite projective dimension in either case. Then contravariant finiteness of $\mathcal{P}^\infty(\Lambda\text{-mod})$ in Λ-mod forces arbitrary modules in $\mathcal{P}^\infty(\Lambda\text{-Mod})$ to be direct limits of objects in $\mathcal{P}^\infty(\Lambda\text{-mod})$. When combined with a theorem of Auslander and Reiten [1], this entails that l fin dim Λ = l Fin dim Λ = $\sup_{1 \le i \le n}$ p dim A_i in this case, where $A_1, \ldots, A_n$ are the minimal right $\mathcal{P}^\infty(\Lambda\text{-mod})$-approximations of the simple left Λ-modules. (Here l fin dim Λ and l Fin dim Λ are the suprema of the projective dimensions attained on $\mathcal{P}^\infty(\Lambda\text{-mod})$ and $\mathcal{P}^\infty(\Lambda\text{-Mod})$, respectively.)

As a byproduct of the described connections, one obtains that contravariant finiteness of $\mathcal{P}^\infty(\Lambda\text{-mod})$ implies <u>co</u>variant finiteness of this category in

---

The first author was partially supported by a grant from the NSF while this work was done, while the second author was supported by the US–Norway Fulbright Foundation. Most of this research was carried out while the second author was visiting the University of California at Santa Barbara, and he wishes to thank his coauthor for her kind hospitality.

Typeset by $\mathcal{A}\mathcal{M}\mathcal{S}$-TEX

Λ-mod. Indeed, by Crawley-Boevey's [4, Theorem 4.2], an additive subcategory $\mathcal{A}$ of Λ-mod is covariantly finite if and only if the closure $\overrightarrow{\mathcal{A}}$ of $\mathcal{A}$ under direct limits is closed under direct products as well. As explained above, contravariant finiteness of $\mathcal{P}^\infty(\Lambda\text{-mod})$ implies $\mathcal{P}^\infty(\Lambda\text{-Mod}) = \overrightarrow{\mathcal{P}^\infty}(\Lambda\text{-mod})$ and $l\operatorname{Fin dim}\Lambda < \infty$, which guarantees that $\overrightarrow{\mathcal{P}^\infty}(\Lambda\text{-mod})$ is closed under direct products.

It is not hard to find examples demonstrating that, in general, contravariant finiteness of $\mathcal{P}^\infty(\Lambda\text{-mod})$ in Λ-mod is properly stronger than covariant finiteness. In fact, the initial example – due to Igusa, Smalø, and Todorov [11] – of a situation where $\mathcal{P}^\infty(\Lambda\text{-mod})$ fails to be contravariantly finite already serves to show this. This leaves one wondering whether $\mathcal{P}^\infty(\Lambda\text{-mod})$ might always be covariantly finite in Λ-mod. The answer is negative, as we show here, but examples are somewhat harder to come by.

The first part of the present paper is devoted to developing a criterion for failure of covariant finiteness of $\mathcal{P}^\infty(\Lambda\text{-mod})$ and to then applying it to a finite dimensional special biserial algebra (for a definition see under 'Notation and Terminology' below). We believe that, in the restricted setting of special biserial algebras Λ, the conditions of this criterion actually provide an equivalent description for failure of covariant finiteness of $\mathcal{P}^\infty(\Lambda\text{-mod})$. The criterion can be considered as a somewhat weaker twin sibling of the sufficient condition for failure of contravariant finiteness developed by Happel and the first author in [7, Criterion 10]. We remark that, while the concepts of contra- and covariant finiteness are mutually dual, the theories relating them to a prescribed subcategory of Λ-mod are of course not; in particular, the argument backing up the criterion presented here differs substantially from that used to prove [7, Criterion 10].

The homological picture available for algebras Λ having the property that $\mathcal{P}^\infty(\Lambda\text{-mod})$ is contravariantly finite in Λ-mod naturally raises the question as to how abundant they are. While this condition is known to 'slice diagonally' through the standard classes of algebras, the authors conjecture that failure can 'often' be fixed in the following sense: Namely, that there exists a sequence $\Lambda_0 = \Lambda, \Lambda_1, \ldots, \Lambda_m$ of Artin algebras such that each Λ_i is a one-point extension of Λ_{i-1} and $\mathcal{P}^\infty(\Lambda_m\text{-mod})$ is contravariantly finite in Λ_m-mod. Since in each passage from Λ_{i-1} to Λ_i both the little and big finitistic dimensions increase by at most 1, this might give a handle on the difference $l\operatorname{Fin dim} - l\operatorname{fin dim}$ for special classes of algebras. Moreover, the existence of a sequence as above guarantees that $l\operatorname{Fin dim}\Lambda < \infty$. In general, this process will force one to leave a given 'nice' class of algebras, however. The second part of this article is devoted to the explicit construction of such a sequence $\Lambda_0 = \Lambda, \Lambda_1, \ldots, \Lambda_m$ such that Λ is a monomial relation algebra while Λ_m cannot be chosen from this class of algebras.

Terminology and Notation.

In the following, Λ will be a split finite dimensional algebra over a field

K, i.e., Λ will be of the form $K\Gamma/I$ for some quiver Γ and an admissible ideal I of the path algebra $K\Gamma$. The Jacobson radical of Λ will be denoted by J. For us, the primitive idempotents of Λ will be those which naturally correspond to the vertices of Γ; in fact, we will identify a complete set of primitive idempotents of Λ with the vertices of the quiver.

According to [2], a full subcategory $\mathcal{A}$ of Λ-mod is said to be *contravariantly finite* in Λ-mod if, for each module M in Λ-mod, there exists a homomorphism $f : A \to M$ with $A \in \mathcal{A}$ such that the following sequence of contravariant functors is exact:

$$\operatorname{Hom}_\Lambda(-, A)|_\mathcal{A} \xrightarrow{\operatorname{Hom}_\Lambda(-,f)} \operatorname{Hom}_\Lambda(-, M)|_\mathcal{A} \to 0;$$

in other words, exactness of this sequence means that each $g \in \operatorname{Hom}_\Lambda(B, M)$ with $B \in \mathcal{A}$ factors through f. In that case, A is called a *right $\mathcal{A}$-approximation* of M. By [2], the $\mathcal{A}$-approximations of M that have minimal dimension are pairwise isomorphic. This justifies reference to *the* minimal right $\mathcal{A}$-approximation of M, whenever $\mathcal{A}$ is contravariantly finite. The terms *covariantly finite* and *left $\mathcal{A}$-approximation* are defined dually.

An algebra Λ is said to be *special biserial* in case it is isomorphic to a path algebra modulo relations, $K\Gamma/I$, with the following properties: Given any vertex e of Γ, at most two arrows enter e and at most two arrows leave e; moreover, for any arrow α in Γ there is at most one arrow β with $\alpha\beta \notin I$ and at most one arrow γ with $\gamma\alpha \notin I$.

Recall, moreover, that given two algebras $\Lambda = K\Gamma/I$ and $\Lambda' = K\Gamma'/I'$, the second is called a *one-point extension* of the first in case Γ' results from Γ through addition of a single vertex which is a *source* of Γ' such that $I' \cap K\Gamma = I$. For importance and properties of one-point extensions, we refer to [13].

Given paths p and q of Γ, we say that q is a *subpath* of p if $p = p_2 q p_1$ in $K\Gamma$ for suitable paths p_1 and p_2. We call q a *right (left) subpath* of p in case $p = p_2 q$ (respectively, $p = q p_1$) for suitable paths p_2 (respectively, p_1). The path p is said to *start (end)* in the arrow α if α is a right (left) subpath of p. Furthermore, we call an element x of a left Λ-module M a *top element* in case $x \in M \setminus JM$ and $ex = x$ for a primitive idempotent e from our distinguished set.

Finally, we refer the reader to previous work of the authors for their graphing conventions (see, e.g., [7,8,9,10]). The graphs most crucial to the present note are zigzags of the type

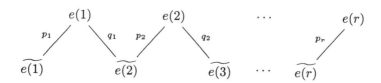

where the $e(i)$ and $\widetilde{e(i)}$ denote vertices and, for each i, the p_i, q_i denote paths of positive length starting in distinct arrows. That a module $M \in \Lambda\text{-mod}$ has the shown graph relative to a sequence $x_1, \ldots, x_r$ of top elements in particular encodes the following information: $x_i = e(i)x_i$, the x_i are K-linearly independent modulo JM, each p_i has starting point $e(i)$ and endpoint $\widetilde{e(i)}$, each q_i has starting point $e(i)$ and endpoint $e(i+1)$, and the multiples $q_i x_i = p_{i+1} x_{i+1}$, $1 \leq i \leq r-1$, are K-linearly independent elements of the socle of M. The information encoded in the graph guarantees, moreover, that there are no 'other' nonzero multiples of the x_i apart from those shown; more precisely, the only paths in $K\Gamma$ not annihilating the element $x_i \in M$ are the right subpaths of p_i and q_i.

2. Covariant finiteness of $\mathcal{P}^\infty(\Lambda\text{-mod})$

The main goal of this section will be the development and application of conditions which guarantee that a simple left Λ-module fails to have a left $\mathcal{P}^\infty(\Lambda\text{-mod})$-approximation. We will start by recalling a result of Auslander and Reiten, including a short alternate proof akin to the arguments of the introduction.

Proposition 1. [1, Proposition 4.2] *If* l fin dim $\Lambda \leq 1$, *then* $\mathcal{P}^\infty(\Lambda\text{-mod})$ *is covariantly finite in* $\Lambda\text{-mod}$.

Proof. Suppose l fin dim $\Lambda \leq 1$. By [4, Theorem 4.2], it suffices to prove that each direct product of objects in $\mathcal{P}^\infty(\Lambda\text{-mod})$ belongs to $\overrightarrow{\mathcal{P}^\infty}(\Lambda\text{-mod})$. Clearly, each direct product M of objects from $\mathcal{P}^\infty(\Lambda\text{-mod})$ has projective dimension ≤ 1 in $\Lambda\text{-Mod}$, and hence [9, Observation 5] shows that M is a direct limit of finitely generated modules of finite projective dimension as required. □

We will see later in this section that the conclusion of Proposition 1 breaks down for algebras of finitistic dimension 2.

Example 2. [11] This is the example exhibited by Igusa, Smalø and Todorov to show that $\mathcal{P}^\infty(\Lambda\text{-mod})$ may fail to be contravariantly finite, even in the case of a special biserial algebra Λ.

Let Γ be the quiver

and $I \subseteq K\Gamma$ such that, for $\Lambda = K\Gamma/I$, the indecomposable projective left Λ-modules have the following graphs:

CO– VERSUS CONTRAVARIANT FINITENESS

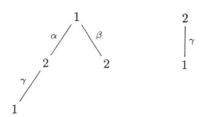

Here l Fin dim $\Lambda = 1$, and hence $\mathcal{P}^\infty(\Lambda\text{-mod})$ is covariantly finite in Λ-mod by Proposition 1. □

In the sequel, $\mathcal{A}$ will denote a full subcategory of Λ-mod.

Criterion 3. *Let $e(1), \ldots, e(r)$ be vertices of Γ, and let $p_1, \ldots, p_r, q_1, \ldots, q_r$ be $2r$ paths of positive length in $K\Gamma$, none of which is a subpath of any of the others. Moreover, suppose that the following conditions (1) and (2) are satisfied:*

(1) For each natural number n, there exists a module $M_n \in \mathcal{A}$ having graph

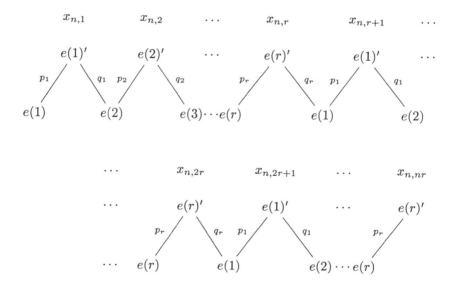

relative to a suitable sequence of top elements $x_{n,1}, \ldots, x_{n,nr}$ of M_n which are K-linearly independent modulo JM_n.

(2) Each module $A \in \mathcal{A}$ has the following properties:
  (i) $e(1)(\operatorname{Soc} A) \subseteq p_1 A$;
  (ii) $q_i A \cap (\operatorname{Soc} A) \subseteq p_{i+1} A$ for $i < r$, and $q_r A \cap (\operatorname{Soc} A) \subseteq p_1 A$;

(iii) *If $x \in A$ with $p_i x \in \operatorname{Soc} A$, then $q_i x \in \operatorname{Soc} A$.*
Then $S = \Lambda e(1)/Je(1)$ fails to have a left $\mathcal{A}$-approximation.

Proof. For $i > r$, let $s(i)$ be the integer in $\{1, \ldots, r\}$ with $i \equiv s(i) \pmod{r}$, and define $p_i := p_{s(i)}$, $q_i := q_{s(i)}$.

We assume that, to the contrary of our claim, $S = \Lambda e(1)/Je(1)$ does have a left $\mathcal{A}$-approximation $f : S \to A$ with $A \in \mathcal{A}$. Choose $n > \dim_K A$ and write $x_1, \ldots, x_{nr}$ for the elements $x_{n,1}, \ldots, x_{n,nr}$ of M_n. Moreover, let $g : S \to M_n$ be the embedding which sends $e(1) + Je(1)$ to $p_1 x_1$, and choose a homomorphism $h : A \to M_n$ with $g = h \circ f$. Since $f(e(1) + Je(1)) \in e(1) \operatorname{Soc} A$, condition 2(i) permits us to pick an element $a_1 \in A$ such that $f(e(1) + Je(1)) = p_1 a_1$. In view of the equality $hf(e(1) + Je(1)) = p_1 x_1$, we see that $h(a_1) = x_1 + y_1$ with $y_1 \in \sum_{j \not\equiv 1 \pmod{r}} \Lambda x_j + \sum_{j \equiv 1 \pmod{r}} J x_j$. Keep in mind that p_1 equals $p_{r+1} = \cdots = p_{lr+1}$, but is not a subpath of the other p_i or any of the q_i. Since q_1 is not a subpath of any of the paths $p_1, \ldots, p_r, q_2, \ldots, q_r$ either, we infer that $q_1 h(a_1) = q_1 x_1$. Due to our choice of a_1 such that $p_1 a_1 \in \operatorname{Soc} A$, condition 2(iii) guarantees that $q_1 a_1 \in \operatorname{Soc} A$ as well, in other words, $q_1 a_1 \in q_1 A \cap \operatorname{Soc} A$. Next, condition 2(ii) yields $a_2 \in A$ with $p_2 a_2 = q_1 a_1$. In view of $h(p_2 a_2) = q_1 h(a_1) = q_1 x_1$, we obtain $h(a_2) = x_2 + y_2$ with $y_2 \in \sum_{j \not\equiv 2 \pmod{r}} \Lambda x_j + \sum_{j \equiv 2 \pmod{r}} J x_j$; indeed this follows at once from the nature of the graph of M_n and the hypothesis that p_2 is not a subpath of any of $p_1, p_3, \ldots, p_r, q_1, \ldots, q_r$. Now the non-occurrence of q_2 as a subpath of $p_1, \ldots, p_r, q_1, q_3, \ldots, q_r$ allows us to deduce $q_2 h(a_2) = q_2 x_2$, and since $p_2 a_2 = q_1 a_1 \in \operatorname{Soc} A$, we observe that also $q_2 a_2 \in \operatorname{Soc} A$ by 2(iii). Thus 2(ii) in turn provides us with an element $a_3 \in A$ such that $p_3 a_3 = q_2 a_2$. As above, we argue that $q_3 h(a_3) = q_3 x_3$, and proceeding inductively, we thus obtain a sequence $a_1, \ldots, a_{nr}$ of elements in A with the property that $q_i h(a_i) = q_i x_i$ for $1 \leq i \leq nr$. Since the elements $q_1 x_1, \ldots, q_{nr} x_{nr}$ of M_n are K-linearly independent by hypothesis, so are $a_1, \ldots, a_{nr}$. But this contradicts our choice of n exceeding $\dim_K A$ and proves the criterion. □

As our argument makes clear, the requirement that none of the p_i, q_i is a subpath of any other as called for in the criterion can certainly be weakened. Other than that, the chain reaction exhibited in the proof of Criterion 3 appears prototypical for the failure of a simple module S to have a left $\mathcal{P}^\infty(\Lambda\text{-mod})$-approximation. In fact, specializing to special biserial algebras, we believe the answer to the following question to be positive.

Problem 4. Suppose $S = \Lambda e(1) Je(1)$ is a simple left module over a finite dimensional special biserial algebra $\Lambda = K\Gamma/I$. If S does not have a left $\mathcal{P}^\infty(\Lambda\text{-mod})$-approximation, do there exist paths $p_1, \ldots, p_r, q_1, \ldots, q_r$ in $K\Gamma$ satisfying the conditions (1) and (2) of Criterion 3?

On the other hand, we point out that it is not known whether failure of covariant finiteness of $\mathcal{P}^\infty(\Lambda\text{-mod})$ in $\Lambda\text{-mod}$ implies that one of the simple left Λ-modules is devoid of a left $\mathcal{P}^\infty(\Lambda\text{-mod})$-approximation. This is in contrast with our level of information on contravariant finiteness of $\mathcal{P}^\infty(\Lambda\text{-mod})$. Indeed, as was shown by Auslander and Reiten in [1], $\mathcal{P}^\infty(\Lambda\text{-mod})$ is contravariantly finite in Λ-mod provided that all simple Λ-modules have right $\mathcal{P}^\infty(\Lambda\text{-mod})$-approximations. We therefore propose

Problem 5. Is $\mathcal{P}^\infty(\Lambda\text{-mod})$ covariantly finite in Λ-mod in case each simple left Λ-module has a left $\mathcal{P}^\infty(\Lambda\text{-mod})$-approximation?

Criterion 3 is particularly suited to the situation where $\mathcal{A} = \mathcal{P}^\infty(\Lambda\text{-mod})$ and Λ is a finite dimensional special biserial algebra (see §1 for a definition). The representation theory of the finitely generated modules over such algebras is exceptionally well understood (see [3,5,6,12,14]). In fact, the indecomposable objects of Λ-mod fall into two classes, 'bands' and 'strings'. All we presently need to know about bands is the following: If $B \in \Lambda$-mod is a band, $e \in \Lambda$ a primitive idempotent and $x \in e(\operatorname{Soc} B)$, then $x \in \sum_i u_i B \cap v_i B$ for pairs (u_i, v_i) of paths of positive length ending in e such that, moreover, $u_i = \alpha_i u_i'$ and $v_i = \beta_i v_i'$ with distinct arrows α_i and β_i. The strings, on the other hand, are precisely the objects in Λ-mod having graphs of the form

relative to a suitable sequence of top elements, such that, for each i, the paths u_i, v_i start in distinct arrows, and the partners of each pair (v_i, u_{i+1}) end in distinct arrows.

Example 6. A finite dimensional special biserial algebra Λ for which $\mathcal{P}^\infty(\Lambda\text{-mod})$ fails to be covariantly finite in Λ-mod: Let $\Lambda = K\Gamma/I$, where Γ is the quiver

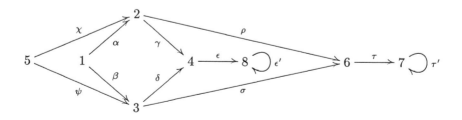

and I is generated by

$$\gamma\alpha - \delta\beta,\ \rho\chi - \sigma\psi,\ \gamma\chi,\ \rho\alpha,\ \delta\psi,\ \sigma\beta,\ \epsilon\gamma,\ \epsilon'\epsilon,\ (\epsilon')^2,\ \tau'\tau,\ (\tau')^2,\ \tau\sigma.$$

Then the graphs of the indecomposable projective left Λ-modules are

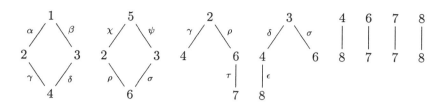

We will use Criterion 3 to show that $S = \Lambda e_3/Je_3$ does not have a left $\mathcal{P}^\infty(\Lambda\text{-mod})$-approximation. For that purpose, let $r = 2$, $e(1) = e_3$, $e(2) = e_5$, $p_1 = \beta$, and $p_2 = \chi$, $q_1 = \alpha$, $q_2 = \psi$. Then condition 1 of our criterion is satisfied by the modules M_n with graphs

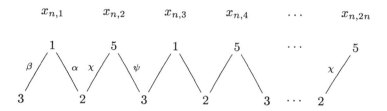

Indeed, one readily verifies that the first syzygy $\Omega^1(M_n)$ of M_n has graph

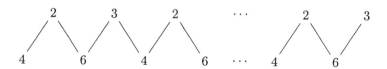

and the second syzygy $\Omega^2(M_n)$ is a direct sum of modules with graphs

This shows that $\mathrm{p\,dim}\, M_n = 2$; in particular, $M_n \in \mathcal{P}^\infty(\Lambda\text{-mod})$.

It is obvious that for every module $A \in \Lambda\text{-mod}$ and every element $x \in A$ the implications $(\beta x \in \mathrm{Soc}\, A \implies \alpha x \in \mathrm{Soc}\, A)$ and $(\chi x \in \mathrm{Soc}\, A \implies \psi x \in \mathrm{Soc}\, A)$ hold. So condition 2(iii) of Criterion 3 is met as well.

To check the remaining conditions under (2), we start by observing that the only nontrivial paths of $K\Gamma$ ending in e_3 are the arrows β and ψ, while the only nontrivial paths ending in e_2 are the arrows α and χ. Given our comments on bands, this implies that, for each band $B \in \Lambda\text{-mod}$, we have $e_3(\operatorname{Soc} B) \subseteq \beta B \cap \psi B$ and, in particular, $\psi B \cap (\operatorname{Soc} B) \subseteq \beta B$. Analogously, $\alpha B \cap (\operatorname{Soc} B) \subseteq e_2(\operatorname{Soc} B) \subseteq \chi B$. Hence, we may focus our attention on the situation where $A \in \mathcal{P}^\infty(\Lambda\text{-mod})$ is a string with $S_3 \subseteq \operatorname{Soc} A$. Note that A cannot be simple, since $\operatorname{p\,dim} S_3 = \infty$. The only possibility for a copy of $S_3 \subseteq \operatorname{Soc} A$ not to belong to βA is that of a string A having graph

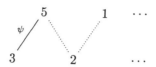

In that case $\Omega^1(A)$ would have a graph of one of the following types

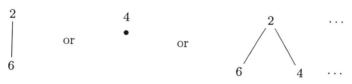

and $\Omega^2(A)$ would be a copy of $S_4 \oplus S_7$ in the first case, a copy of S_8 in the second, and have a direct summand isomorphic to S_7 in the third. In all of these cases, we would have $\operatorname{p\,dim}\Omega^2(A) = \infty$ contradicting our choice of A in $\mathcal{P}^\infty(\Lambda\text{-mod})$. This proves $e_3(\operatorname{Soc} A) \subseteq \beta A$ and thus 2(i).

For 2(ii) it suffices to observe that any $A \in \mathcal{P}^\infty(\Lambda\text{-mod})$ has the stronger property that $e_2(\operatorname{Soc} A) \subseteq \chi A$. The argument is analogous to the one we just completed for e_3 and β.

Thus, by Criterion 3, $S = \Lambda e_3 / J e_3$ does not have a left $\mathcal{P}^\infty(\Lambda\text{-mod})$-approximation. □

We remark that $\operatorname{l\,Fin\,dim}\Lambda = 2$ in the preceding example, which shows that Auslander-Reiten's Proposition 1 cannot be extended to the case of finitistic dimensions exceeding 1.

3. Curing failure of contravariant finiteness of $\mathcal{P}^\infty(\Lambda\text{-mod})$

We conjecture that, for any monomial relation algebra and for any special biserial algebra Λ, there exists a sequence of one-point extensions $\Lambda = \Lambda_0, \ldots, \Lambda_m = \Delta$ such that $\mathcal{P}^\infty(\Delta\text{-mod})$ is contravariantly finite in $\Delta\text{-mod}$. At this point, our conviction is based mainly on a long list of examples. One of our most interesting examples shows that, in general, one cannot expect Δ to retain the 'good' properties of Λ in this process, in other words, a 'cure' for failure of contravariant finiteness of $\mathcal{P}^\infty(\Lambda\text{-mod})$ by successive one-point extensions, will usually involve a trade-off. Here we will construct an example of a monomial relation algebra Λ, together with a sequence $\Lambda = \Lambda_0, \Lambda_1, \Lambda_2 = \Delta$ as above such that Δ cannot be chosen within the class of monomial relation algebras. Simultaneously, this example will illustrate the potential intricacy of the structure of the minimal right $\mathcal{P}^\infty(\Delta\text{-mod})$-approximations of the simple modules.

Example 7. Let $\Lambda = K\Gamma/I$, where Γ is the quiver

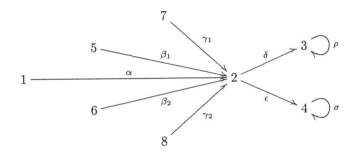

and the ideal $I \subseteq K\Gamma$ is generated by

$$\delta\alpha, \; \epsilon\alpha, \; \epsilon\beta_i \; (i=1,2), \; \delta\gamma_i \; (i=1,2), \; \rho\delta, \; \sigma\epsilon, \; \rho^2, \; \sigma^2.$$

This yields indecomposable projective left Λ-modules with graphs

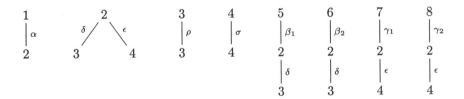

To see that $\mathcal{P}^\infty(\Lambda\text{-mod})$ fails to be contravariantly finite in $\Lambda\text{-mod}$, more precisely, that $\Lambda e_1/Je_1$ fails to have a right $\mathcal{P}^\infty(\Lambda\text{-mod})$-approximation, we exhibit a $\mathcal{P}^\infty(\Lambda\text{-mod})$-phantom of infinite K-dimension for $\Lambda e_1/Je_1$ (see [7, Definition 5 and Theorem 9]). The routine check that the following module H is indeed such a phantom is left to the reader: $H = \varinjlim H_n$ where

$$H = \left(\Lambda z \oplus \bigoplus_{i=1}^n \Lambda x_i \oplus \bigoplus_{i=1}^n \Lambda y_i \right) \bigg/ U_n,$$

with $z = e_1$, $x_{2m-1} = e_5$, $x_{2m} = e_6$, $y_{2m-1} = e_7$, $y_{2m} = e_8$ for $m \geq 1$, and

$$U_n = \Lambda(\alpha z - \beta_1 x_1 - \gamma_1 y_1) + \sum_{i=1}^{n-1} \Lambda(\widetilde{\gamma}_i y_i - \widetilde{\beta}_{i+1} x_{i+1} - \widetilde{\gamma}_{i+1} y_{i+1}),$$

with $\widetilde{\gamma}_i$ equal to γ_1 or γ_2, depending on whether i is odd or even, and $\widetilde{\beta}_i$ equal to β_1 or β_2, depending on whether i is odd or even. The modules H_n can be pictured via graphs of the form

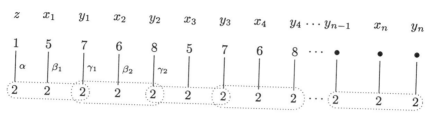

relative to the top elements $z, x_1, \ldots, x_n, y_1, \ldots, y_n$, where we extend our graphing conventions as follows: The dotted loop around the vertices labeled '2' which represent $\alpha z, \beta_1 x_1, \gamma_1 y_1$ indicates that any two of the three listed vectors are K-linearly independent while $\dim_K(K\alpha z + K\beta_1 x_1 + K\gamma_1 y_1) = 2$. The same holds for the additional triples $\widetilde{\gamma}_i y_i, \widetilde{\beta}_i x_{i+1}, \widetilde{\gamma}_{i+1} y_{i+1}$ surrounded by loops. Note that $\Omega^1(H_n) \cong (\Lambda e_2)^n$ for $n \in \mathbb{N}$; in particular, $H_n \in \mathcal{P}^\infty(\Lambda\text{-mod})$ for $n \in \mathbb{N}$.

Let Γ_1 be the quiver obtained from Γ by adding a single vertex, labeled 9, and two arrows leaving 9, namely $\chi_1 : 9 \to 5$ and $\chi_2 : 9 \to 6$. Moreover, let $\Lambda_1 = K\Gamma_1/I_1$, where the ideal $I_1 \subseteq K\Gamma_1$ is generated by I and the relation $\beta_1\chi_1 - \beta_2\chi_2$.

Next let Γ_2 be the quiver obtained from Γ_1 by adding a single vertex, 10, and two arrows leaving 10, namely $\psi_1 : 10 \to 7$ and $\psi_2 : 10 \to 8$. Now $\Delta = \Lambda_2 = K\Gamma_2/I_2$, where I_2 is generated by I_1 and the relation $\gamma_1\psi_1 - \gamma_2\psi_2 \in K\Gamma_2$.

Clearly, Λ_1 is a one-point extension of $\Lambda = \Lambda_0$, and $\Delta = \Lambda_2$ is a one-point extension of Λ_1. The 'new' indecomposable projective left Δ-modules have graphs

Note that $\Delta e_i = \Lambda e_i$ for $1 \leq i \leq 8$.

One can verify that $\mathcal{P}^\infty(\Delta\text{-mod})$ is contravariantly finite in $\Delta\text{-mod}$ by exhibiting right $\mathcal{P}^\infty(\Delta\text{-mod})$-approximations of the simple left Δ-modules

$S_i = \Delta e_i/J(\Delta)e_i$ for $1 \leq i \leq 10$. It is comparatively easy to see that the following are the (minimal) right $\mathcal{P}^\infty(\Delta\text{-mod})$-approximations of $S_2, \ldots, S_{10}$: Namely, Δe_i for $i = 2, 3, 4$, and

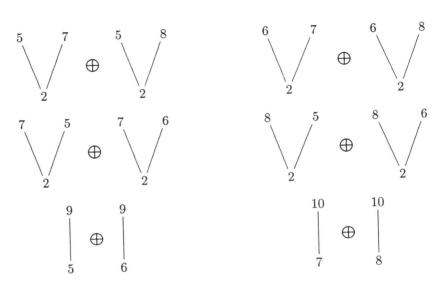

for $i = 5, 6, \ldots, 10$, respectively. We will sketch an argument backing the claim that the canonical epimorphism $f : A_1 \to S_1$ with

$$A_1 = (\Delta e_1 \oplus \Delta e_9 \oplus \Delta e_{10})/\Delta(\alpha, \beta_1\chi_1, \gamma_1\psi_1)$$

is a right $\mathcal{P}^\infty(\Delta\text{-mod})$-approximation of S_1. To buttress intuition, start by noting that A_1 has graph

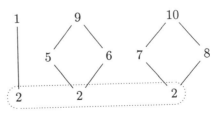

with the above convention for loops. Observe, moreover, that $\Omega^1_\Delta(A_1) \cong \Delta e_2$, whence $A_1 \in \mathcal{P}^\infty(\Delta\text{-mod})$.

We will sketch an argument showing that each epimorphism $g : M \to S_1$ with $M \in \mathcal{P}^\infty(\Delta\text{-mod})$ factors through f. It is clearly harmless to assume that M is indecomposable. Moreover, it suffices to consider the case where $\ker(g)$ *does not contain any nonzero submodules in* $\mathcal{P}^\infty(\Delta\text{-mod})$; indeed, given $U \subseteq M$ with $U \in \mathcal{P}^\infty(\Delta\text{-mod})$, it is enough to factor the induced map $\bar{g} : M/U \to S_1$ through f. In particular, this means that M does not contain any submodules isomorphic to Δe_i, with $i \geq 2$, nor any submodules with graphs of type

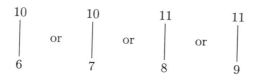

As a consequence, we can zero in on the structure of M as follows: Let $\mathcal{A}$ be the class of Δ-modules isomorphic to Δe_1, $\mathcal{B}$ the class of modules isomorphic to one of the Δ-modules with graphs

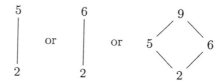

and, finally, $\mathcal{C}$ the class of those Δ-modules which have one of the graphs

Our assumptions on M guarantee that, up to isomorphism, $M = X/Y \in \mathcal{P}^\infty(\Delta\text{-mod})$, where

$$X = \bigoplus_{1 \leq i \leq m_1} A_i \oplus \bigoplus_{1 \leq i \leq m_2} B_i \oplus \bigoplus_{1 \leq i \leq m_3} C_i$$

with $A_i \in \mathcal{A}$, $B_i \in \mathcal{B}$, $C_i \in \mathcal{C}$ and $Y \subseteq \operatorname{Soc} X \cong S_2^{m_1+m_2+m_3}$.

Let a_i, b_i, c_i be top elements of A_i, B_i, C_i, respectively, and let $a'_i = \alpha a_i$, $b'_i = (\beta_1 + \beta_2 + \beta_1\chi_1)b_i$, $c'_i = (\gamma_1 + \gamma_2 + \gamma_1\psi_1)c_i$. Moreover, let

$$\bigoplus_{1 \leq i \leq m_1} \Delta \bar{a}_i \oplus \bigoplus_{1 \leq i \leq m_2} \Delta \bar{b}_i \oplus \bigoplus_{1 \leq i \leq m_3} \Delta \bar{c}_i \longrightarrow X/Y$$

be the obvious projective cover of X/Y mapping $\bar{a}_i$, $\bar{b}_i$, $\bar{c}_i$ to a_i, b_i, c_i respectively. Then

$$\operatorname{Soc} X = \bigoplus_{1 \leq i \leq m_1} \Delta a'_i \oplus \bigoplus_{1 \leq i \leq m_2} \Delta b'_i \oplus \bigoplus_{1 \leq i \leq m_3} \Delta c'_i,$$

and we can therefore write Y in the form $\bigoplus_{1\leq h\leq t}\Delta y_h$ with

$$y_h = \sum_{1\leq i\leq m_1} k_{hi}a'_i + \sum_{1\leq i\leq m_2} l_{hi}b'_i + \sum_{1\leq i\leq m_3} m_{hi}c'_i,$$

where $k_{hi}, l_{hi}, m_{hi} \in K$ such that $\Delta y_h \cong \Delta e_2 = \Lambda e_2$ for each h. Let $\mathbf{k}_h = (k_{hi})_{1\leq i\leq m_1} \in K^{m_1}$, $\mathbf{l}_h = (l_{hi})_{1\leq i\leq m_2} \in K^{m_2}$, and $\mathbf{m}_h = (m_{hi})_{1\leq i\leq m_3}$ in K^{m_3}. Then the vectors $\mathbf{l}_1,\ldots,\mathbf{l}_t \in K^{m_2}$ are K-linearly independent; indeed if we had $\sum_{1\leq h\leq t} d_h \mathbf{l}_h = 0$ with $d_h \in K$ not all zero, we would obtain a top element z of $\Omega^1_\Lambda(M)$ with the property that $\delta z = 0$, namely $z = \sum_{1\leq h\leq t} d_h z_h$, where

$$z_h = \sum_{1\leq i\leq m_1} k_{hi}\alpha\bar{a}_i + \sum_{1\leq i\leq m_2} l_{hi}(\beta_1 + \beta_2 + \beta_1\chi_1)\bar{b}_i$$
$$+ \sum_{1\leq i\leq m_3} m_{hi}(\gamma_1 + \gamma_2 + \gamma_1\psi_1)\bar{c}_i$$

in $\Omega^1_\Lambda(M)$. This would place a direct summand isomorphic to S_3 into $\Omega^1_\Lambda(M)$, which – in view of $\text{p}\dim_\Lambda \Delta e_3/J(\Delta e_3) = \text{p}\dim \Lambda e_3/Je_3 = \infty$ – is incompatible with $\text{p}\dim_\Lambda(M) < \infty$. Similarly $\mathbf{m}_1,\ldots,\mathbf{m}_t$ in K^{m_3} are linearly independent, since otherwise we would obtain a direct summand isomorphic to S_4 in $\Omega^1_\Lambda(M)$. If we set $\widetilde{g}(a_i) = \overline{(r_i e_1, 0, 0)} \in A_1$, where $f(a_i) = \overline{r_i e_1}$ with $r_i \in K$, the above independence information allows us to extend the assignment $\widetilde{g}$ to a homomorphism $\widetilde{g} : M \to A$. Any such homomorphism clearly satisfies $f \circ \widetilde{g} = g$.

Now let $\Lambda = R_0, R_1, \ldots, R_m = R$ be successive one-point extensions such that R is a monomial relation algebra. We leave the justification of our claim that $\mathcal{P}^\infty(R\text{-mod})$ fails to be contravariantly finite in R-mod as an exercise, but provide hints of the underlying ideas.

1.) Due to the fact that R is a monomial relation algebra obtained from Λ via one-point extensions, the following holds: Whenever a left R-module M has a submodule N with graph

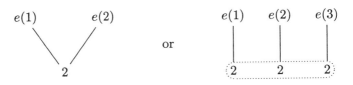

where $e(1)$, $e(2)$, $e(3)$ are distinct vertices in $\{e_1, e_5, e_6, e_7, e_8\}$, the first syzygy $\Omega^1_R(M)$ of M has a top element x of type e_2; moreover, if $e(1)$, $e(2)$, $e(3)$ belong to $\{e_1, e_5, e_6\}$, we can choose x such that Rx has graph

and if $e(1), e(2), e(3) \in \{e_1, e_7, e_8\}$, we can choose x such that Rx has graph

In each of these two situations, $\operatorname{p dim}_R(M) = \infty$.

2.) If there were a $\mathcal{P}^\infty(R\text{-mod})$-approximation B_1 of $Re_1/J(R)e_1 = \Lambda e_1/Je_1 = S_1$, then all homomorphisms in $\operatorname{Hom}_R(H_n, S_1) = \operatorname{Hom}_\Lambda(H_n, S_1)$ with H_n as above would factor through B_1, because $H_n \in \mathcal{P}^\infty(R\text{-mod})$. Using the first part, one would deduce the existence of a submodule of B_1 with graph

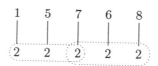

and then proceed to show that $H = \varinjlim H_n$ would still be a $\mathcal{P}^\infty(R\text{-mod})$-phantom for S_1.

References

1. M. Auslander and I. Reiten, *Applications of contravariantly finite subcategories*, Advances in Math. **86** (1991), 111-152.
2. M. Auslander and S.O. Smalø, *Preprojective modules over Artin algebras*, J. Algebra **66** (1980), 61-122.
3. M.C.R. Butler and C.M. Ringel, *Auslander-Reiten sequences with few middle terms and applications to string algebras*, Comm. Algebra **15** (1987), 145-179.
4. W. Crawley-Boevey, *Locally finitely presented additive categories*, Comm. Algebra **22** (1994), 1641-1674.
5. P.W. Donovan and M.R. Freislich, *The indecomposable representations of certain groups with dihedral Sylow subgroups*, Math. Ann. **238** (1978), 207-216.
6. I.M. Gelfand and V.A. Ponomarev, *Indecomposable representations of the Lorentz group*, Uspehi Mat. Nauk **23** (1968), 3-60; English Transl., Russian Math. Surveys **23** (1969), 1-58.
7. D. Happel and B. Huisgen-Zimmermann, *Viewing finite dimensional representations through infinite dimensional ones*, manuscript.
8. B. Huisgen-Zimmermann, *Predicting syzygies over monomial relation algebras*, manuscripta math. **70** (1991), 157-182.

9. _____, *Homological assets of positively graded representations of finite dimensional algebras*, in Representations of Algebras (Ottawa 1992) (V. Dlab and H. Lenzing, eds.), Canad. Math. Soc. Conf. Proc. Series 14, 1993, pp. 463-475.
10. B. Huisgen-Zimmermann and S.O. Smalø, *A homological bridge between finite and infinite dimensional representations of algebras*, manuscript.
11. K. Igusa, S.O. Smalø, and G. Todorov, *Finite projectivity and contravariant finiteness*, Proc. Amer. Math. Soc. **109** (1990), 937-941.
12. C.M. Ringel, *The indecomposable representations of the dihedral 2-groups*, Math. Ann. **214** (1975), 19-34.
13. _____, *Tame Algebras and Integral Quadratic Forms*, Lecture Notes in Math. 1099, Springer-Verlag, Berlin, 1984.
14. B. Wald and J. Waschbüsch, *Tame biserial algebras*, J. Algebra **95** (1985), 480-500.

DEPARTMENT OF MATHEMATICS, UNIVERSITY OF CALIFORNIA, SANTA BARBARA, CA 93106, USA
E-mail address: birge@math.ucsb.edu

DEPARTMENT OF MATHEMATICS, THE NORWEGIAN UNIVERSITY FOR SCIENCE AND TECHNOLOGY, 7055 DRAGVOLL, NORWAY
E-mail address: sverre@matstat.unit.no

MONOMIALS AND THE LEXICOGRAPHIC ORDER

Heather Hulett

ABSTRACT. This paper investigates the relationship between the lexicographic order on monomials and lex-segments of monomials.

1. Introduction

The relationship between monomials in a polynomial ring and the lexicographic ordering of these monomials has been studied for some time. Results involving the extremal properties of initial lexicographic ideals began in 1926 when Macaulay [Ma] used the lexicographic ordering to determine bounds on the Hilbert function of certain ideals. In the 1960's, Kruskal [Kr] and Katona [Ka] generalized Macaulay's theorem using combinatorial arguments. Further interest was spurred by the development of computer algebra programs (e.g., *Macaulay* [BS]) where the ordering of the monomials plays an important role in determining the Gröbner basis of an ideal. Two different proofs by Bigatti [Bi] and this author [Hu] in 1993 showed the relationship between initial lexicographic ideals and size of the Betti numbers of monomial ideals. More recently, Martin and this author [HM] studied the resolution of the more general lexicographic ideal.

The purpose of this note is to discuss some basic properties of the relationship between a monomial of degree d in $R = k[x_1, \ldots, x_n]$ and its position in the lexicographic ordering. Specifically, there are two goals: 1) to write the pth monomial explicitly given only p, n, and d and conversely, given the monomial, immediately tell where it is in the lexicographic ordering and 2) predict what the pth monomial of degree d will generate in degree $d+1$. The ultimate objective is to predict when a string of consecutive monomials in degree d will generate consecutive monomials in the next degree. This is of interest because, while an initial lex-segment in degree d will necessarily generate an initial lex-segment in degree $d+1$, this is not true for a general lex-segment. Thus, the second goal will help determine when an ideal is a lexicographic ideal.

1991 *Mathematics Subject Classification*. Primary 13F20 ; Secondary 05A10.

Typeset by $\mathcal{A}_{\mathcal{M}}\mathcal{S}$-TEX

2. The Lexicographic Order

We begin with some definitions and basic facts.

Recall that for two monomials $u = x_1^{a_1} \cdots x_n^{a_n}$ and $v = x_1^{b_1} \cdots x_n^{b_n}$, the lexicographic ordering specifies that $u > v$ if either $\deg u > \deg v$ or $\deg u = \deg v$ and $a_i - b_i > 0$ the first time it is non-zero. Thus the degree d monomials in R in the lex-order can be written by grouping them as follows:

$$x_1^d,\ x_1^{d-1} \cdot (x_2, \ldots, x_n),\ \ldots,\ x_1^{d-i} \cdot (x_2, \ldots, x_n)^i,\ \ldots,\ x_1^0 \cdot (x_2, \ldots, x_n)^d$$

where $(x_2, \ldots, x_n)^i$ is short-hand notation for the monomials of degree i in $x_2, \ldots, x_n$ written in lexicographic order. For a given degree d we write m_p for the pth monomial in this order. Since the number of monomials of degree d in n variables is $\binom{n+d-1}{d}$, the number of monomials having $d - i$ as the highest power of x_1 that occurs is $\binom{n-1+i-1}{i}$. Moreover, it is an easy exercise to show that

$$\sum_{k=0}^{i-1} \binom{n-1+k-1}{k} = \binom{n+i-2}{i-1} = \binom{n+i-2}{n-1}.$$

Therefore, the degree of x_1 in m_p is $d - i$ if and only if $\binom{n+i-2}{n-1} < p \leq \binom{n+i-1}{n-1}$.

In the spirit of Macaulay's work, given positive integers n and p we can write a unique representation of p in the form

$$p = 1 + \binom{b_1}{n-1} + \binom{b_2}{n-2} + \cdots + \binom{b_j}{n-j}$$

where $n - k \leq b_k$ and $b_{k+1} < b_k$ for all $1 \leq k \leq n - 1$. (For notational convenience, we let $b_{j+1} = \cdots = b_{n-1} = 0$.) Because the lexicographic order is a total order, we can find a unique relationship between the b_i's in the expansion of p above and the exponents e_i in the pth monomial $m_p = x_1^{e_1} \cdots x_n^{e_n}$.

Theorem 2.1. Let $R = k[x_1, \ldots, x_n]$ with $n \geq 2$ and let $m_p = x_1^{e_1} \cdots x_n^{e_n}$ be a degree d monomial in R with $1 < p \leq \binom{n+d-1}{d}$. Then

$$p = 1 + \binom{b_1}{n-1} + \binom{b_2}{n-2} + \cdots + \binom{b_j}{n-j}$$

iff $e_i = n + d - (e_1 + \cdots + e_{i-1}) - b_i - (i+1)$ for $1 \leq i \leq j$, $e_{j+1} = d - (e_1 + \cdots + e_j)$ and $e_k = 0$ for $j + 2 \leq k \leq n$. For $2 \leq i \leq j$, this simplifies to $e_i = b_{i-1} - b_i - 1$.

Proof. This is easy in $k[x_1, x_2]$. The degree d monomials in lexicographic order are

$$x_1^d,\ x_1^{d-1} x_2,\ x_1^{d-2} x_2^2,\ \ldots,\ x_2^d$$

so it is clear that the exponent of x_2 is one less that the position of the monomial in the ordering. Thus, if $m_p = x_1^{e_1} x_2^{e_2}$ has degree d and $p = 1 + \binom{b_1}{1}$, then $e_2 = b_1$ and $e_1 = d - e_2 = d - b_1 = n + d - b_1 - 2$.

Now, let the dimension n be greater than 2. By the above discussion about the degree of x_1, since $\binom{b_1}{n-1} < p < \binom{b_1+1}{n-1}$, we must have $e_1 = n + d - b_1 - 2$. Let $w = x_2^{e_2} \cdots x_n^{e_n}$. Then by the grouping of the monomials in lex-order given above, w is the $1 + \binom{b_2}{n-2} + \cdots + \binom{b_j}{n-j}$-th monomial of degree $d - e_1$ in the variables $x_2, \ldots, x_n$. If $b_2 = 0$, then $w = x_2^{d-e_1}$. Otherwise, by induction, for $2 \leq i \leq j$ we have $e_i = (n-1) + (d - e_1) - (e_2 + \cdots + e_{i-1}) - b_i - i = n + d - (e_1 + \cdots + e_{i-1}) - b_i - (i+1)$ and $e_{j+1} = (d - e_1) - (e_2 + \cdots + e_j)$. We are done since the ordering gives a one-to-one correspondence between a monomial and its place in the lexicographic order.

Using this theorem we can predict what the pth monomial in degree d will generate in degree $d+1$, or more precisely, where those monomials will occur.

Corollary 2.2. *Let* $p = 1 + \binom{b_1}{n-1} + \binom{b_2}{n-2} + \cdots + \binom{b_r}{n-r}$ *as above and let* $w_{p_i} = x_i m_p$ *for* $1 \leq i \leq n$. *Then* $p_1 = p$; *if* $1 < i \leq r+1$, *then*

$$p_i = 1 + \binom{b_1+1}{n-1} + \cdots + \binom{b_{i-1}+1}{n-(i-1)} + \binom{b_i}{n-i} + \cdots + \binom{b_r}{n-r};$$

and if $r+1 < i \leq n$, *then*

$$p_i = 1 + \binom{b_1+1}{n-1} + \cdots + \binom{b_r+1}{n-r} + \binom{n-r-1}{n-r-1} + \cdots + \binom{n-i+1}{n-i+1}.$$

Proof. If $i = 1$, $w_{p_1} = x_1^{e_1+1} x_2^{e_2} \cdots x_n^{e_n}$ where the e_i are written in terms of the b_i in the expansion of p above as in the theorem. Using b'_j's in the unique expansion of p_i, we see that $e_1 + 1 = n + (d+1) - b'_1 - 2$ and hence $b'_1 = b_1$. The 1 added to the degree offsets the added 1 in the new exponent of x_1. These will continue to cancel throughout the b'_j's so that

$$b'_j = n + (d+1) - ((e_1+1) + e_2 + \cdots + e_j) - (j+1) = b_j.$$

For $1 < i \leq r+1$, the increase in the degree again necessitates the increase in the b'_j's so that the exponents will stay the same until the ith variable: $b'_j = n + (d+1) - (e_1 + e_2 + \cdots + e_j) - (j+1) = b_j + 1$ for $1 \leq j < i$. In w_{p_i}, we have $b'_i = n + (d+1) - (e_1 + e_2 + \cdots + e_i + 1) - (i+1)) = b_i$ and similarly $b'_j = n + (d+1) - (e_1 + e_2 + \cdots + e_i + 1 + e_{i+1} + \cdots + e_j) - (j+1) = b_j$ for $i \leq j \leq r$.

Finally, suppose $r+1 < i \leq n$. As argued above, $b'_j = n + (d+1) - (e_1 + e_2 + \cdots + e_j) - (j+1) = b_j + 1$ for $1 \leq j \leq r$. Since $e_{r+2} = \cdots = e_n = 0$ in

m_p, all of the corresponding exponents except the ith will still equal 0 in w_{p_i}. Thus, we need $b'_j = n - j$ for $r + 1 \leq j \leq (i-1)$. This will keep the exponents of x_1 through x_{i-1} the same so we are left with the exponent of x_i equal to $(d+1) - (e_1 + \cdots + e_{i-1}) = (d+1) - d = 1$. Also, for $j > i$ we have $n + (d+1) - (e_1 + \cdots + e_i + 1 + \cdots + e_j) - j + 1 = n - j - 1 < n - j$ so $b'_j = 0$.

3. Lex-Segments

We recall the definitions introduced in several of the papers mentioned in the introduction.

Definition 3.1. *A string of consecutive monomials of degree d in the lexicographic order is called a lex-segment. If the first term of a lex-segment is x_1^d, then we call it an initial lex-segment and if the last term in a lex-segment is x_n^d, then it is called a final lex-segment.*

It is easy to check that every initial lex-segment in degree d generates an initial lex-segment in degree $d+1$ (see [Hu]). We now consider when a lex-segment will generate a lex-segment in view of our results above.

Proposition 3.2. *Let $m_i, \ldots, m_{i+r-1}$ be a lex-segment of degree d monomials with $1 < i = 1 + \binom{b_1}{n-1} + \binom{b_2}{n-2} + \cdots + \binom{b_t}{n-t}$. If this lex-segment generates a lex-segment in degree $d+1$, then $r \geq \binom{b_1}{n-2}$.*

Proof. As above, let $w_{j_k} = x_k m_j$. By the corollary and the fact that the lexicographic ordering is multiplicative, the first $r + 1$ monomials of degree $d+1$ generated are (in order) $w_i, \ldots, w_{i+r-1}, w_{i_2}$. Now, the first r monomials are clearly consecutive so if the degree $d+1$ monomials form a lex-segment, we must have $i < i_2 \leq i + r$. Since $i_2 = 1 + \binom{b_1+1}{n-1} + \binom{b_2}{n-2} + \cdots + \binom{b_t}{n-t}$, then $r \geq i_2 - i = \binom{b_1+1}{n-1} - \binom{b_1}{n-1} = \binom{b_1}{n-2}$.

Note that if $m_i = x_2^d$, then $i = 1 + \binom{n+d-2}{n-1}$ and so $r = \binom{n+d-2}{n-2}$ which makes $m_{i+r-1} = x_n^d$ and hence the segment was a final lex-segment. This agrees with Deery's result that a final lex-segment generates a final lex-segment if and only if x_2^d is in the generating set.

As noted above if $m_i, \ldots, m_k$ is a lex-segment, then so is $x_1 \cdot (m_i, \ldots, m_k)$ and in fact, these are the first monomials that the lex-segment generates in the next degree.

Proposition 3.3. *Let $k^i = 1 + \binom{a}{n-1} + \binom{a-1}{n-2} + \cdots + \binom{a-(i-2)}{n-(i-1)}$ for $i \leq n-1$ and $\alpha = \binom{a+1}{n-1} = 1 + \binom{a}{n-1} + \binom{a-1}{n-2} + \cdots + \binom{a-n+2}{1}$. Then $x_i \cdot (m_{k^i}, \ldots, m_\alpha)$ is a lex-segment, as is $x_{i-1} m_\alpha, x_i m_{k^i}$.*

Proof. The first claim is clear since, if $m_{k^i} < w_p < m_\alpha$, then $p = 1 + \binom{a}{n-1} + \cdots + \binom{a-(i-2)}{n-(i-1)} + \binom{b_i}{n-i} + \cdots + \binom{b_s}{n-s}$ for some $b_j \leq a - (j-2)$ with strict inequality holding for at least one b_j. Then by the corollary above, $p_i = 1 + \binom{a+1}{n-1} + \binom{a}{n-2} + \cdots + \binom{a-(i-2)+1}{n-i+1} + \binom{b_i}{n-i} + \cdots + \binom{b_s}{n-s}$ and clearly this is the monomial preceding $x_i w_{p+1}$.

To show the second claim we again use the corollary. If $w_p = x_{i-1} m_\alpha$, then $p = 1 + \binom{a+1}{n-1} + \cdots + \binom{a-i+4}{n-i+2} + \binom{a-i+2}{n-i+1} + \cdots + \binom{a-n+2}{1}$. Note that the last terms of p have a constant difference between the "numerator" and the "denominator" of $a - n + 1$ so adding one makes that last sum "roll over" and so $p + 1 = 1 + \binom{a+1}{n-1} + \cdots + \binom{a-i+4}{n-i+2} + \binom{a-i+3}{n-i+1}$. If $w_q = x_i m_{k^i}$, then $q = 1 + \binom{a+1}{n-1} + \binom{a}{n-2} + \cdots + \binom{a-(i-2)+1}{n-(i-1)}$ which equals $p + 1$.

We use this proposition to determine when a lex-segment $m_i, \ldots, m_k$ generates a lex-segment in the next degree.

Theorem 3.4. *Let* $\dim R > 2$ *and let* $m_i, \ldots, m_k$, *with* $i > 1$, *be a lex-segment of monomials of degree* d *in* R *such that* $x_1 | m_i$. *Let* $k = 1 + \binom{a_1}{n-1} + \cdots + \binom{a_s}{n-s}$ *and let* $1 < j$ *be the first index such that* $a_j + 1 \neq a_{j-1}$ *(j could equal* $s+1$*). If* $p = 1 + \binom{a_1-1}{n-1} + \cdots + \binom{a_{j-1}-1}{n-(j-1)} + \binom{a_j}{n-j} + \cdots + \binom{a_s}{n-s} + 1 \geq i$, *then* $m_i, \ldots, m_k$ *generates a lex-segment in degree* $d + 1$.

Proof. Note m_p is in the given lex-segment since $i \leq p < k$.
Claim: $x_1(m_i, \ldots, m_k), x_j m_p$ is a lex-segment.
As before, we write $x_i m_j = w_{j_i}$. We only need to check the last two monomials in the sequence. We have $p_j = 1 + \binom{a_1}{n-1} + \cdots + \binom{a_{j-1}}{n-(j-1)} + \binom{a_j}{n-j} + \cdots + \binom{a_s}{n-s} + 1 = k + 1$ so the monomials are consecutive.

Now, let $\alpha = \binom{a_1}{n-1} = 1 + \binom{a_1-1}{n-1} + \binom{a_1-2}{n-2} + \cdots + \binom{a_1-(n-1)}{1}$. Note $i < \alpha < k$.
Claim: $x_j(m_p, \ldots, m_\alpha), x_{j+1} m_{k^{j+1}}$ is a lex-segment.
The proof that the first terms are consecutive proceeds just as in the previous proposition so we need only check $\alpha_j + 1 = k_{j+1}^{j+1}$.
As above, $k^{j+1} = 1 + \binom{a_1-1}{n-1} + \binom{a_1-2}{n-2} + \cdots + \binom{a_1-j}{n-j}$ so $k_{j+1}^{j+1} = 1 + \binom{a_1}{n-1} + \binom{a_1-1}{n-2} + \cdots + \binom{a_1-j+1}{n-j}$. But $\alpha_j + 1 = 1 + \binom{a_1}{n-1} + \binom{a_1-1}{n-2} + \cdots + \binom{a_1-(j-1)+1}{n-(j-1)} + \binom{a_1-j}{n-j} + \cdots + \binom{a_1-(n-1)}{1} + 1 = 1 + \binom{a_1}{n-1} + \binom{a_1-1}{n-2} + \cdots + \binom{a_1-j+1}{n-j}$ so the monomials are consecutive. Again, $m_{k^{j+1}}$ is in the original lex-segment.

Continuing in this fashion we get the following lex-segment where all m_j's are between m_i and m_k:

$$x_1(m_i, \ldots, m_k), x_j(m_p, \ldots, m_\alpha), x_{j+1}(m_{k^{j+1}} \ldots, m_\alpha), \ldots, x_n m_{k^n} = x_n m_\alpha.$$

Now, by induction on the ring $k[x_2, \ldots, x_n]$, we can finish the argument after noting that $x_n m_\alpha, x_2 m_{\alpha+1}$ is a lex-segment. But this is clear since $\alpha_n + 1 = 1 + \binom{a_1}{n-1} + \binom{a_1-1}{n-2} + \cdots + \binom{a_1-(n-1)+1}{1} + 1 = 1 + \binom{a_1+1}{n-1} = (\alpha+1)_2$. The induction gives the next chunk of lex-segment as $x_2(m_{\alpha+1}, \ldots, m_\beta)$ where $\beta = \binom{a_1}{n-1} + \binom{a_2}{n-2}$ and eventually the last element in the lex-segment will be given by $x_n m_\gamma$ where $\gamma = \binom{a_1}{n-1} + \binom{a_2}{n-2} + \cdots + \binom{a_s}{n-s} = k$. Thus we have described every monomial between $x_1 m_i$ and $x_n m_k$ as a product of $x_j m_l$ for some $1 \leq j \leq n$ and some $i \leq l \leq k$, therefore $m_i, \ldots, m_k$ generates a lex-segment.

This is a slight improvement of the result in [HM] that says if x_j is the highest indexed variable in m_k and $w_q = m_k/x_j$ then $m_i, \ldots, m_k$ will generate a lex-segment if $i < q$.

References

[BS] D. Bayer, and M. Stillman, *Macaulay: A system for computation in algebraic geometry and commutative algebra*, Download via anonymous ftp from zariski.harvard.edu, 1989.

[Bi] A. Bigatti, *Upper bounds for the Betti numbers of a given Hilbert function*, Comm. Algebra **21** (1993), 2317–2334.

[De] T. Deery, *Rev-lex segment ideals and minimal Betti numbers*, Queen's Papers in Pure and Applied Math **102**, 1996.

[EK] S. Eliahou and M. Kervaire, *Minimal resolutions of some monomial ideals*, J. Algebra **129** (1990), 1–25.

[Hu] H. Hulett, *Maximum Betti numbers for a given Hilbert function*, Comm. Algebra **21** (1993), 2335–2350.

[HM] H. Hulett and H. Martin, *Betti Numbers of Lex-Segment Ideals*, preprint, 1996.

[Ka] G. Katona, *A Theorem of Finite Sets*, Theory of Graphs (Proc. Colloq., Tihany, Hungary, 1966) (P. Erdös & G. Katona, ed.), Academic Press, New York and London, 1968.

[Kr] J. B. Kruskal, *The Number of Simplices in a Complex*, Mathematical Optimization Techniques (R. Bellman, ed.), University of California Press, Berkeley, CA, 1963, pp.251-278.

[Ma] F. S. Macaulay, *Some Properties of Enumeration in the Theory of Modular Systems*, Proc. London Math. Soc. **26** (1927), 531–555.

Department of Mathematics and Statistics, Miami University, Oxford, OH 45056

RINGS OVER WHICH DIRECT SUMS OF CS MODULES ARE CS

DINH VAN HUYNH AND BRUNO J. MÜLLER

ABSTRACT. A module M is defined to be a CS module, if every submodule of M is essential in a direct summand of M. In this paper we show, among other results, that for a right nonsingular ring R, all direct sums of CS right R-modules are CS if and only if R is a right artinian ring and every indecomposable injective right R-module has length at most two.

1. Introduction

A module M is called a CS module (or an extending module, or a module with (C_1)), if every submodule of M is essential in a direct summand of M.

CS modules generalize quasi-continuous modules, which in turn generalize continuous, quasi-injective and injective modules, in this order. It is well-known that a ring R is semisimple artinian if and only if every right (left) R-module is injective. In analogy, a ring R is said to be a *CS-semisimple* ring if every right R-module is CS (cf. [9]). By [3, 13.5], a ring R is CS-semisimple if and only if R is a right and left artinian, right and left serial ring with $J(R)^2 = 0$, where $J(R)$ is the Jacobson radical of R.

A ring R is right noetherian if and only if every direct sum of injective right R-modules is injective (see e.g. [1, 18.13]). On the other hand, by [8, Corollary 2], a ring R is semisimple artinian if and only if each direct sum of any two quasi-continuous right R-modules is quasi-continuous. Motivated by these results, we consider here rings for which direct sums of CS modules are CS, and show that such a ring is semiprimary with Jacobson radical square zero. For nonsingular rings, we obtain the following result:

Theorem 1. *For a right nonsingular ring R the following conditions are*
equivalent:
(i) R is a right artinian ring and every indecomposable injective right R-module
has length at most two;
(ii) Every direct sum of CS right R-modules is CS.

1991 *Mathematics Subject Classification*. Primary 16D70. Secondary 16D50, 16D40, 16D60..

We also give an example which shows that a ring of Theorem 1 is not necessarily CS-semisimple. However the following result holds.

Theorem 2. *Let R be any ring over which every direct sum of CS right R-modules is CS. If R is right CS, then every right R-module is CS, i.e. R is CS-semisimple.*

2. Preliminary Results

Throughout this paper, all modules are unitary right modules over associative rings with identity. For modules over a ring R we study, besides the *Eigenschaft*

(E) every direct sum of CS modules is CS,

the two weaker versions

(E_1) every direct sum of uniform modules is CS, and

(E_2) every direct sum of uniform modules and *one* injective module is CS.

For a module M, $J(M), Soc(M)$ and $E(M)$ denote the Jacobson radical, the socle and the injective hull of M, respectively. A submodule N of M is called *closed* if it has no proper essential extension in M. The notations $N \subseteq' M$, $N \subseteq_{cl} M$, $N \subseteq^{\oplus} M$ signify that N is an essential, closed, direct summand submodule of M, respectively. If applied to a ring R, terms like nonsingular, artinian, etc. mean that R has the respective property as a right module over itself. For a direct sum $\oplus_{i \in I} M_i$ and a subset K of the index set I, the subsum $\oplus_{i \in K} U_i$ is abbreviated by $U(K)$. For background references we refer to [1], [3], [10] and [14].

We note that the properties $(E_1), (E_2)$ and (E), being purely lattice theoretical, are inherited by factor rings, and by Morita equivalent rings.

Lemma 3 ([2, Proposition 2.2]). *Being closed is transitive, i.e. $A \subseteq_{cl} B \subseteq_{cl} C$ implies $A \subseteq_{cl} C$.*

Harada [5] (cf. [10, 2.25]) has shown the following important supplement to the Krull-Schmidt-Azumaya Theorem: *A direct sum of modules with local endomorphism rings satisfies lsTn, if and only if it complements direct summands, if and only if every local direct summand is a direct summand.* Here *lsTn* (locally semi-T-nilpotent) means that the composition of any non-isomorphisms between distinct summands, is eventually zero on any particular element. We call such a module a *Harada module*.

Lemma 4. *If $U = \oplus U_i$ is a direct sum of uniform modules of length at most two, then U is a Harada module.*

Proof. The endomorphism rings of each U_i is obviously local, and $lsTn$ holds since non-isomorphisms decrease lengths. □

Next we discuss the category $\sigma[M]$ generated by a module M. It contains all subfactors of direct sums of copies of M. Every module X has a unique largest submodule $X' \in \sigma[M]$, namely the sum of all images of maps from M to X. In consequence, $\sigma[M]$ is a reflective subcategory of Mod-R; colimits and finite limits in $\sigma[M]$ coincide with those in Mod-R, while arbitrary limits are the reflections of those in Mod-R. (For more details cf. [14].)

The category $\sigma[M]$ is well suited for the discussion of M-injectivity: A module $X \in \sigma[M]$ is $\sigma[M]$-injective if and only if X is M-injective. $\sigma[M]$ is closed under M-injective hulls. The analogue for projectivity is not quite true: $\sigma[M]$-projectivity is usually stronger than M-projectivity (except if the module is finitely generated or has a projective cover).

Lemma 5 ([14, 27.3 and 27.5]). *A module M is locally noetherian (i.e. every finitely generated submodule of M is noetherian), if and only if any direct sum of M-injective modules is M-injective. If so, then every module in $\sigma[M]$ contains a maximal M-injective submodule, and every M-injective module in $\sigma[M]$ is a direct sum of uniform submodules.*

We shall say that a module M has property $(E_1), (E_2)$ or (E) if the category $\sigma[M]$ has the property. We find it convenient to prove our results in this section in the general setting of $\sigma[M]$. The reader who doesn't want to be bothered with these technicalities, should always read $M = R$ hence $\sigma[M] = $ Mod-R. But we emphasize that the use of $\sigma[M]$ is needed in the proof of Theorem 8.

Lemma 6 (cf. [3, 13.1]). *For any module M the following conditions are*

equivalent:
(1) M satisfies (E_1);
(2) Each direct sum of any two uniform modules in $\sigma[M]$ is CS;
(3) Every uniform module in $\sigma[M]$ has length at most two.

Although the proof of Lemma 6 can be found in [3, 13.1], we remark that the equivalence (1) ⇔ (3) was also proved by us in an earlier preprint (announced in [7]), using different techniques.

Recall that a module X is called a V-module if every submodule of X is the intersection of maximal submodules. It is easily checked that this holds if and only if every simple module is X-injective.

A module M is said to satisfy (L) (or to be a module with (L)) if for M

one of the equivalent conditions (1), (2) or (3) in Lemma 6 holds.

Proposition 7. *If (L) holds for M, then for each $X \in \sigma[M]$, $J(X)$ is a V-module.*

Proof. For an arbitrary module $X \in \sigma[M]$, every submodule A is the intersection of completely meet-irreducible submodules A_i. The factor modules X/A_i are uniform, hence of length at most 2: $A_i \subseteq B_i \subset X$. If A is a submodule of $J(X)$, then $A = \cap A_i = \cap(A_i \cap J(X))$. As $J(X) \subseteq B_i$, the intersection $A_i \cap J(X)$ either equals $J(X)$ or it is maximal in $J(X)$. □

Theorem 8. *If a module M satisfies (E_2), then every module in $\sigma[M]$ is semiartinian.*

Proof. It suffices to show that every nonzero module $N \in \sigma[M]$ has a simple submodule. Suppose this is not true for some N. Without loss of generality, we may assume that N is M-injective.

Consider an arbitrary direct sum $U = \oplus U_i$ of N-injective uniform modules $U_i \in \sigma[N] \subseteq \sigma[M]$. By Lemma 6, each U_i is of length at most two. We claim that U is N-injective.

Let $f : H \to U$ be a homomorphism from an essential submodule H of N to U. Define $H_1 = \{x - f(x) | \ x \in H\}$. As $N \oplus U$ is CS by (E_2), there exists $H_1 \subseteq' H_2$ with $N \oplus U = H_2 \oplus K$. Let $p_1 : N \oplus U \to U$ and $p_2 : H_2 \oplus K \to K$ be the projections. Since $H_2 \cap U = 0$, p_2 embeds U into K. In fact, $p_2(U)$ is essential in K, since H is essential in N. As U is essential over its socle, due to (L), so is K. But N has zero socle by supposition, so $N \cap K = 0$. Therefore p_1 embeds K into U.

Together we have $p_1 p_2(U) \subseteq' p_1(K) \subseteq U$. Since each U_i is N-injective, and hence U-injective, $p_1 p_2(U)$ is a local direct summand in U. By Lemma 4, it is a direct summand, in U hence in $p_1(K)$. But then, $p_1 p_2(U) = p_1(K)$, therefore $p_2(U) = K$. We conclude that $N \oplus U = H_2 \oplus p_2(U) = H_2 \oplus U$. Hence, the restriction $(p_3|N)$ of the projection $p_3 : H_2 \oplus U \to U$ extends f. This proves our claim.

So far we have shown that any direct sum of N-injective uniform modules in $\sigma[N]$ is N-injective. By Lemma 5, N is locally noetherian, and a direct sum of uniform modules. Then (L) implies that N has a simple submodule, contrary to the initial supposition. □

A module in $\sigma[M]$ is M-*singular* if it is of the form A/B with $B \subseteq' A \in \sigma[M]$. This is equivalent to the usual "singular" if $M = R$. In generalization of the well-known concept for rings, one calls a module M an SI-module if every M-singular module is M-injective (cf. [3, Section 17]). A module $N \in \sigma[M]$ is called M-*nonsingular* if N does not contain any nonzero M-singular submodule.

Theorem 9. *An M-nonsingular module M satisfies (E_2) if and only if*

M is an SI-module with (L).

Proof. ($\Rightarrow$): We have to show that any M-singular module X is M-injective. By Theorem 8, X is semiartinian, hence essential over its socle. We repeat the proof of the claim, within the proof of Theorem 8, with U replaced by $Soc(X)$, and N by the M-injective hull M^* of M. The crucial observation, $M^* \cap K = 0$, holds here because M^* is M-nonsingular, while K is essential over $Soc(K) \cong Soc(X)$ which is M-singular. We conclude that $Soc(X)$ is M^*-(hence M-) injective. We deduce $X = Soc(X)$, and so X is indeed M-injective.

($\Leftarrow$): Let E, $U = \oplus_{i \in I} U_i \in \sigma[M]$, where E is M-injective and the U_i ($i \in I$) are uniform, and hence of length at most 2. We have to show that every closed submodule V of $E \oplus U$ is a direct summand, i.e. $E \oplus U$ is CS.

We prove this first under the additional assumption that all U_i are M-nonsingular and M-injective. Then by Lemma 5, U is quasi-injective, as U is locally noetherian. Let $p : E \oplus U \to U$ be the projection. As U is M-nonsingular, $V \cap E = ker(p|V)$ is closed in V, hence in $E \oplus U$, hence in E. Therefore, $E = (V \cap E) \oplus E_1$ and $E \oplus U = (V \cap E) \oplus W$, where $W = E_1 \oplus U$. By modularity, $V = (V \cap E) \oplus (V \cap W)$.

The projection p induces an isomorphism between $V \cap W$ and $p(V)$ ($p(V) \subseteq U$). In particular, $V \cap W$ is locally noetherian. By Lemma 5, $V \cap W$ contains a maximal U-injective submodule V^*, which splits in $V \cap W$, say $V \cap W = V^* \oplus V_1$. If we can prove that $V_1 = 0$, then we are done, because this will imply that $E_1 \oplus (V \cap W) \subseteq^\oplus W = E_1 \oplus U$, and hence $V \subseteq^\oplus E \oplus U$, as desired.

Assume on the contrary that $V_1 \neq 0$. Let K be a subset of I such that $k \in K$ if and only if U_k has length 2. Set $T = (E \oplus U(K)) \cap V_1$. If $T \neq 0$, then there is a minimal submodule S in T. Hence there exists a finite subset F of K such that $S \subseteq E \oplus U(F)$. Clearly, $E \oplus U(F)$ is M-injective. If S is closed in $E \oplus U(F)$, then S is M-(and hence U-) injective, a contradiction to the maximality of V^*. If S is not closed in $S \oplus U(F)$, then S is not closed in V_1, since V_1 is closed in V and hence in $E \oplus U$ (cf. Lemma 3). Therefore the closure S^* of S in V_1 has length two, implying the U-injectivity of S^*, again a contradiction to the maximality of V^*. Thus $T = 0$, which shows that V_1 is embedded in $U(K')$ where $K' = I - K$. Hence $U(K') \neq 0$. Moreover, each submodule of $U(K')$ is U-injective. In particular, V_1 is U-injective, a contradiction to the maximality of V^*. Hence $V_1 = 0$, as desired.

Now we consider the general case. First of all, if $Soc(U_i)$ is M-singular, then it is M-injective by assumption, and so $Soc(U_i) = U_i$ is M-singular. The direct sum of all such U_i is therefore M-singular, hence M-injective since M is an SI-module, and can be incorporated into E. Thus we may assume that U is M-nonsingular.

We decompose $U = U' \oplus U''$, where U' is the direct sum of those U_i which are M-injective. The remaining U_j are simple, due to (L), and $\sigma[M]$-projective. (If some epimorphism in $\sigma[M]$ onto a simple module does not split, then the kernel is essential, and consequently the simple module is M-injective.) Therefore, $U"$ is semisimple and $\sigma[M]$-projective.

Consider the projection $q : E \oplus U' \oplus U" \to U"$. Then $(q|V)$ splits, say $V = V' \oplus V"$ where $V' = ker(q|V) = (E \oplus U') \cap V$ and $q(V) = q(V") \subseteq^\oplus U"$. The module V' is closed in V, hence in $E \oplus U$, hence in $E \oplus U'$. The previous dealt with special case applies here to obtain $V' \subseteq^\oplus E \oplus U'$. We conclude that $V = V' \oplus V" \subseteq^\oplus (E \oplus U') \oplus V" = (E \oplus U') \oplus q(V") \subseteq^\oplus (E \oplus U') \oplus U"$. □

The following result is an immediate consequence of Lemma 6 and Theorem 9.

Corollary 10. *An M-nonsingular module M is an SI-module with (L) if every direct sum of any two CS modules in $\sigma[M]$ is CS.*

3. Proofs of Theorems 1 and 2

By Oshiro [12, 3.18] if R is a ring such that every direct sum of copies of the R-module R is CS, then R is semiprimary and satisfies the ascending chain condition (briefly, ACC) on right annihilators. We use this to prove the following lemma.

Lemma 11. *Let R be a right nonsingular ring such that every direct sum of copies of $E(R)$ is CS. Then R has finite right uniform dimension.*

Proof. The maximal right quotient ring Q of R is von Neumann regular and self-injective, and isomorphic to $E(R)$ as an R-module. Thus $\oplus Q_R$ is CS. Hence it is easy to check that any closed R-submodule of $\oplus Q$ is a closed Q-module. It follows that $\oplus Q$ is a CS Q-module. Hence the ring Q has ACC on right annihilators (cf.[12, 3.18]). This implies that Q is a semisimple ring, and therefore R has finite right uniform dimension. □

Lemma 12. *A ring R with (E) is semiprimary with $J(R)^2 = 0$ and satisfies (L).*

Proof. Let R be a ring that satisfies (E). Hence R is semiartinian with (L) and $J(R)^2 = 0$, by the results in Section 2. A minimal right ideal of R/J ($J = J(R)$) cannot be nilpotent. Therefore, it is idempotent and nonsingular in Mod-(R/J). Thus R/J is a right nonsingular ring. The injective hull E of R/J as an (R/J)-module is quasi-injective as an R-module. Hence any direct sum of copies of E is a CS R-module. By Lemma 11, R/J has finite right uniform dimension, and is semisimple. □

Proof of Theorem 1. Let R be a right nonsingular ring.

$(i) \Rightarrow (ii)$. By [11], every CS R-module is a direct sum of uniform modules, and by (i), each of length at most two. Therefore, any direct sum of CS R-modules is a direct sum of uniform modules, which is CS by Lemma 6.

$(ii) \Rightarrow (i)$. By Theorem 9, R is right SI with (L). Then $R/Soc(R_R)$ is semisimple artinian by [4, 3.11]. Hence R is right artinian by Lemma 11. □

Proof of Theorem 2. Let R be a ring that satisfies (E). By Lemma 12, $R/Soc(R_R)$ is a semisimple artinian ring. If R is right CS, then it implies that $Soc(R_R)$ has finite length by [3, 9.11]. Hence R is right artinian. Therefore, $R_R = R_1 \oplus \cdots \oplus R_n$ where each R_i is a uniform right ideal. By Lemma 12, R satisfies (L), this means that each R_i has length at most two, and if an R_j has length two, R_j must be injective. Hence R is CS-semisimple by [3, 13.5]. □

4. Some Examples

First we give an example which shows that a ring in Theorem 1 is not necessarily CS-semisimple. In particular, this shows that [3, 13.3 $(e) \Rightarrow (a)$] is incorrect. Consider the matrix ring

$$R = \begin{bmatrix} \mathbb{C} & \mathbb{C} \\ 0 & \mathbb{R} \end{bmatrix}.$$

Then R is right and left artinian, right and left hereditary, in particular R is right and left nonsingular. Moreover R is left serial, and $R/J(R) \cong \mathbb{C} \times \mathbb{R}$. Hence, by [6], for each simple R-module S, $E(S)$ is uniserial. Assume that $E(S) \neq S$. As R is a right (and left) SI-ring, every singular R-module is semisimple (cf. [4, 3.1]). Hence $E(S)/S$ is semisimple. But $E(S)/S$ is also uniserial, therefore, $E(S)/S$ is simple. This proves that any indecomposable injective R-module is of length at most two. It follows that R is a ring of Theorem 1, i.e. every direct sum of CS R-modules is CS. Furthermore, since the local R-module $e_{11}R$ has uniform dimension 2, the ring R is not right CS. Thus R is not CS-semisimple.

Next we provide examples to separate the conditions (E_1), (E_2) and (E). Consider an arbitrary commutative regular ring R.

(1) A cyclic uniform R-module is isomorphic to a factor ring, and is therfore a field. It follows that every uniform module is simple. Thus (L) holds, and R satisfies (E_1) by Lemma 6. In particular, there are rings with zero socle which satisfy (E_1), for instance the ring $(\Pi_{i \in \mathbb{N}} F_i)/(\oplus_{i \in \mathbb{N}} F_i)$ has no minimal ideals, where F_i are fields.

(2) If R satisfies (E_2), then it is semiartinian by Theorem 8. Moreover, there are non-semisimple rings with (E_2), e.g. the SI-ring in [4, 3.2]: $R = F + (\oplus_{i \in \mathbb{N}} F_i)$ where $F = F_i$, a field.

(3) If R satisfies (E), then it is artinian by Theorem 1, hence a finite product of fields.

Now we discuss the structure of rings which appear in Theorem 9: Goodearl ([4, 3.11]) has classified SI-rings as products of a ring which is semisimple modulo its socle, and finitely many rings Morita equivalent to SI-domains. Since our ring is semiartinian, by Theorem 8, the domains must be division rings. It follows that the rings in Theorem 9 (with $M = R$), are just the products of a semisimple ring, and a right nonsingular ring semisimple modulo its socle and satisfying (L). The example mentioned in (2) shows that a ring of Theorem 9 is not necessarily right (or left) artinian.

Finally we consider a semiprimary ring R with $J^2 = 0$ $(J = J(R))$. The following details are elementary but a bit messy, and we present them informally. Without loss of generality, we assume that R is basic and ring-directly indecomposable.

Such a ring has a matrix representation, $R = (e_i R e_i)$, from a decomposition $1 = \Sigma_{i=1}^n e_i$ into indecomposable orthogonal idempotents. We use the notations $A_i = e_i R e_i$ (local rings with radical square zero), $X_{ii} = J(A_i) = e_i J e_i$, $D_i = A_i / X_{ii}$ (division rings), and $X_{ij} = e_i R e_j = e_i J e_j$ for $i \neq j$. Due to $J^2 = 0$, the X_{ij} are D_i-D_j-bispaces, and multiply to zero. Thus R is completely determined by the knowledge of the A_i and X_{ij}.

Set $\overline{e_j R} = e_j R / e_j J$ and $E_j = E(\overline{e_j R})$. By [12], $E_j / \overline{e_j R}$ is isomorphic to $Hom_R(J, \overline{e_j R})$. This module can be identified with the row $(X_{1j}^*, \cdots, X_{nj}^*)$ built from the duals $X_{ij}^* = Hom_{D_j}(X_{ij}, D_j)$. Therfore, E_j has length at most 2, as required by (L), if and only if $X_{ij} = 0$ for all but at most one i, and this exceptional X_{ij}^* is one-dimentional over D_i.

Thus each column of the matrix $(X_{ij}) \subseteq R$ has at most one nonzero entry. But indecomposability forces at least one nonzero off-diagonal entry into every column but at most one (which we may take to be the first one).

In summary: The basic indecomposable semiprimary rings with (L) are, up to isomorphism, precisely the ones whose matrix representation has, in the first column, either only D_1, or only A_1 with $X_{11} \neq 0$, or D_1 and one nonzero off-diagonal X_{i1}; and in all other columns, D_j and one nonzero off-diagonal X_{ij}; moreover the nonzero X_{ij}^* are one-dimentional.

Over such a ring, the uniform injectives E_j all have length two, except that E_1 has length one if the first column carries only D_1.

(Some readers might find it illuminating to think of R in terms of its quiver: It is a "rooted" tree, where the "root" can be either a vertex, or a vertex with self-arrow, or a directed cycle.)

It remains to discuss the condition that X^* is one-dimentional over T, for a bispace $_T X_D$. Assume first that $dim X_D < \infty$. This is true for all X_{ij} if R is right artinian. Then $X_T^* \cong T_T$ leads to a division ring embedding $D \subset T$ such that $dim_D T = dim X_D$. There results a bispace isomorphism $X \cong {}^*(X^*) \cong {}^*T$. Conversely, any bispace of the form ${}^*T$ satisfies

$(^*T)^* \cong T$. Thus, in the artinian case, all our bispaces come from left-finite division ring embeddings.

If $dim X_D = \infty$, we still have the embeddings $D \subset T$ and $X \to^* T$ as before, but the latter is definitely not surjective. It would be interesting and useful to characterize X as a subspace of $^*T$.

Acknowledgment. Dinh Van Huynh gratefully acknowledges the support of the Department of Mathematics and Statistics, McMaster University, and the NSERC (Canada) for his visit.

References

1. A. W. Anderson and K.R. Fuller, *Rings and Categories of Modules*, Springer-Verlag 1974.
2. A.W. Chatters and C.R. Hajarnavis, *Rings in which every complement right ideal is a direct summand*, Quart. J. Math. Oxford 28 (1977), 61-80.
3. N.V. Dung, D.V. Huynh, P.F. Smith and R. Wisbauer, *Extending Modules*, Research Notes in Mathematics Series 313, Pitman, London 1994.
4. K.R. Goodearl, *Singular Torsion and Splitting Properties*, Memoirs Amer. Math. Soc. 124 (1972).
5. M. Harada, *Factor Categories with Applications to Direct Decomposition of Modules*, Marcel Dekker 1983.
6. D.H. Hill, *Rings whose indecomposable injective modules are uniserial*, Canad. J. Math. 34 (1982), 797-805.
7. D.V. Huynh and B.J. Müller, *Rings for which direct sums of CS modules are CS*, Vietnam J. Math. 22 (No. 3&4) (1994), 120-123 (An announcement without proofs).
8. D.V. Huynh and S.T. Rizvi, *An approach to Boyle's Conjecture*, Proc. Edinburgh Math. Soc., to appear.
9. D.V. Huynh and S.T. Rizvi, *On some classes of artinian rings*, to appear.
10. S. H. Mohamed and B.J. Müller, *Continuous and Discrete Modules*, London Math. Soc. Lecture Note Series No 147, Cambridge Univ. Press 1990.
11. M. Okado, *On decomposition of extending modules*, Math. Japonica 29 (1984), 939-941.
12. K. Oshiro, *Lifting modules, extending modules and their application to QF-rings*, Hokkaido Math. J. 13 (1984), 310-338.
13. A. Rosenberg and D. Zelinsky, *Finiteness of the injective hull*, Math. Z. 70 (1959), 372-380.
14. R. Wisbauer, *Foundations of Module and Ring Theory*, Gordon and Breach 1991.

Institute of Mathematics, P.O.Box 631 Boho, Hanoi, Vietnam

Department of Mathematics and Statistics, McMaster University
1280 Main Street West, Hamilton, Ontario, Canada L8S 4K1

EXCHANGE PROPERTIES AND THE TOTAL

FRIEDRICH KASCH AND WOLFGANG SCHNEIDER

Dedicated to our dear colleague and friend Bruno Mueller

ABSTRACT. We study several exchange properties for modules for their behavior under the formation of direct sums and their relationship with the total of a module.

1. INTRODUCTION

In the study of the total of the endomorphism ring of a module, certain conditions are naturally involved, which can be considered as exchange properties (abbreviated by EPs). These EPs are "similar" to but weaker than the classical $2-EP$. In our prior work ([2],[5]), we were mainly interested in studying the total of a module and the EPs were only used to get information about the total. Here we study the EPs in their own right. Mainly we show that the EPs have, for the most part, the expected behavior with respect to direct sums. Only one of the expected properties cannot be demonstrated and is left as an open question. By transferring these results to the total of a module, we get some new results and new proofs for known properties of the total.

2. EXCHANGE PROPERTIES

Our starting point is the classical $2-EP$. W. Schneider ([2]) defined a weaker EP, which he denoted by $D2-EP$ (where D stands for direct summand). By studying the total of a module, we were inspired to define a further EP, which is situated properly between the $2-EP$ and the $D2-EP$ and which we call $B2-EP$ (where "B" stands for "between"). To give the definitions for these EPs, we need some notation. Let R be a ring with $1 \in R$ and let $A, B, C, \ldots$ be unitary right R-modules.

2.1 Definition.

1) The module A has the $2-EP :\Leftrightarrow$ in every situation

$$M = A \oplus B = C \oplus D ,\qquad(1)$$

there exist $C' \hookrightarrow C, D' \hookrightarrow D$ such that

$$M = A \oplus C' \oplus D' .\qquad(2)$$

1991 Mathematics Subject Classification. 16D10.

2) The module A has the $B2 - AE$:$\Leftrightarrow$ in every situation (1), at least one of the following conditions is satisfied:

   (i) There exists $C' \hookrightarrow C$ such that
   $$M = A \oplus C' \oplus D \ ; \tag{3}$$

   (ii) There exists $D' \hookrightarrow D$ such that
   $$M = A \oplus C \oplus D' \ ; \tag{4}$$

   (iii) There exists $0 \neq A' \hookrightarrow A, 0 \neq A^* \hookrightarrow A, C' \hookrightarrow C, D^* \hookrightarrow D$ such that
   $$M = A' \oplus C' \oplus D = A^* \oplus C \oplus D^* \ . \tag{5}$$

3) The module A has the $D2 - EP$:$\Leftrightarrow$ for every $A_0 \stackrel{\oplus}{\hookrightarrow} A, A_0 \neq 0$ and in every situation
$$M = A_0 \oplus B = C \oplus D \ . \tag{6}$$
there exists $A_0' \hookrightarrow A_0, A_0' \neq 0, C' \hookrightarrow C, D' \hookrightarrow D$ such that
$$M = A_0' \oplus C' \oplus D' \ . \tag{7}$$

We remark that all the submodules which occur in these definitions are direct summands (by the modular law). Further, it is obvious that these definitions are preserved under module isomorphisms.

To get the reader interested in the new EPs, we state some results from the literature which we reprove in part here.

1) We have the following implications:
$$2 - EP \stackrel{\Rightarrow}{\neq} B2 - EP \stackrel{\Rightarrow}{\neq} D2 - EP \ ;$$

2) A has the $B2 - EP \Leftrightarrow RAD(End_R(A)) = TOT(End_R(A))$;

3) A has the $D2 - EP \Leftrightarrow TOT(End_R(A))$ is an ideal of $End_R(A)$.

As a guide, we consider first some well-known properties of the $2 - EP$.

2.2 Properties of the $2-EP$.

1) If A has the $2-EP$ and if
$$M = U \oplus A \oplus B = U \oplus C \oplus D,$$
then there exists $C' \hookrightarrow C, D' \hookrightarrow D$ such that
$$M = U \oplus A \oplus C' \oplus D'.$$

2) If $A = A_1 \oplus A_2$, then A has the $2-EP$ if and only if A_1 and A_2 have the $2-EP$.

3) If A has the $2-EP$, then A has the $n-EP$ for each $n \in N$ $(=$ finite $EP)$.

It is appropriate to ask whether the new EPs satisfy the same properties. First, this is true for the property 2.2.1. Since the proof is just the same as for the $2-EP$, we will not give it here. It means that we can always add in (1), ..., (7) a direct summand U. The properties 2.2.2 and 2.2.3 for the $B2-$ and the $D2-EP$ will be considered later. For this, we need the following lemma, which is as simple as well as useful.

2.3 Lemma. Let
$$M = C \oplus D = H \oplus C' \oplus D', \quad C' \hookrightarrow C, \quad D' \hookrightarrow D. \qquad (8)$$
Then there exists $H_0 \stackrel{\oplus}{\hookrightarrow} H$, such that
$$M = H_0 \oplus C \oplus D' : \quad \text{and} \qquad (9)$$
if $D' \neq D$, then $H_0 \neq 0$.

Proof. By the modular law C', resp. D', are direct summands in C, resp. $D: C = C' \oplus C^* \cdot D = D' \oplus D^*$.

Denote by π the projection onto $C^* \oplus D^*$ belonging to the decomposition $M = C' \oplus C^* \oplus D' \oplus D^*$. By applying π on (8), it induces an isomorphism
$$\pi_0 : H \longrightarrow C^* \oplus D^*. \qquad (10)$$

Define $H_0 := \pi_0^{-1}(D^*)$, then we show that with this H_0, (9) is satisfied. First we have
$$\pi(H_0 + C + D') = D^* + C^* = Im(\pi).$$
Since further,

$$Ke(\pi) = C' \oplus D' \hookrightarrow H_0 + C + D' ,$$

we get $M = H_0 + C + D'$. It remains to show that this sum is direct. From (8), the following sum is direct: $H_0 \oplus C' \oplus D'$. Consider now $a + c' + d' = c^*$ with $a \in H_0, c' \in C', d' \in D', c^* \in C^*$. Then $\pi(a + c' + d') = \pi(a) = \pi(c^*) = c^*$, and since $\pi(a) \in D^*$, this implies $\pi(a) = c^* = 0$, and then also $a - c' = d' = 0$. Therefore, (9) is satisfied. That $H_0 \neq 0$ for $D' \neq D$ is obvious.

Now we formulate two corollaries which are useful for later applications.

2.4 Corollary 1. Assume

$$M = U \oplus C \oplus D = U \oplus H \oplus C' \oplus D' , \quad C' \overset{\leftrightarrow}{\neq} C , \quad D' \overset{\leftrightarrow}{\neq} D .$$

Then there exists $0 \neq A_1 \hookrightarrow H, 0 \neq A_2 \hookrightarrow H$ such that

$$M = u \oplus A_1 \oplus C' \oplus D = U \oplus A_2 \oplus C \oplus D' . \tag{11}$$

Proof. We prove the second equation in (11). For this, one has to substitute D' by $U + D'$ and H_0 by A_2 in (8).

2.5 Corollary 2. Assume

$$M = u \oplus \left(\bigoplus_{i=1}^{n} C_i \right) = u \oplus H \oplus \left(\bigoplus_{i=1}^{n} C'_i \right) , \quad c'_i \hookrightarrow C_i ,$$

and for at least one i.e., $\{1, \cdots, n\}$ assume further $C'_{i_0} \neq C_{i_0}$. Then there exists $0 \neq A_0 \hookrightarrow H$ such that

$$M = U \oplus A_0 \oplus C'_{i_0} \oplus \left(\bigoplus_{\substack{i=1 \\ i \neq i_0}}^{n} C_i \right) . \tag{12}$$

Proof. By 2.3, each $C'_i, \neq i_0$ can be filled up to C_i.

Now, we consider the $B2 - AE$.

2.6 Theorem.

1) If A has the $2 - EP$, then A has the $B2 - EP$.
2) for $A = A_1 \oplus A_2$, it follows that A has the $B2 - EP$ if and only if A_1 and A_2 have the $B2 - EP$.
3) If A has the $B2 - EP$, then A has the $D2 - EP$.
4) Arbitrary direct sums of modules with the $B2 - EP$ have the $D2 - EP$.

Proof.

1) If A has the $2-EP$, we can assume (1) and (2). If in (2) already $D' = D$ or $C' = C$, then one of the conditions (i) or (ii) in the definition of the $B2-EP$ is satisfied. But if $C' \neq C$ and $D' \neq D$, then by 2.4 (with $U = 0$), (iii) follows.

2) We assume first that A has the $B2-EP$ and that $A_1 \neq 0, A_2 \neq 0$. We will show that A_1 has the $B2-EP$. For this, we consider the situation

$$M = A_1 \oplus B = C \oplus D .$$

From the (outer) direct sum of M and A_2, it follows (up to an isomorphism) that

$$M \oplus A_2 = A \oplus B = A_2 \oplus C \oplus D . \qquad (13)$$

Now we apply the assumption about A on the case (i) in the definition and on the decomposition

$$A \oplus B = (A_2 \oplus C) \oplus D . \qquad (14)$$

It then follows that

$$A \oplus B = A \oplus (A_2 \oplus C)' \oplus D , \quad (A_2 \oplus C)' \hookrightarrow A_2 \oplus C . \qquad (15)$$

Let π be the projection onto M along $M \oplus A_2$. Then (15) implies

$$M = A_1 + C' + D , \quad C' : \pi((A_2 \oplus C)') . \qquad (16)$$

We still have to show that this sum is direct. Let $a_1 + c' + d = 0$ with $a_1 \in A_1, c' \in C', d \in D$. For c' there exists $a_2 \in A_2$ with $a_2 + c' \in (A_2 + C)'$. With this a_2 we have $a_1 - a_2 + (a_2 + c') + d = 0$, and then (15) implies $a_1 - a_2 = a_2 + c' = d = 0$ and, finally, $a_1 = a_2 = 0, c' = d = 0$. Hence (16) is really a direct sum. The case (ii) in the definition of $B2-EP$ is not possible for (14) (since $A_2 \hookrightarrow A$). Similarly, we have for the decomposition:

$$M \oplus A_2 = A \oplus B = C \oplus (A_2 \oplus D) \qquad (17)$$

in the case (ii)

$$M = A_1 \oplus C \oplus D' , \quad D' \hookrightarrow D . \qquad (18)$$

Now the case (i) is not possible. If we have (16) or (18), the proof is finished. If for both decompositions (14) and (17), only (iii) is satisfied, then this implies that

$$M \oplus A_2 = A' \oplus (A_2 + C)' \oplus D , \quad 0 \neq A' \hookrightarrow A \qquad (19)$$

$$= A'' \oplus (A_2 \oplus C) \oplus D'' , \quad 0 \neq A'' \hookrightarrow A ,$$

$$M \oplus A_2 = A^* \oplus C^* \oplus (A_2 \oplus D) , \quad 0 \neq A^* \hookrightarrow A \qquad (20)$$

$$= A^{**} \oplus C \oplus (A_2 \oplus D)^{**} , \quad 0 \neq A^{**} \hookrightarrow A .$$

In (19), $A'' \neq 0$; hence (19) implies $A'' \not\subset A_2$ and, therefore, $\pi(A'') \neq 0$. By applying π on (19), we get

$$M = \pi(A^*) \oplus C \oplus D , \quad \pi(A'') \neq 0 ,$$

Similarly (20) implies

$$M = \pi(A^*) \oplus C^* \oplus D , \quad \pi(A^*) \neq 0 , \quad \pi(A^*) \hookrightarrow A_1 .$$

These equations show that (iii) is satisfied for A_1. Altogether we have proved that A_1 has the $B2 - EP$.

Now we assume that $A_1 \neq 0, A_2 \neq 0$ have the $B2 - EP$, and we consider the situation

$$M = A \oplus B = A_1 \oplus A_2 \oplus B = C \oplus D . \qquad (21)$$

Now we move A_2 to the right side, where for A_2 all three cases $(i), (ii), (iii)$ are possible; and then we move A_1, where also all three cases have to be considered. We begin with (i) for A_2 and (i) for A_1.

(i), (i):

$$M = A_2 \oplus C' \oplus D = A_1 \oplus A_2 \oplus C'' \oplus D . \qquad (22)$$

The first equation follows immediately from (21) in the case (i) for A_2. To get the second equation, we use 2.2.1 with A_2 in the place of U. The second equation is the condition (i) for A.

In the next step, we combine (i) for A_2 with (ii) for A_1. In the following, 2.2.1 will be used without citation.

(i), (ii):

$$M = A_2 \oplus C' \oplus D = A_1 \oplus A_2 \oplus C' \oplus D'.$$

From $A_2 \neq 0, A_1 \neq 0$, it follows that $C' \neq C, D' \neq D$. Then 2.4 can be applied to the second equation with $U = 0$ and $H = A_1 \oplus A_2$; this gives (iii) for A.
(i), (iii):

$$M = A_2 \oplus C' \oplus D = A_1' \oplus A_2 \oplus C'' \oplus D \tag{23}$$

$$= A_1^* \oplus A_2 \oplus C' \oplus D'.$$

Now we apply 2.4 on the third equation to get again (iii) for A.
Now we consider (only) (iii) for A_2.
(iii)

$$M = A_2' \oplus C' \oplus D = A_2^* \oplus C \oplus D^*, \tag{24}$$

whereby $A_2' \neq 0, A_2^* \neq 0, C' \neq C, D^* \neq D$. If now (ii) is satisfied for A_1 and the first equation, then it follows that

$$M = A_1 \oplus A_2' \oplus C' \oplus D', \quad C' \neq C, \quad D' \neq D.$$

Then by 2.4, (iii) follows for A. If (i) is satisfied for A_1 and the second equation in (24), then (iii) follows for A. If, however, (i) is satisfied for A_1 and the first equation in (24) and (ii) is true for A_1 and the second equation in (24), then

$$M = A_1 \oplus A_2' \oplus C'' \oplus D = A_1 \oplus A_2^* \oplus C \oplus D^{**},$$

which is again (iii) for A. If finally (iii) is true for A_1 and the first, resp. the second equation in (24), then it follows that

$$M = A_1' \oplus A_2' \oplus C'' \oplus D = A_1^* \oplus A_2' \oplus C' \oplus D' \tag{25}$$

resp.,

$$M = A_1' \oplus A_2^* \oplus C^* \oplus D^* = A_1^* \oplus A_2^* \oplus C \oplus D^{**} \tag{26}$$

with $C' \neq C, D' \neq D, C^* \neq C, D^* \neq D$. Now we apply 2.4 on the second decomposition in (25), resp. the first in (26) to get again (iii) for A.

The still missing cases follow by exchanging C and D, which means exchanging (i) and (ii).

3) If A has the $B2-EP$ and if $0 \neq A_0 \overset{\oplus}{\hookrightarrow} A$ then A_0 also has, by (2), the $B2-EP$. All conditions $(i),(ii),(iii)$ then imply the condition for the $D2-EP$ to hold.

4) We consider a decomposition

$$A = \bigoplus_{i \in I} B_i , \quad \text{all } B_i \text{ have the } B2-EP$$

and assume that $A = A_0 \oplus A_1, A_0 \neq 0$. Let $a \in A_0, a \neq 0$. Then there exists a finite subset $J \subset I$ such that

$$a \in F = \bigoplus_{j \in J} B_j$$

By 2.1, F also has the $B2-EP$. Then we apply to the situation

$$A = A_0 \oplus A_1 = F \oplus G \quad \text{with} \quad G := \bigoplus_{i \in I \setminus J} B_i$$

the $B2-EP$ of F. Then one of the following conditions must be satisfied:

(i)
$$A = F \oplus A_0' \oplus A_1 , \quad A_0' \neq A_0 \tag{27}$$

(ii)
$$A = F \oplus A_0 \oplus A_1' ,$$

(iii)
$$A = F' \oplus A_0' \oplus A_1 = F^* \oplus A_0 \oplus A_1^* \tag{28}$$

Since $a \in A_0 \cap F$, condition (ii) is not possible. Let $A_0 = A_0' \oplus A_0''$ and denote by π the projection of the decomposition $A = A_0' \oplus A_0'' \oplus A_1$ onto A_0''. If (27) is true, then π induces an isomorphism of F onto A_0'' and, therefore, A_0'' also has the $B2-EP$. Then in the situation

$$A_0 \oplus B = A_0'' \oplus (A_0' \oplus B) = C \oplus D ,$$

one of the conditions (i), (ii), (iii) for A_0'' is satisfied, and this includes the definition for the $D2-EP$ for A. If (28) is true, we apply π on the first decomposition of (28). Then π induces an isomorphism $F' \longrightarrow A_0''$ and since F' is a direct summand of F, A_0'' has the $B2-EP$. As before, it follows that the $D2-EP$ is satisfied for A.

An arbitrary direct sum of modules with the $B2 - EP$ need not again have the $B2 - EP$. A module with a LE-decomposition (LE stands for local endomorphism ring) has the $B2 - EP$ if and only if the module is a Harada-module. An example of a module which is not Harada is the following:

$$M := \bigoplus_{n \in \mathbf{N}} \mathbf{Z}/p^n \mathbf{Z} \text{ with } p, \text{ a prime number}.$$

It is possible to generalize the definition of the $B2 - EP$ to the definition of the $Bn - EP$ and to show that the $B2 - EP$ implies the $Bn - EP$. With respect to the high expenditure (of words) we will not give these demonstrations here. Later, as an example, we formulate the $Dn - EP$.

Now we study the $D2 - EP$. First we intend to show that the original definition by W. Schneider ([2]) is equivalent to our formulation in 2.1.3. The original definition is A has the $D2 - EP :\Leftrightarrow$. For every $A_0 \stackrel{\oplus}{\hookrightarrow} A, A_0 \neq 0$ and every situation $M = A_0 \oplus B = C \oplus D$, there exist $A_0' \hookrightarrow, A_0, A_0' \neq 0$ and $C' \hookrightarrow C$ such that

$$M = A_0' \oplus C \oplus D$$

or there exists $A_0'' \hookrightarrow A, A_0'' \neq 0$ and $D'' \hookrightarrow D$ such that

$$M = A_0'' \oplus C \oplus D''.$$

It is obvious that this formulation implies our definition 2.1.3. But if this is satisfied and if in (7) $C' = C$ or $D' = D$, then the foregoing definition is satisfied. If, however, $C' \neq C, D' \neq D$, then we get this by 2.4.

From the definition, it follows immediately that direct summands of modules with the $D2 - EP$ have also this property. The question is open whether direct sums of $D2 - EP$ modules have the $D2 - EP$. The best result in this direction is the following theorem ([2]), for which we give a new proof.

2.7 Theorem. If $A = A_1 \oplus A_2$ and if A_1 has the $B2 - EP$ and A_2 has the $D2 - EP$, then A has the $D2 - EP$.

Proof. In the following, we assume $A_1 \neq 0, A_2 \neq 0$; for otherwise the statement is true. Let $U \stackrel{\oplus}{\hookrightarrow} A, U \neq 0$ and consider

$$U \oplus B = C \oplus D. \tag{29}$$

We have to prove that

$$U \oplus B = U_0 \oplus C' \oplus D', \quad 0 \neq U_0 \hookrightarrow U, \quad C' \hookrightarrow C, \quad D' \hookrightarrow D. \quad (30)$$

By assumption on U, there exists $V \hookrightarrow A$ with

$$A = A_1 \oplus A_2 = U \oplus V. \quad (31)$$

We now apply the $B2-EP$ on A_1 on (31); then one of the following cases is true:

(i) $A_1 \oplus A_2 = A_1 \oplus U' \oplus V$, $U' \hookrightarrow U$;

(ii) $A_1 \oplus A_2 = A_1 \oplus U \oplus V'$, $V' \hookrightarrow V$;

(iii) $A_1 \oplus A_2 = A'_1 \oplus U' \oplus V = A^*_1 \oplus U \oplus V^*$,
$0 \neq A'_1 \hookrightarrow A_1$, $0 \neq A^*_1 \hookrightarrow A_1$, $U' \hookrightarrow U$, $V^* \hookrightarrow V$.

We consider the three cases.

(i) By (31) and (i), it follows that U has a direct summand U_1 which is isomorphic to A_1. Let $U = U_1 \oplus U_2$. Since A_1 has the $B2-EP$, U_1 also has the $B2-EP$, and hence the $D2-EP$. With this (29) implies that

$$U \oplus B = U_1 \oplus (U_2 \oplus B) = C \oplus D = U'_1 \oplus C' \oplus D'$$

with $U'_1 \hookrightarrow U, U'_1 \neq 0$. This is (30), what we had to show.

(ii) By (ii), $U \oplus V'$ and A_2 are isomorphic. Therefore, U has the $D2-EP$ (since A_2 has it) and then again (29) implies (30).

(iii) As in case (i), U has a direct summand which is isomorphic to A'_1. Then, as in case (i), (30) follows.

2.8 Remark. Assume that

$$M = A \oplus B = \bigoplus_{i=1}^{n} C_i, \quad n \geq 2. \quad (32)$$

Then the following conditions are equivalent:

1) There exist $0 \neq A' \hookrightarrow A$ and $C'_i \hookrightarrow C_i, i = 1, ..., n$, such that

$$M = A' \oplus \left(\bigoplus_{i=1}^{n} C'_i \right). \quad (33)$$

2) There exist $0 \neq A^* \hookrightarrow A$ and $j \in \{1,...n\}$, such that

$$M = A^* \oplus C'_j \oplus \left(\bigoplus_{\substack{1 \leq i \leq n \\ i \neq j}} C_i \right), \quad C'_j \neq C_j. \quad (34)$$

Proof. 2) $\Rightarrow$ 1): Obvious, since $C'_i \neq C_i$ is not assumed.
1) $\Rightarrow$ 2): Since $A' \neq 0$, there exists at least one j with $C'_j \neq C_j$. Then apply 2.5.

2.9 Definition. If for every direct summand $A_0 \neq 0$ of A and for every situation (32) the conditions (33) and (34) are satisfied, then A has the $Dn - EP$.

2.10 Theorem. The $D2 - EP$ implies the $Dn - EP$.

Proof. Proof by induction. We assume that A has the $Di - EP$ for $i = 2, ..., n$. We show that A has the $D(n+1)-EP$. For $0 \neq A_0 \hookrightarrow A$, we consider the situation

$$M = A_0 \oplus B = \bigoplus_{i=1}^{n+1} C_i = C_1 \oplus \left(\bigoplus_{i=2}^{n+1} C_i \right).$$

From the $D2 - EP$ for A, it follows that

$$M = A'_0 \oplus C'_1 \oplus \left(\bigoplus_{i=2}^{n+1} C_i \right)', \quad A'_0 \neq 0.$$

1) Case: $C'_1 \neq C_1$. Then in the foregoing equation, we can fill up the last summand by 2.3 to $\bigoplus_{i=2}^{n+1} C_i$. This is what we had to show.

2) Case: $C'_1 = C_1$. Then we have

$$M = C_1 \oplus A'_0 \oplus \left(\bigoplus_{i=2}^{n+1} C_i \right)' = C_1 \oplus \left(\bigoplus_{i=2}^{n+1} C_i \right).$$

Now we consider $C_1 (= U)$ fixed in the sense of 2.2. Then we can apply the induction step for n of A'_0 and the sum $\bigoplus_{i=2}^{n+1} C_i$. This gives the induction for $n + 1$.

If we have a module which satisfies a chain condition for direct summands, then all three EP are equivalent, which we now intend to prove.

2.11 Remark. The maximum and the minimum condition for direct summands of a module are equivalent. If these are satisfied, we say that the module satisfies the *MD-condition*. Artinian and Noetherian modules obviously satisfy the MD-condition.

2.12 Theorem. For a module with the MD-condition, the $2 - EP$, the $B2 - EP$ and the $D2 - EP$ are equivalent.

Proof. We have only to show that the $2 - EP$ and the $D2 - EP$ are equivalent. We already proved that in general the $2 - EP$ implies the $D2 - EP$. Now let A have the $D2 - EP$ and we consider the situation

$$M = A \oplus B = C \oplus D, \quad A \neq 0.$$

Then there exists a decomposition

$$A = A_1 \oplus K_1, \quad A_1 \neq 0$$

such that

$$M = A \oplus B = A_1 \oplus K_1 \oplus B = A_1 \oplus C_1 \oplus D_1, \quad (35)$$

with $C_1 \hookrightarrow C, D_1 \hookrightarrow D$. Since K_1 has the $D2 - EP$, then in the case $K_1 \neq 0$, we have again a decomposition

$$K_1 = A_2 \oplus K_2, \quad A_2 \neq 0.$$

For 2.2 with $U = A_1$, it then follows that

$$M = A_1 \oplus A_2 \oplus C_2 \oplus D_2, \quad C_2 \hookrightarrow C_1, \; D_2 \hookrightarrow D_1. \quad (36)$$

Now we assume that A_1 is a maximal direct summand of A with (35). Then with respect to (36), K_1 must be 0, hence $A_1 = A$ and the $2 - EP$ is satisfied.

3. Characterization of Exchange Properties by the Total

In several papers we have defined and studied partially invertible elements and the total in a ring R (see the references). There are several possible ways to define the total of $R(= TOT(R))$, for example:

$$TOT(R) := \{r \in R \mid \forall\, s \in R\, [rs = (rs)^2 \longrightarrow rs = 0]\}\,.$$

We mention here two useful properties of the total:

1) The total is a semi-ideal in R; that means that for all $s \in R$,
$$sTOT(R) \subset TOT(R)\,,\quad TOT(R)s \subset TOT(R)\,.$$

2)
$$RAD(R) + TOT(R) = TOT(R)\,,$$
that means: For all $u \in RAD(R)$ and all $t \in TOT(R), u + t \in TOT(R)$. Since $0 \in TOT(R)$, it follows that $RAD(R) \subset TOT(R)$.

With respect to these properties, there are two interesting questions.

1) Question: Under which conditions is $TOT(R)$ additively closed; i.e., is an ideal of R?
2) Question: Under which conditions is $RAD(R) = TOT(R)$?

If A is an R-module, we denote $S := End_R(A)$. Then we call A a *total module* = TO-module, resp. a *radical total module* = RT-module, iff $TOT(S)$ is an ideal in S, resp. $RAD(S) = TOT(S)$. Then an RT-module is also a TO-module. Examples of RT-modules are discrete and continuous modules, to the study of which Bruno Mueller made very many interesting contributions. All the properties of RT-modules can be applied to these modules. As an example, we mention that any decomposition of these modules is locally semi-T-nilpotent.

As we mentioned already in the beginning, we have the following conditions.

3.1 Theorem.

1) A has the $B2 - EP \Leftrightarrow A$ is RT-module.
2) A has the $D2 - EP \Leftrightarrow A$ is TO-module.

Proof. See [2] and [5].

3.2 Corollary. If A is a TO-module and has the MD-condition, then A is an RT-module.

Proof. This follows by 2.12 and 3.1.

Examples where the $D2 - EP$ is satisfied but not the $B2 - EP$, and where the $B2 - EP$ is satisfied and not the $2 - EP$ are in the literature (see [2] and [5]).

REFERENCES

[1] Kasch, F. *Moduln mit LE-Zerlegungen und Harada- Moduln.* Lecture Notes, University of Munich, 1982.

[2] Schneider, W. *Das Total von Moduln und Ringen.* Algebra Berichte **55**, 1987, Verl. R. Fischer, München.

[3] Kasch, F. *Partiell Invertierbare Homomorphismem und das Total.* Algebra Berichte **60**, 1988, Verl. R. Fischer, München.

[4] Kasch, F. *The total in the category of modules.* General Algebra, 1988, Elsevier Sc. Publisher B.V., 1990.

[5] Kasch F. and W. Schneider. *The total of modules and rings.* Algebra Berichte **69**, 1992. Verlag R. Fischer, München.

[6] Kasch, F. *Verallgemeinerung der Sätze von Johnson und Utumi auf Hom*, Periodica Mathematic Humgaria **31** (3), 1995.

[7] Kasch, F. it Regularität in Hom, Algebra Berichte **75**, 1996, Verlag R. Fischer, München.

Current address: Friedrich Kasch, Department of Mathematics, University of Munich, Germany.

Current address: Wolfgang Schneider, Department of Mathematics, University of Augsburg, Germany.

LOCAL BIJECTIVE GABRIEL CORRESPONDENCE AND TORSION THEORETIC FBN RINGS

Paul Kim and Günter Krause

ABSTRACT. A right noetherian ring R that has local bijective Gabriel correspondence with respect to a torsion theory τ need in general not be right fully τ-bounded, but it is, if and only if the τ-closed prime ideals satisfy a version of the second layer condition. Rings with local bijective Gabriel correspondence are characterized by the tameness of their τ-torsionfree modules, and this implies that their relative (Gabriel–Rentschler) Krull and classical Krull dimensions coincide.

1. Introduction

Let R be a right noetherian ring, and let τ be a hereditary torsion theory on the category mod-R of right R-modules. Theorems about R-modules can often be relativized with respect to τ, the corresponding result is usually obtained by suitably reformulating the original one. However, this procedure is not without pitfalls, there are cases where the most tempting reformulation fails. For example, trying to relativize the well-known result ([11], [16]) that R is right fully bounded if and only if R has bijective Gabriel correspondence (that is, the map $[E] \to \mathrm{Ass}(E)$ from the set of isomorphism classes $[E]$ of injective indecomposable right R-modules E to the set $\mathrm{Spec}(R)$ of prime ideals of R is a bijection), one might be led to conjecture that R is right fully τ-bounded (that is, every essential τ-closed right ideal of every prime factor ring of R contains a nonzero two-sided ideal) if and only if R has local bijective Gabriel correspondence with respect to τ (that is, the restriction of the above map to isomorphism classes of injective indecomposable τ-torsionfree right R-modules is a bijection onto the set $\mathrm{Spec}_\tau(R)$ of τ-closed prime ideals). While this is true in one direction (see the proof of (ii) $\Rightarrow$ (i) in [1, Theorem 17]), Example 3.1 below provides a noetherian ring that has local bijective Gabriel correspondence with respect to a torsion theory τ, but it is not right fully τ-bounded.

The failure of the above conjecture poses the problem of finding a necessary and sufficient condition for a right noetherian ring R with local bijective Gabriel correspondence to be right fully τ-bounded. This is achieved

---

1991 *Mathematics Subject Classification*. Primary 16P40; Secondary 16P60, 16S90.
Support by a research grant from NSERC Canada is gratefully acknowledged.

Typeset by $\mathcal{AMS}$-TEX

in Theorem 5.2, where it is shown that the right restricted strong second layer condition for $\operatorname{Spec}_\tau(R)$ serves the purpose. We also show (Theorem 3.3) that R has local bijective Gabriel correspondence with respect to τ, if and only if nonzero τ-torsionfree right R-modules are tame, so that statements (1) and and (3) of [12, Theorem 2.4] remain equivalent. Finally, relativizing the well-known result that the classical Krull dimension of a right noetherian ring with bijective Gabriel correspondence is equal to its right Krull dimension (in the sense of Gabriel and Rentschler), we prove (Theorem 4.5) that $\operatorname{cl.K}_\tau\dim(R) = \kappa_\tau(R_R)$ whenever R is right noetherian and has local bijective Gabriel correspondence with respect to τ. Here κ_τ denotes the relative Krull dimension as introduced in [5], and $\operatorname{cl.K}_\tau\dim$ is defined like the classical Krull dimension, using the set $\operatorname{Spec}_\tau(R)$ in place of $\operatorname{Spec}(R)$.

Note that our results are stated and proved for right noetherian rings. Although some details have to be worked out, we are almost certain that they still hold for rings that are merely right τ-noetherian, that is, rings with ACC for τ-closed right ideals.

2. Definitions and Notations

All rings considered are associative with unit element 1, modules are unitary. For standard terminology the reader is referred to [4] and [13]. Basic torsion-theoretic results can be found in [14] and [16].

Let M be a right R-module, and let X and Y be subsets of M and R, respectively. Then

$$\ell_M(Y) = \text{annihilator of } Y \text{ in } M = \{m \in M \mid mY = 0\},$$
$$r_R(X) = \text{annihilator of } X \text{ in } R = \{r \in R \mid Xr = 0\}.$$

Subscripts are deleted if there is no danger of ambiguity. A prime ideal P of R is *associated* with the right R-module M if there exists a submodule $0 \neq N \subseteq M$ such that $P = r(N')$ for all submodules $0 \neq N' \subseteq N$.

$$\text{mod-}R = \text{category of right } R\text{-modules}$$
$$\operatorname{Ass}(M) = \text{set of associated primes of the } R\text{-module } M$$
$$\operatorname{Spec}(R) = \text{set of all prime ideals of } R$$
$$\operatorname{Spec}_\tau(R) = \{P \in \operatorname{Spec}(R) \mid R/P \text{ is } \tau\text{-torsionfree}\}$$
$$N \subseteq_{\text{ess}} M = N \text{ is an essential submodule of } M$$
$$\operatorname{E}_R(M) = \operatorname{E}(M) = \text{injective envelope of the right } R\text{-module } M$$
$$E_P = \text{injective indecomposable direct summand of } \operatorname{E}_R(R/P)$$
$$\rho(M) = \text{reduced rank of the module } M$$
$$\operatorname{dev}(\mathcal{S}) = \text{deviation of the partially ordered set } \mathcal{S}$$
$$|M| = |M|_R = \text{Krull dimension of the right } R\text{-module } M$$

$\kappa_\tau(M)$ = relative Krull dimension of M with respect to the hereditary torsion theory τ.

cl.K.dim(R) = classical Krull dimension of the ring R

$\mathcal{C}(I) = \{c \in R \mid c + I \text{ is regular in the ring } R/I\}$

Given a hereditary torsion theory τ on mod-R, a submodule N of a right R-module M is τ-dense if M/N is τ-torsion, τ-closed if M/N is τ-torsionfree. The submodule $\overline{N}^\tau = \bigcap\{C \mid C \supseteq N, C \text{ } \tau\text{-closed}\}$ is called the τ-closure of N (in M), it is the smallest τ-closed submodule of M that contains N and also the largest submodule of M in which N is τ-dense.

A right R-module M is called P-primary if Ass$(M) = P$; it is called P-prime if Ass$(M) = P = r(M)$. A uniform P-primary right R-module U is P-tame, or simply tame, if the P-prime submodule $\ell_M(P)$ is torsionfree as a right R/P-module, that is, no nonzero element of $\ell_M(P)$ is annihilated by an element of $\mathcal{C}(P)$; it is called P-wild, or simply wild, if $\ell_M(P)$ is $\mathcal{C}(P)$-torsion. A right R-module is tame (wild) if all its uniform submodules are tame (wild). A right R-module M is $\mathfrak{X}$-tame for a set $\mathfrak{X}$ of prime ideals, if it is tame and Ass$(M) \subseteq \mathfrak{X}$. Note that submodules, essential extensions and direct sums of tame modules are tame. It is also easy to see that extensions of tame modules by tame modules are tame.

A right R-module M is *finitely annihilated* if there exist finitely many elements $m_1, m_2, ..., m_n \in M$ such that $r(M) = r(m_1, ..., m_n) = \bigcap_{i=1}^n r(m_i)$, it is called a Δ-*module* if R satisfies the descending chain condition for right annihilators of subsets of M. Note that a Δ-module is tame.

Let R be a ring, and let $Q, P \in \mathrm{Spec}(R)$. Then Q is *linked* to P (via A), denoted by $Q \rightsquigarrow P$, if A is an ideal with $QP \subseteq A \subsetneq Q \cap P$, such that $Q \cap P/A$ is torsionfree as a right R/P-module and fully faithful (that is, has no nonzero unfaithful submodules) as a left R/Q-module.

A subset $\mathfrak{X}$ of $\mathrm{Spec}(R)$ is said to be *right link closed* if $P \in \mathfrak{X}$ and $Q \rightsquigarrow P$, implies that $Q \in \mathfrak{X}$.

3. Local Bijective Gabriel Correspondence

In [11], the following statements were shown to be equivalent for a right noetherian ring R.

(a) R is right fully bounded.
(b) R has *bijective Gabriel correspondence*, that is, the map $E \to \mathrm{Ass}(E)$ induces a 1–1 correspondence between the set of isomorphism classes of injective indecomposable right R-modules E and the set $\mathrm{Spec}(R)$ of all prime ideals of R.
(c) For any injective indecomposable right R-module E with $\mathrm{Ass}(E) = P$ there exists a P-prime cyclic submodule eR with $|eR| = |R/P|_R$ (Note that this is a somewhat more elaborate way of saying that all nonzero right R-modules are tame).

The above statements can be formulated relative to a hereditary torsion theory τ on mod-R, where, following [12], a ring R is called *right τ-bounded* if every τ-closed essential right ideal of R contains a nonzero two-sided ideal, and R is *right fully τ-bounded* if every prime factor ring of R is right τ-bounded. (Note that the definition in [12] required the prime factor ring R/P to be τ-torsionfree. This is unecessary, since a τ-closed right ideal A of R that contains a two-sided ideal I also contains its τ-closure $\overline{I}^\tau$ which is easily seen to be a two-sided ideal as well. Obviously, $\overline{I}^\tau \supsetneq I$ if I is not τ-closed. Thus, any factor ring of R that is not τ-torsionfree is automatically (more than) right τ-bounded.) Expressing the above statements (a)–(c) relative to the torsion theory τ one gets

(a_τ) R is right fully τ-bounded.
(b_τ) R has *local bijective Gabriel correspondence with respect to τ*, that is, the map $E \to \operatorname{Ass}(E)$ induces a 1–1 correspondence between the set of isomorphism classes of τ-torsionfree injective indecomposable right R-modules E and the set $\operatorname{Spec}_\tau(R)$ of τ-closed prime ideals of R.
(c_τ) For any τ-torsionfree injective indecomposable right R-module E with $\operatorname{Ass}(E) = P$ there exists a P-prime cyclic submodule eR with $\kappa_\tau(eR) = \kappa_\tau(R/P)$ (Note, as above for (c), that this is equivalent to saying that τ-torsionfree right R-modules are tame).

Note that the statements (a_τ)–(c_τ) become the statements (a)–(c) when τ is the 0-torsion theory where all nonzero modules are torsionfree. It is stated in [12, Theorem 2.4] that the above three statements are equivalent for a τ-noetherian ring R, that is, a ring that satisfies ACC for τ-closed right ideals. That result expands (and its proof uses) Theorem 17 of [1], which claims the equivalence of (a_τ) and (b_τ). Unfortunately, there is a flaw in the proof of that result (on the third last line of page 160 of [1], where J/P is claimed to be a two-sided ideal). It turns out that, while the implication (a_τ) $\Rightarrow$ (b_τ) remains true for any hereditary torsion theory τ for a right noetherian ring (see Corollary 3.6 below), (b_τ) does not always imply (a_τ).

Example 3.1. Let k be a field of characteristic zero, and let A_1 denote the first Weyl algebra over k, $A_1 = k[x][y; \partial/\partial x]$. Let $R = k + xA_1$, the idealizer of the maximal right ideal xA_1. Then R is a noetherian domain, and it is easy to see (cf. [13, Example 1.3.10]) that R, $P = xA_1$ and 0 are the only two-sided ideals of R. Let τ be the torsion theory that is cogenerated by $\operatorname{E}(R_R) \oplus \operatorname{E}_R(R/P)$. We first show that there is a 1–1 corespondence between the set $\{0, P\}$ and the set of isomorphism classes of τ-torsionfree injective indecomposable right R-modules. Let E be such a module. First, assume that $\operatorname{Ass}(E) = P$, and let eR be a cyclic P-prime submodule of E. Since P is a maximal right ideal of R, $P = r(e)$, so $E \simeq \operatorname{E}(eR) = \operatorname{E}(R/r(e)) = \operatorname{E}_R(R/P)$. Next, consider the case when $\operatorname{Ass}(E) = 0$, and assume that $E \not\simeq \operatorname{E}(R_R)$, so $E \simeq \operatorname{E}(R/A)$ for some right ideal $A \neq 0$. Since R has no zero-divisors, $|R/A| < |R_R|$, and since $|R_R| = |A_1| = 1$ by [13, 6.5.2 and 6.6.8], $M = R/A$ is artinian, so it may be assumed to be simple. Now, M is τ-torsionfree, so

$$0 \neq \text{Hom}(M, \text{E}(R_R) \oplus \text{E}_R(R/P)) \simeq \text{Hom}(M, \text{E}(R_R)) \oplus \text{Hom}(M, \text{E}_R(R/P)).$$

Since R has no nonzero minimal right ideals, $\text{Hom}(M, \text{E}(R_R)) = 0$, so M must be isomorphic to a submodule of $\text{E}_R(R/P)$, contradicting the fact that $\text{Ass}(M) = \text{Ass}(E) = 0 \neq P$.

Now, consider the essential right ideal xP of R. Note that $P = xA_1 \supsetneq x^2A_1 = xP$, so $P \not\subseteq r(R/xP)$. Since P is the only proper nonzero two-sided ideal, $r(R/xP) = 0$. It remains to show that R/xP is τ-torsionfree. For this, observe that $P/xP = xA_1/x^2A_1 \simeq A_1/xA_1 = A_1/P$ as right A_1-modules, and hence also as right R-modules. It is easy to see that $(A_1/P)_R$ is uniform, hence an essential extension of R/P, so that P/xP is τ-torsionfree. Thus R/xP is τ-torsionfree.

The failure of (a$_\tau$) $\Rightarrow$ (b$_\tau$) necessitates an examination of two results in the literature where this implication has been used. The first of these is [12, Theorem 2.4] which is incorrect as stated, although parts of it can be recovered, as is shown in in this section. The other is the equivalence of statements (i) and (ii) of [2, Theorem 2.9] which remains true, as can be seen from what follows.

Definition. A torsion theory τ on mod-R is *ideal invariant* if I/DI is τ-torsion for every two-sided ideal I and every τ-dense right ideal D of R.

Lemma 3.2. *Let R be a right τ-noetherian ring where τ is an ideal invariant hereditary torsion theory on* mod-R. *If R has local bijective Gabriel correspondence with respect to τ, then R is right fully τ-bounded.*

Proof. Let P be a prime ideal, and let E/P be an essential right ideal of R/P such that R/E is τ-torsion free. We have to show that $r(R/E) \supsetneq P$. If P is τ-dense, then $E = R$, and there is nothing to show. Thus, assume that P is τ-closed and suppose that $r(R/E) = P$. We may assume that E is maximal in the set $\mathcal{S} = \{X \mid X/P \subseteq_{\text{ess}} R/P, X \tau\text{-closed}, r(R/X) = P\}$. It is easy to see that R/E is uniform. We claim that R/E is fully faithful as an R/P-module. For this, let $0 \neq F/E \subseteq R/E$. Now, F/E is τ-dense in its τ-closure $\overline{F/E}^\tau = \overline{F}^\tau/E$, so $r(\overline{F}^\tau/E) = r(F/E) \supseteq r(R/E) = P$ by [2, Prop. 2.5]. As $r(R/\overline{F}^\tau) \cdot r(\overline{F}^\tau/E) \subseteq r(R/E) = P$ and $r(R/\overline{F}^\tau) \supsetneq P$ by the maximality of E in $\mathcal{S}$, we have $r(F/E) = r(\overline{F}^\tau/E) \subseteq P$, so $r(F/E) = P$. Consequently $\text{Ass}(R/E) = P$, and the local bijective Gabriel correspondence with respect to τ implies that $\text{E}(R/E) \simeq E_P$. Since R/P is a right Goldie prime ring by [14, Prop. 7.3.9], and since $E/P \subseteq_{\text{ess}} R/P$, this is impossible. □

While, thus, in a right τ-noetherian ring R with local bijective Gabriel correspondence the ideal invariance of τ is a sufficient condition for R to be right fully τ-bounded, it remains open whether it is necessary as well. In Section 5 we investigate this problem for right noetherian rings and show that the restricted strong second layer condition for $\text{Spec}_\tau(R)$ is a necessary and sufficient condition for the equivalence of fully τ-boundedness and local bijective Gabriel correspondence with respect to τ.

We proceed to show that, in spite of the above example, statements (1) and (3) of [12, Theorem 2.4], that is, our statements (b_τ) and (c_τ) are equivalent, at least when R is right noetherian. However, note that statement (3) of [12, Theorem 2.4] should be amended to include that the module eR is P-prime, for otherwise there is no guarantee that $P \subseteq r(e)$ (Incidentally, the same remark applies to statement (3) in [11, Theorem]).

Note that for a hereditary torsion theory τ the set $\text{Spec}_\tau(R)$ consists of all prime ideals Q that are contained in some $P \in \text{Ass}(E)$, where E is any injective cogenerator of τ. For the following, let τ^* denote the torsion theory that is cogenerated by $\bigoplus_{P \in \text{Spec}_\tau(R)} \text{E}_R(R/P)$.

Theorem 3.3. *Let R be a right noetherian ring and let τ be a hereditary torsion theory. Then the following statements are equivalent.*
  (i) *R has local bijective Gabriel correspondence with respect to τ.*
  (ii) *$\tau = \tau^*$ and direct products of $\text{Spec}_\tau(R)$-tame right R-modules are tame.*
  (iii) *Any τ-torsionfree right R-module $M \neq 0$ is tame.*
  (iv) *For any τ-torsionfree injective indecomposable right R-module E with $\text{Ass}(E) = P$ there exists a P-prime cyclic submodule eR of E with $\kappa_\tau(eR) = \kappa_\tau(R/P)$.*

Proof. (i) $\Rightarrow$ (ii): As a hereditary torsion theory τ is cogenerated by an injective right R-module E. Since R is right noetherian, $E = \bigoplus_{i \in I} E_i$ with injective indecomposable right R-modules E_i. By hypothesis, $E_i \simeq E_P$ for each $i \in I$, where $P = \text{Ass}(E_i)$. Thus, τ is cogenerated by $\bigoplus_{P \in \text{Spec}_\tau(R)} \text{E}_R(R/P)$, that is, $\tau = \tau^*$. Since any $\text{Spec}_\tau(R)$-tame module is τ-torsionfree, a direct product M of $\text{Spec}_\tau(R)$-tame modules is certainly τ-torsionfree. Let U be a uniform submodule of M. Then $\text{E}(U)$ is an injective indecomposable τ-torsionfree right R-module. If $\text{Ass}(U) = \text{Ass}(\text{E}(U)) = P$, then $\text{E}(U) \simeq E_P$ by (i), so U is tame.

(ii) $\Rightarrow$ (iii): Let $M \neq 0$ be τ-torsionfree. Since $\tau = \tau^*$, M embeds in a direct product of copies of $\bigoplus_{P \in \text{Spec}_\tau(R)} \text{E}_R(R/P)$. Since each $\text{E}_R(R/P)$ is $\text{Spec}_\tau(R)$-tame, and since direct products of $\text{Spec}_\tau(R)$-tame modules are tame by hypothesis, M is thus tame.

(iii) $\Rightarrow$ (iv): Let E be a τ-torsionfree injective indecomposable right R-module with $\text{Ass}(E) = P$, and let eR be a cyclic P-prime submodule of E. Since E is tame by hypothesis, $r(e)/P$ is not an essential right ideal of R/P. Since R/P is κ_τ-homogeneous, $\kappa_\tau(eR) = \kappa_\tau(R/r(e)) = \kappa_\tau(R/P)$.

(iv) $\Rightarrow$ (i): Let E be a τ-torsionfree injective indecomposable right R-module with $\text{Ass}(E) = P$. Note that P is τ-closed, so $\kappa_\tau(R/P) > -1$. If

eR is a P-prime submodule of E with $\kappa_\tau(R/r(e)) = \kappa_\tau(eR) = \kappa_\tau(R/P)$, then $r(e)/P$ cannot be an essential right ideal of R/P, for otherwise, by Goldie's Theorem, it would contain a regular element of R/P, implying that $\kappa_\tau(R/r(e)) < \kappa_\tau(R/P)$. Consequently, there exists a right ideal $A \supsetneq P$ such that $r(e) \cap A = P$, whence

$$E \simeq \mathrm{E}(A + r(e)/r(e)) \simeq \mathrm{E}(A/A \cap r(e)) = \mathrm{E}_R(A/P) \subseteq \mathrm{E}_R(R/P),$$

so $E \simeq E_P$, proving the claim. □

In order to pinpoint the gap (see Remark (b) after Theorem 5.2) between fully τ-boundedness and the existence of a local bijective Gabriel corespondence with respect to τ, we remark that the condition "Direct products of $\mathrm{Spec}_\tau(R)$-tame modules are tame" in Theorem 3.3(ii) can be replaced by seemingly weaker conditions.

Lemma 3.4. *The following conditions are equivalent for a set $\mathfrak{X}$ of prime ideals of the right noetherian ring R.*

(i) *Direct powers of $\mathfrak{X}$-tame right R-modules are tame.*
(ii) *Direct products of $\mathfrak{X}$-tame right R-modules are tame.*
(iii) *Direct products of finitely generated $\mathfrak{X}$-tame right R-modules are tame.*
(iv) *Direct products of cyclic $\mathfrak{X}$-tame right R-modules are tame.*

Proof. (i) ⇒ (ii): Let $\{M_i \mid i \in I\}$ be a set of $\mathfrak{X}$-tame right R-modules. Then $\bigoplus_{i \in I} M_i$ is $\mathfrak{X}$-tame, so $(\bigoplus_{i \in I} M_i)^I$ is tame by hypothesis. Since obviously $\prod_{i \in I} M_i \hookrightarrow (\bigoplus_{i \in I} M_i)^I$, and since submodules of tame modules are tame, $\prod_{i \in I} M_i$ is tame.

The implications (ii) ⇒ (iii) and (iii) ⇒ (iv) are trivial.

(iv) ⇒ (i): Let M be an $\mathfrak{X}$-tame right R-module and let $I \neq \emptyset$. In order to show that M^I is tame, it is obviously sufficient to show that xR is tame for any $0 \neq x \in M^I$. Let $x = (x_i)_{i \in I}, x_i \in M$. Since each $x_iR \subseteq M$ is $\mathfrak{X}$-tame and cyclic, the module $\prod_{i \in I} x_iR$ is tame by hypothesis, so its submodule xR is also tame. □

The class of right noetherian rings with local bijective Gabriel correspondence is closed under homomorphic images in the following sense. Let $\overline{R} = R/I$ be a homomorphic image of R. Given the hereditary torsion theory τ on mod-R, it induces a hereditary torsion theory $\overline{\tau}$ on mod-$\overline{R}$ by specifying $\{A/I \mid A \text{ is } \tau\text{-closed in } R\}$ as the set of $\overline{\tau}$-closed right ideals of $\overline{R}$.

Corollary 3.5. *Let R be a right noetherian ring that has local bijective Gabriel correspondence with respect to the hereditary torsion theory τ, and let $\overline{R} = R/I$ be a homomorphic image of R. Then $\overline{R}$ has local bijective Gabriel correspondence with respect to $\overline{\tau}$.*

Proof. Let M be a $\bar{\tau}$-torsionfree right $\bar{R}$-module. As an R-module, M is τ-torsionfree, hence tame by (iv) of Theorem 3.3, so M is also tame as an $\bar{R}$-module. The claim thus follows from Theorem 3.3. □

Corollary 3.6. *Let R be a right noetherian ring, and let τ be a hereditary torsion theory on mod-R. If R is right fully τ-bounded, then R has local bijective Gabriel correspondence with respect to τ.*

Proof. Using Theorem 3.3, we proceed to show that every τ-torsionfree right R-module is tame. It is obviously sufficient to establish this for any uniform prime right R-module that is τ-torsionfree. Let U be such a module, and let $\mathrm{Ass}(U) = P$. For any element $u \in U$, $uR \simeq R/r(u)$ is τ-torsionfree, so $r(u)$ is τ-closed. If $r(u)/P$ were an essential right ideal of R/P, then there would exist a two-sided ideal I, $r(u) \supseteq I \supsetneq P = r(R/r(u))$, which is impossible. Thus U is $\mathcal{C}(P)$-torsionfree and hence tame. □

4. RELATIVE KRULL DIMENSIONS

In [11, Theorem 2.4] it was shown that $\mathrm{cl.Kdim}(R) = |R_R|$ for every right fully bounded right noetherian ring. This has been generalized to right fully τ-bounded rings in [12], where a classical Krull dimension relative to a torsion theory was introduced as follows.

Definition. Let R be a right noetherian ring, let τ be a hereditary torsion theory on mod-R, and let $\mathrm{Spec}_\tau(R)$ be the set of τ-closed prime ideals of R. Set $\mathrm{Spec}_\tau^{-1}(R) = \emptyset$, and for an ordinal $\alpha > -1$ define

$$\mathrm{Spec}_\tau^\alpha(R) = \{P \in \mathrm{Spec}_\tau(R) \mid P \subsetneq Q \in \mathrm{Spec}_\tau(R) \Rightarrow Q \in \bigcup_{\beta < \alpha} \mathrm{Spec}_\tau^\beta(R)\}.$$

The smallest ordinal α with $\mathrm{Spec}_\tau^\alpha(R) = \mathrm{Spec}_\tau(R)$ is called the *classical τ-Krull dimension* of R and denoted by $\mathrm{cl.K}_\tau\mathrm{dim}(R)$.

Theorem 3.5 of [12] states that $\mathrm{cl.K}_\tau\mathrm{dim}(R) = \kappa_\tau(R_R)$ for a right fully τ-bounded ring R with Krull dimension that is not τ-torsion. We proceed to establish this equality for a right noetherian ring R that has local bijective Gabriel correspondence with respect to τ. Simultaneously, for such a ring, we establish the equality $\mathrm{cl.K}_\tau\mathrm{dim}(R) = \gamma_\tau(R_R)$, where γ_τ is defined as follows.

Definition. Let R be a right noetherian ring, and let τ be a hereditary torsion theory on mod-R. For a right R-module M let

$$\Gamma_\tau(M) = \{T \mid T \subseteq M, M/T \text{ is } \mathrm{Spec}_\tau(R)\text{-tame}\},$$

and define $\gamma_\tau(M) = \mathrm{dev}(\Gamma_\tau(M))$. Thus $\gamma_\tau(M) = -1$ if no homomorphic image of M is $\mathrm{Spec}_\tau(R)$-tame and $\gamma_\tau(M) = \alpha$ for an ordinal $\alpha > -1$ if $\gamma_\tau(M) \not< \alpha$ and, given any infinite descending chain $M = T_0 \supsetneq T_1 \supsetneq \cdots \supsetneq$

$T_i \supsetneq T_{i+1} \supsetneq \ldots$ with $T_i \in \Gamma_\tau(M)$, then $\gamma_\tau(T_i/T_{i+1}) < \alpha$ for all but finitely many indices i.

Note that $\gamma_\tau(M)$ is defined for any right R-module M that has Krull dimension. This concept is a relativization of the *classical Krull dimension* Cl.dim(M) of a module M introduced by Kosler in [9]. Note that $\gamma_\tau =$ Cl.dim whenever τ is a torsion theory with Spec$_\tau(R) = $ Spec(R). It was proved in [9, Theorem 2.9] that for a right noetherian ring R satisfying the right second layer condition, Cl.dim(R_R) coincides with the usual classical Krull dimension cl.Kdim(R) (denoted by dim(R) in [9]). Specializing Theorem 4.5 below to a torsion theory τ with Spec$_\tau(R) = $ Spec(R), the same equality is obtained for a right noetherian ring R with the property that direct products of tame right R-modules are tame. Note, however, that the ring described in our Example 3.1 has this latter property, yet it does not satisfy the right second layer condition (see [13, Example 4.3.15]). Thus, Kosler's equality holds for rings other than just those satisfying the second layer condition.

Lemma 4.1. *Let R be a right noetherian ring, and let τ be a hereditary torsion theory for* mod-R. *Then $\gamma_\tau(M) \leq \kappa_\tau(M)$ for any finitely generated right R-module M.*

Proof. Since the set $\Gamma_\tau(M) = \{T \mid T \subseteq M, M/T \text{ is Spec}_\tau(R)\text{-tame}\}$ is a subposet of $K_\tau(M) = \{S \mid S \subseteq M, M/S \text{ is } \tau\text{-torsionfree}\}$, it follows that $\gamma_\tau(M) = \text{dev}(\Gamma_\tau(M)) \leq \text{dev}(K_\tau(M)) = \kappa_\tau(M)$ by [13, Lemma 6.1.5]. □

Corollary 4.2. *Let R be a right noetherian ring that has local bijective Gabriel correspondence with respect to a hereditary torsion theory τ. Then $\gamma_\tau(M) = \kappa_\tau(M)$ for every finitely generated right R-module M.*

Proof. Obvious, since $\Gamma_\tau(M) = K_\tau(M)$ by Theorem 3.3. □

Lemma 4.3. *Let R be a right noetherian prime ring with $\kappa_\tau(R_R) > -1$, and let E be an essential right ideal of R. Then*

(i) *If R/E is* Spec$_\tau(R)$-*tame, then $\gamma_\tau(R/E) < \gamma_\tau(R_R)$.*
(ii) $\kappa_\tau(R/E) < \kappa_\tau(R_R)$.

Proof. (i) Set $E_0 = R$, $E_1 = E$, and assume that for $k \geq 1$ essential right ideals $E_i, 0 \leq i \leq k$ have been found, such that $E_{i-1} \supsetneq E_i$, R/E_i is Spec$_\tau(R)$-tame and $\gamma_\tau(E_{i-1}/E_i) \geq \beta = \gamma_\tau(R/E)$ for $1 \leq i \leq k$. Let c be a regular element in E_k, and choose a submodule $E_{k+1}/cE_k \subseteq E_k/cE_k$, maximal with respect to $cR/cE_k \oplus E_{k+1}/cE_k \subseteq_{\text{ess}} E_k/cE_k$. Then

$$R/E_k \simeq cR/cE_k \simeq \frac{cR + E_{k+1}}{E_{k+1}} \subseteq_{\text{ess}} E_k/E_{k+1}.$$

Since R/E_k is Spec$_\tau(R)$-tame, its essential extension E_k/E_{k+1} is Spec$_\tau(R)$-tame. Since extensions of Spec$_\tau(R)$-tame modules by Spec$_\tau(R)$-tame modules are Spec$_\tau(R)$-tame, R/E_{k+1} is Spec$_\tau(R)$-tame. Note that

$$\gamma_\tau(E_k/E_{k+1}) \geq \gamma_\tau(R/E_k) \geq \gamma_\tau(E_{k-1}/E_k) \geq \beta.$$

Since $c^2R \subseteq cE_k \subseteq E_{k+1}$, E_{k+1} is an essential right ideal. Thus, there esists an infinite descending chain $R \supsetneq E_1 \supsetneq ... \supsetneq E_i \supsetneq E_{i+1} \supsetneq ...$, such that each R/E_i is $\text{Spec}_\tau(R)$-tame and $\gamma_\tau(E_i/E_{i+1}) \geq \beta$ for all i. Consequently, $\gamma_\tau(R_R) > \beta = \gamma_\tau(R/E)$.

(ii) Let $c \in E$ be regular and consider the infinite descending chain

$$R \supsetneq cR \supsetneq c^2R \supsetneq ... \supsetneq c^iR \supsetneq c^{i+1}R \supsetneq ... .$$

For all but finitely many i, $\kappa_\tau(c^iR/c^{i+1}R) < \kappa_\tau(R_R)$. Since $c^iR/c^{i+1}R \simeq R/cR$, it follows that $\kappa_\tau(R/E) \leq \kappa_\tau(R/cR) < \kappa_\tau(R_R)$. □

Lemma 4.4. $\text{cl.K}_\tau\dim(R) \leq \gamma_\tau(R_R) \leq \kappa_\tau(R_R)$ *for any right noetherian ring R and any hereditary torsion theory τ.*

Proof. Only the first inequality has to be established, the second one follows from Lemma 4.1. Set $\gamma_\tau(R_R) = \alpha$ and proceed by induction on α. Assume first that $\alpha = -1$. Then no homomorphic image of R_R is $\text{Spec}_\tau(R)$-tame, so no prime ideal belongs to $\text{Spec}_\tau(R)$, meaning that $\text{Spec}_\tau(R) = \emptyset$, whence $\text{cl.K}_\tau\dim(R) = -1$. Next, let $\alpha > -1$ and assume that the result holds for all ordinals $\beta < \alpha$. Let $P \in \text{Spec}_\tau(R)$. It has to be shown that $P \in \text{Spec}_\tau^\alpha(R)$. For this, let $P \subsetneq Q \in \text{Spec}_\tau(R)$. We have to show that $Q \in \text{Spec}_\tau^\beta(R)$ for some $\beta < \alpha$. Note that Q/P is an essential right ideal of R/P and that R/Q is $\text{Spec}_\tau(R)$-tame, so that $\beta = \gamma_\tau(R/Q) < \gamma_\tau(R/P) \leq \alpha$ by Lemma 4.3(i). By the inductive hypothesis, $\text{cl.K}_\tau\dim(R/Q) \leq \gamma_\tau(R/Q) = \beta$, whence $\text{Spec}_\tau^\beta(R/Q) = \text{Spec}_\tau(R/Q)$, so $Q \in \text{Spec}_\tau^\beta(R)$. □

Theorem 4.5. *Let R be a right noetherian ring that has local bijective Gabriel correspondence with respect to a hereditary torsion theory τ. Then* $\text{cl.K}_\tau\dim(R) = \gamma_\tau(R_R) = \kappa_\tau(R_R)$.

Proof. In view of Lemma 4.4, only $\alpha = \kappa_\tau(R_R) \leq \text{cl.K}_\tau\dim(R)$ has to be established. Proceed by induction on α, observing that the case $\alpha = -1$ is dealt with by the above lemma. Let $\alpha > -1$ and note that R may be assumed to be a prime ring, as $\kappa_\tau(R) = \kappa_\tau(R/Q)$ for some (minimal) prime ideal Q by [5, Proposition 1.8]. Since $\alpha > -1$, the prime ring R is not τ-torsion, so it must be τ-torsionfree by [14, Lemma 7.3.10]. In order to establish the inequality $\kappa_\tau(R_R) \leq \text{cl.K}_\tau\dim(R)$, it has to be shown that, given any infinite descending chain $R = A_0 \supsetneq A_1 \supsetneq ... \supsetneq A_i \supsetneq A_{i+1} \supsetneq ...$ of τ-closed right ideals A_i, $\kappa_\tau(A_i/A_{i+1}) < \text{cl.K}_\tau\dim(R)$ for all but finitely many indices i. Now, since $\kappa_\tau(R_R) = \alpha$, $\kappa_\tau(A_i/A_{i+1}) < \alpha$ for almost all i. Let $M \neq 0$ be a τ-torsionfree subfactor of R with $\kappa_\tau(M) = \beta < \alpha$, and let N be a β-critical (with respect to κ_τ) factor module of M. It is easy to see that N is τ-torsionfree, hence tame by Theorem 3.3. Let $\text{Ass}(N) = P$ and let nR be a P-prime cyclic submodule of N. Since $r(n)/P$ is not an essential right ideal of R/P, and since R/P is κ_τ-homogeneous, it follows that $\kappa_\tau(R/P) = \kappa_\tau(nR) = \kappa_\tau(N) = \beta < \alpha$. By the inductive hypothesis, $\text{cl.K}_\tau\dim(R/P) = \beta$. Since R is a τ-torsionfree prime ring, and since obviously $0 \neq P \in \mathrm{Spec}_\tau(R)$, it follows that $\kappa_\tau(M) = \beta = \mathrm{cl.K}_\tau\dim(R/P) < \mathrm{cl.K}_\tau\dim(R)$. □

5. Torsion-theoretic Fully Boundedness

As has been demonstrated in Example 3.1, a right noetherian ring R that has local bijective Gabriel correspondence with respect to a hereditary torsion theory τ need not be right fully τ-bounded. As remarked earlier, the set $\mathrm{Spec}_\tau(R)$ (= $\mathrm{Spec}(R)$ for the particular torsion theory τ) of that example does not satisfy the right second layer condition, so it does not satisfy the stronger *right restricted strong second layer condition*, defined below.

Definition. A prime ideal P of a right noetherian ring R satisfies the *right restricted strong second layer condition* if for every prime ideal $Q \subsetneq P$, every finitely generated P-tame right R/Q-module M is unfaithful over R/Q. A set $\mathfrak{X}$ of prime ideals satisfies the *right restricted strong second layer condition* if every $P \in \mathfrak{X}$ satisfies this condition. The ring R satisfies the *right restricted strong second layer condition* if $\mathrm{Spec}(R)$ does so.

We have chosen this particular name for the condition, because it is formally defined like the right strong second layer condition (cf. [7, p. 220]), the class of right modules being restricted from P-primary to P-tame. It was first introduced by Jategaonkar, labelled $(*)_r$ in [6], and (†) in [4, Exercise 11L]. A prime ideal P with this property satisfies the usual *right second layer condition* which requires the second layer of E_P to be tame (see [8, Lemma 2.5]), but prime ideals exist that satisfy the latter condition but not the former one (see [4, Exercise 11M]). However, it follows from [8, Prop. 2.6] that any right link closed set of prime ideals of a right noetherian ring R that satisfies the right second layer condition also satisfies the right restricted strong second layer condition. Thus, in particular, R satisfies one of these conditions if and only if it satisfies the other one.

The following characterization of the right restricted strong second layer condition will be used in the proof of Theorem 5.2.

Lemma 5.1. *The following conditions are equivalent for a prime ideal P of a right noetherian ring R.*

(i) *P satisfies the right restricted strong second layer condition.*
(ii) *If M is a finitely generated P-tame right R-module, then M^I is P-tame for any set $I \neq \emptyset$.*

Proof. (i) $\Rightarrow$ (ii): Let M be a finitely generated P-tame right R-module, and let $Q \in \mathrm{Ass}(M^I)$. Choose $n = (n_i)_{i \in I} \in M^I$, such that $Q = r(nR) = \bigcap_{i \in I} r(n_i R) = r(\sum_{i \in I} n_i R)$. Since M is noetherian, $N = \sum_{i \in I} n_i R \subseteq M$ is a finitely generated P-tame module. As P is assumed to satisfy the right restricted strong second layer condition, $Q = P$ follows. Now M^I is also P-tame, since otherwise $r(n)/P$ would have to be an essential right ideal of R/P for some $n = (n_i)_{i \in I} \in \ell_{M^I}(P)$, making each $r(n_i)/P$ an essential right ideal of R/P, which contradicts the tameness of M.

(ii) $\Rightarrow$ (i): Let M be a finitely generated P-tame right R-module, let Q be a prime ideal, $Q \subsetneq P$, and assume that $r(M) = Q$. Then $R/Q \hookrightarrow M^M$. Since M^M is P-tame by hypothesis, R/Q would also be P-tame, an obvious contradiction. □

Now, the equivalence of statements (i) and (ii) in the following result shows that it is precisely the right restricted strong second layer condition that is needed to close the gap between the two conditions (a_τ) and (b_τ). Note that when the theorem below is specialized to the 0-torsion theory (where all nonzero modules are torsionfree), then it gives the various characterizations of right fully boundedness for a right noetherian ring. The last one is perhaps less well-known, it appears in [10].

Theorem 5.2. *Let R be a right noetherian ring, and let τ be a hereditary torsion theory on mod-R. Then the following statements are equivalent.*

(i) *R is right fully τ-bounded.*
(ii) (a) *$\mathrm{Spec}_\tau(R)$ has right restricted strong second layer condition.*
    (b) *R has local bijective Gabriel correspondence with respect to τ.*
(iii) *Finitely generated τ-torsionfree right R-modules are Δ-modules.*
(iv) *If σ is a hereditary torsion theory with $\mathrm{Spec}_\tau(R) \subseteq \mathrm{Spec}_\sigma(R)$, then $\kappa_\sigma(M) = \kappa_\sigma(R/r(M))$ for every finitely generated τ-torsionfree right R-module M.*
(v) *Every factor ring of R is right τ-bounded.*
(vi) *$\kappa_\tau(M) = \mathrm{cl.K}_\tau\dim(R/r(M))$ for any finitely generated τ-torsionfree right R-module M.*

Proof. (i) $\Rightarrow$ (ii): Let $M = \sum_{i=1}^n m_i R$ be a finitely generated P-tame right R-module, where $P \in \mathrm{Spec}_\tau(R)$, and let $Q \subseteq r(M)$ be a prime ideal, $Q \subsetneq P$. Note that M, and hence also each $m_i R$ is τ-torsionfree. Note also that each $r(m_i)/Q$ is an essential right ideal of R/Q, for otherwise $Q \in \mathrm{Ass}(m_i R) = P$. Since R is assumed to be right fully τ-bounded, it follows that $r(m_i R) \supsetneq Q$ for each i, hence that $r(M) = \bigcap_{i=1}^n r(m_i R) \supsetneq Q$. Thus P satisfies the right restricted strong second layer condition. Part (b) of (ii) is the assertion of Corollary 3.6.

(ii) $\Rightarrow$ (iii): Let M be a finitely generated τ-torsionfree right R-module, and let $\mathrm{Ass}(M) = \{P_1, ..., P_n\} \subseteq \mathrm{Spec}_\tau(R)$. Since M is noetherian, there exist submodules N_i with $\mathrm{Ass}(M/N_i) = P_i$, such that M embeds as an essential submodule in $\bigoplus_{i=1}^n M/N_i$. Each M/N_i is τ-torsionfree, and since submodules and finite direct sums of Δ-modules are Δ-modules, we may assume that M is P-primary for some $P \in \mathrm{Spec}_\tau(R)$. By Theorem 3.3, M is P-tame. Now let $S \neq \emptyset$ be a subset of M. Since P satisfies the right restricted strong second layer condition, $\mathrm{Ass}(M^S) = P$ by Lemma 5.1. As $R/r(S) \hookrightarrow M^S$, $\mathrm{Ass}(R/r(S)) = P$, so $L/r(S) = \ell_{R/r(S)}(P) \subseteq_{\mathrm{ess}} R/r(S)$. Now let $s_1 \in S$. If $A_1 = L \cap r(s_1) = r(S)$, it follows that $r(s_1) = r(S)$, and it is done. Otherwise there exists an element $s_2 \in S$ such that $A_1 \not\subseteq r(s_2)$. Set $A_2 = L \cap r(s_1) \cap r(s_2) \subsetneq A_1$. If $A_2 = r(S)$, it is done, otherwise continue in this fashion to obtain a strictly descending chain of right ideals $A_i = L \cap r(s_1, s_2, ..., s_i)$. Note that $A_i/A_{i+1} = A_i/A_i \cap r(s_{i+1}) \simeq s_{i+1}A_i$ and that $s_{i+1}A_iP \subseteq s_{i+1}LP \subseteq s_{i+1}r(S) = 0$, so $s_{i+1}A_i \subseteq \ell_M(P)$. Since M is P-tame, it follows that $\rho_{R/P}(A_i/A_{i+1}) = \rho_{R/P}(s_{i+1}A_i) > 0$ whenever $A_i \supsetneq A_{i+1}$. Since R is right noetherian, $\rho_{R/P}(L/r(S)) < \infty$, so by the additivity of the reduced rank, the chain of the A_i's can have at most $\rho_{R/P}(L/r(S))$ proper inclusions. Thus $A_i = r(S)$ for some index i, whence $r(S) = r(s_1, s_2, ..., s_i)$.

(iii) $\Rightarrow$ (iv): By [5, Corollary 1.6], $\kappa_\sigma(M) \leq \kappa_\sigma(R/r(M))$ in any case. Since M is supposed to be a Δ-module, $R/r(M)$ embeds in a finite direct sum of copies of M, so the reverse inequality holds as well.

(iv) $\Rightarrow$ (i): Let P be a prime ideal, and let E/P be an essential right ideal of R/P, such that R/E is τ-torsionfree. If $\kappa_\sigma(R/P) = -1$, then $P \notin \text{Spec}_\sigma(R)$, so $P \notin \text{Spec}_\tau(R)$. Thus R/P is τ-torsion by [14, Lemma 7.3.10], whence R/E is τ-torsion, implying that $E = R$, and there is nothing to prove. If $\kappa_\sigma(R/P) > -1$, then $\kappa_\sigma(R/r(R/E)) = \kappa_\sigma(R/E) < \kappa_\sigma(R/P)$ by Lemma 4.3(ii), so $r(R/E) \supsetneq P$.

(iii) $\Rightarrow$ (v): It is obviously sufficient to prove that R is right τ-bounded. For this, let E be an essential right ideal, such that R/E is τ-torsionfree. Then the annihilator of each element of R/E is also an essential right ideal. By hypothesis, $r(R/E)$ is the intersection of a finite number of these, whence $r(R/E) \supsetneq 0$.

(v) $\Rightarrow$ (i): Trivial.

(iv) & (v) $\Rightarrow$ (vi): Let M be a finitely generated τ-torsionfree right R-module. By (iv), $\kappa_\tau(M) = \kappa_\tau(R/r(M))$. By (v), $R/r(M)$ is right fully τ-bounded, so $\kappa_\tau(R/r(M)) = \text{cl.K}_\tau\dim(R/r(M))$ by Corollary 3.6 and Theorem 4.5.

(vi) $\Rightarrow$ (i): Let P be a prime ideal, and let E/P be an essential right ideal of R/P, such that R/E is τ-torsionfree. Then R/P is also τ-torsionfree, so $\kappa_\tau(R/E) < \kappa_\tau(R/P)$ by Lemma 4.3(ii). Also, $R/r(R/E)$ is τ-torsionfree, since it embeds in a direct product of copies of R/E. Applying (vi) twice yields

$$\kappa_\tau(R/r(R/E)) = \text{cl.K}_\tau\dim(R/r(R/E)) = \kappa_\tau(R/E) < \kappa_\tau(R/P),$$

whence $r(R/E) \supsetneq P$. □

Remarks. (a) There is a certain amount of overlap between the preceding result and [2, Theorem 2.9], where the authors prove (among other things) the equivalence of fully τ-boundedness, the local bijective Gabriel correspondence with respect to τ and the finite annihilation of finitely generated τ-torsionfree right R-modules for a τ-noetherian ring, provided τ is ideal invariant. In some sense, our result is stronger since it does not need the blanket hypothesis that τ is ideal invariant. However, Theorem 5.2 is stated for a right noetherian ring, so it would be interesting to know whether it still holds if the ring is merely τ-noetherian. Preliminary work by the authors on this question strongly indicate that this is indeed the case.

(b) Condition (ii) of the preceding theorem shows what is needed to close the gap between the existence of a local bijective Gabriel correspondence with respect to τ and fully τ-boundedness. For, if R has local bijective Gabriel correspondence, then direct powers of a $\mathrm{Spec}_\tau(R)$-tame module M are tame by Theorem 3.3, but they need not be $\mathrm{Ass}(M)$-tame, even when M is finitely generated. Indeed, it is easy to see (cf. [13, Example 4.3.15]) that for the ring R of Example 3.1 the right R-module $\mathrm{E}_R(R/P)$ contains a finitely generated submodule M with $r(M) = 0$, so that $0 = \mathrm{Ass}(R_R) \subsetneq \mathrm{Ass}(M^M)$, hence $\mathrm{Ass}(M^M) \neq \mathrm{Ass}(M) = P$. Thus, in view of Lemma 3.4, the restricted strong second layer condition is precisely what is needed to close the gap.

We formulate the equivalence of (i) and (ii) of Theorem 5.2 in a special instance. As has been noted after the definition of the right restricted strong second layer condition, a right link closed set of prime ideals of a right noetherian ring R satisfies this condition if and only it satisfies the right second layer condition. Since the set $\mathrm{Spec}(R)$ is trivially right link closed, R satisfies the right restricted strong second layer condition if and only if it satisfies the right second layer condition. Recall also (Theorem 3.3) that for a hereditary torsion theory τ satisfying $\tau = \tau *$ (the torsion theory cogenerated by $\bigoplus_{P \in \mathrm{Spec}_\tau(R)} \mathrm{E}_R(R/P)$) the local bijective Gabriel correspondence with respect to τ is equivalent to direct products of $\mathrm{Spec}_\tau(R)$-tame right R-modules being tame. With this in mind, the following is an immediate consequence of Theorem 5.2.

Corollary 5.3. *Let R be a right noetherian ring, and let ω be the torsion theory cogenerated by $\bigoplus_{P \in \mathrm{Spec}(R)} \mathrm{E}_R(R/P)$. Then the following statements are equivalent.*
  (i) *R is right fully ω-bounded.*
  (ii) *R satisfies the right second layer condition, and direct products of tame right R-modules are tame.*

References

1. G. P. Aguilar, M. J. Arroyo and C. J. E. Signoret, *A torsion-theoretic generalization of fully bounded noetherian rings*, Comm. Algebra **17** (1989), 149–164.
2. M. J. Asensio and B. Torrecillas, *The local Gabriel correspondence*, Comm. Algebra **20** (1992), 847–866.
3. G. Cauchon, *Les T-anneaux, la condition (H) de Gabriel et ses conséquences*, Comm. Algebra **4** (1976), 11–50.
4. K. R. Goodearl and R. B. Warfield, Jr., *An Introduction to Noncommutative Noetherian Rings*, London Math. Soc. Student Texts 16, Cambridge, 1989.
5. A. V. Jategaonkar, *Relative Krull dimension and prime ideals in right noetherian rings*, Comm. Algebra **2** (1974), 429–468.

6. _____, *Solvable Lie algebras, polycyclic-by-finite groups and bimodule Krull dimension*, Comm. Algebra **10** (1982), 19–69.
7. _____, *Localization in Noetherian Rings*, London Math. Soc. Lecture Notes, Vol. 98, Cambridge University Press, Cambridge, 1986.
8. K. A. Kosler, *Module extensions and the second layer condition*, Comm. Algebra **20** (1992), 69–91.
9. _____, *Classical Krull dimension and the second layer condition*, Ring Theory (Granville, OH, 1992), World Sci. Publishing, River Edge, NJ, 1993, pp. 224–241.
10. G. Krause, *Zur Krull-Dimension linksnoetherscher Ringe*, Rings, Modules and Radicals (Proc. Colloq., Keszthely, 1971), Colloq. Math. Soc. János Bolyai, Vol. 6, North-Holland, Amsterdam, 1973, pp. 259–269.
11. _____, *On fully left bounded left noetherian rings*, J. Algebra **23** (1972), 88–99.
12. H. Lee and P. L. Vachuska, *On relatively FBN rings*, Comm. Algebra **23** (1995), 2991–3001.
13. J. C. McConnell and J. C. Robson, *Noncommutative Noetherian Rings*, John Wiley, New York, 1987.
14. C. Năstăsescu and F. van Oystaeyen, *Dimensions in Ring Theory*, D. Reidel Publishing, Dordrecht, 1987.
15. J. Shapiro, *Localization and right Krull links*, Comm. Algebra **18** (1990), 1789–1804.
16. B. Stenström, *Rings of Quotients*, Springer-Verlag, Berlin/New York, 1975.

DEPARTMENT OF MATHEMATICS & ASTRONOMY, THE UNIVERSITY OF MANITOBA, WINNIPEG, MANITOBA, CANADA R3T 2N2

E-mail address: pkim@cc.umanitoba.ca, gkrause@cc.umanitoba.ca

NORMALIZING EXTENSIONS AND THE SECOND LAYER CONDITION

Karl A. Kosler

ABSTRACT. We characterize the second layer condition for a link closed subset of Spec(S) where S is a Noetherian normalizing extension of a Noetherian ring R and R satisfies the second layer condition. The second layer condition is shown to depend on the R-module structure of tame injective S-modules that are naturally associated with prime ideals in the link closed set. This is used to demonstrate that certain twisted polynomial rings satisfy the second layer condition when R is the coefficient ring. In case S is a centralized extension, our characterization is applied to show that the strong second layer condition for S amounts to a diluted version of AR-separation for S whenever R is AR-separated.

1. INTRODUCTION

The importance of the second layer condition in both the theory of localization and in module theory for noncommutative Noetherian rings is well documented. For example, in [9], it is shown that each prime ideal in a localizable subset of $Spec(R)$ must satisfy that second layer condition. On the other hand, in [3] the second layer condition is a key ingredient in using a generalization of the Loewy series to describe indecomposable injective modules.

In recent years, considerable work has been devoted to determine what Noetherian ring extensions of a Noetherian ring R must satisfy the second layer condition given that R satisfies the second layer condition. Research on this question essentially started with Brown [2] and Jategaonkar [9] who showed that the ordinary group ring RG satisfies the second layer condition provided R is commutative and G is a polycyclic-by-finite group. Further advances on this problem have been made by Bell [1], Byun [4], Letzter [13] and others.

The aim of this paper is to address this problem for certain normalizing extensions. Here a *normalizing extension* of a ring R is a ring S with $S \supseteq R$ and $S = RX$ where $1 \in X \subseteq S$ such that for all $x \in X$, $Rx = xR$. The ring S is called a *centralizing extension* if $rx = xr$ for all $r \in R$ and $x \in X$.

The type of normalizing extension we are interested in includes those centralizing extensions S that satisfy condition (†) from [4]. There the chief advantage of (†) is that for every prime P of S and "undesirable" prime $Q \subset P$ arising from Jategaonkar's main Lemma [10; 6.1.3], $Q \cap R \subset P \cap R$. If S is a normalizing extension and, in addition, $P \cap R/Q \cap R$ is essential in $R/Q \cap R$, then P is called *restricted*. From Section 4, examples include all prime ideals of the Laurent polynomial ring $R[y, y^{-1}; \sigma]$ as well as those

---

1991 Mathematics Subject Classification, 16P50, 16U20.

primes P of $R[y;\sigma]$ with $y \notin P$ where $\sigma \in Aut(R)$. One goal of this paper is to determine when each collection satisfies the second layer condition since this amounts to the second layer condition for each ring (see 4.6). Our approach is to find an applicable characterization of the second layer condition for a link closed collection Ω of restricted prime ideals of S when the base ring R satisfies the second layer condition.

The technique we use for this investigation is module-theoretic in nature and makes heavy use of an extension of classical Krull dimension to one-sided R-modules that was developed in [12]. In the following, $Cldim(M)$ will denote the classical Krull dimension of a module M. Section 2 includes a summary of the results we need from [12] concerning this dimension function.

Section 3 introduces a general abstract setting that we work in for the rest of the paper. The conditions imposed on S and Ω insure that our results are applicable to centralizing extensions as well as certain Ore extensions. Section 3 then proceeds to examine the R-module structure of the injective hull $E_S(S/P)$ where $P \in \Omega$. It is shown that $E_S(S/P)$ contains a unique largest R-submodule B_P such that B_P is $P \cap R$-tame and every non-zero finitely generated submodule $F \subseteq B_P$ has $Cldim(F) = Cldim(R/P \cap R)$. This result is used to characterize the restricted condition for Ω and, in Section 4, it is used to show that Ω satisfies the second layer condition precisely when $B_P = E_S(S/P)$ for all $P \in \Omega$.

The remainder of Section 4 is devoted to applying our results to determine when Ω satisfies the second layer condition in some special cases such as $S = R[y;\sigma], \sigma \in Aut(R)$ and $\Omega = \{P \in Spec(S) : y \notin P\}$. We show that Ω satisfies the second layer condition if every prime ideal of R is maximal or R is a principal ideal ring or R is a fully bounded Noetherian (FBN) ring and the induced automorphism on $R/P \cap R$ has finite order.

In order to obtain the second layer condition for $S = R[y;\sigma]$, we need to consider the set $\Lambda = \{P \in Spec(S) : y \in P\}$. In fact, in some instances, this is the only set of prime ideals worth considering. For example, let k be an algebraically closed field, let $R = k[x]$ and define $\sigma \in Aut(R)$ by $\sigma(x) = x + 1$. Then $S = R[y;\sigma]$ is a domain and $y \in P$ for all $0 \neq P \in Spec(S)$ by [15; 4.3.8(ii)]. Furthermore, by [10; A.3.6], S satisfies the second layer condition. We show that in the general set-up described above, Λ inherits the second layer condition from R.

The goal of Section 5 is to examine the relationship between AR-separation and the strong second layer condition for S when S is a centralizing extension. In this case, if R is AR-separated and S satisfies the strong second layer condition, then S is AR- separated with respect to all pairs $Q \subset P$ in $Spec(S)$ for which $Q \cap R \subset P \cap R$.

Throughout, a condition unmodified by 'left' or 'right' will mean that the condition holds on both sides. Unless specified otherwise, all modules will be left R-modules. We will use the notation $_RM$ to indicate the appropriate action. If M is a left or right R-module and $T \subseteq M$ is a non-empty subset, then $\ell_R(T)$ and $rt_R(T)$ denote the usual left and right annihilator of T. If I is an ideal and M is a module, $ann_M(I) = \{m \in M : Im = 0\}$ is a submodule. For $N \subseteq M$ a submodule of M, the terms *essential* (denoted $N \leq_e M$) and *closed* are taken as defined in [7].

If U is a uniform module, then $P = ass(U) = \{r \in R \mid rU' = 0$ for some submodule $0 \neq U' \subseteq U\}$ is a prime ideal. A prime ideal P is called an *associated* prime of a module M provided $P = ass(U)$ for some uniform submodule $0 \neq U \subseteq M$. The collection of all associated prime ideals of a module M is denoted $Ass(M)$. The collection of all prime ideals of a ring R will be denoted by $Spec(R)$.

A uniform module U over a left Noetherian ring R is called *tame* provided the injective hull $E(U)$ is isomorphic to a direct summand of $E(R/ass(U))$; equivalently, U contains a non-zero submodule that is torsion free over $R/ass(U)$. If the prime ideal $P = ass(U)$ needs to be specified, then we will call U P-*tame*. If U is not tame, then U is called *wild*. A module M is called *tame* provided M contains an essential direct sum of tame uniform submodules. In case M is tame and every $P \in Ass(M)$ is minimal over a semiprime ideal I, then M is called I-*tame*.

2. The Second Layer Condition and Classical Krull Dimension

In this section, we collect some information about the second layer condition and classical Krull dimension that we need in the sequel. The definitions of second layer condition and *(left) link* between prime ideals P and Q, denoted $P \rightsquigarrow Q$, can be found in [7] or [10]. However, using [6; 1.3], one can think of a link $P \rightsquigarrow Q$ as indicating that $Q = \ell_R(F/ann_F(P))$ where $F \subseteq E(R/P)$ is a finitely generated *left* R-submodule with $PF \neq 0$ and $F/ann_F(P)$ is torsion-free over R/Q. A *right link* (using right R-modules) is defined similarly.

If $\phi \neq \Omega \subseteq Spec(R)$, then Ω is called *left link closed* provided $P \rightsquigarrow Q$ and $P \in \Omega$ implies that $Q \in \Omega$. Replacing $P \rightsquigarrow Q$ with a right link from Q to P in this definition yields the definition of *right link closed*. If R is left and right Noetherian, it is easy to see, using [6; 1.3], that Ω is right link closed if and only if $P \rightsquigarrow Q \in \Omega$ implies that $P \in \Omega$. A module M will be called *finitely annihilated* provided $\ell_R(M) = \ell_R(m_1, \cdots, m_n)$ for some $m_1, \cdots, m_n \in M$. The following is a list of characterizations of the second layer condition taken from [6], [10] and [11].

Proposition 2.1. *The following are equivalent for a left Noetherian ring R and left link closed subset Ω of $Spec(R)$.*

1) Ω *satisfies the left second layer condition.*

2) *Every finitely generated submodule of $E(R/P)$ is finitely annihilated for all $P \in \Omega$.*

3) Let $P \in \Omega$ with $E = E(R/P)$. Let $Q \in Ass(E/ann_E(P))$, say $Q = \ell(F+L/L)$ where $F \subseteq E$ is finitely generated and $F+L/L \neq 0$ is uniform. Then it is *never* the case that $Q = \ell(F) \subset P$.

In this case, for all tame indecomposable injective modules E with $P = ass(E) \in \Omega, Q \in Ass(E/ann_E(P))$ if and only if $P \leadsto Q$.

Proof. (1) $\Leftrightarrow$ (2) is [11; 2.7] while (1) $\Leftrightarrow$ (3) follows from [10; 6.1.3] and the definition of the left second layer condition.

The forward implication of the last statement follows from [10; 6.1.3] and the definition of the left second layer condition. The reverse implication is [6; 1.3].

In instances where the second layer condition is unavailable, we will use the following weaker relationship between tame modules and finitely annihilated modules.

Proposition 2.2. Let R be a left Noetherian ring and $M \neq 0$ a module. If every finitely generated submodule of M is finitely annihilated, then M is tame. In case R is Noetherian and M is a Noetherian $R-R$ bimodule, then every left R-submodule of M is finitely annihilated and M is tame. A non-zero uniform module M is tame iff M contains a non-zero finitely annihilated submodule.

Proof. For the first statement, let $0 \neq U \subseteq M$ be a finitely generated uniform submodule with $P = ass(U)$. Then the submodule $N = ann_U(P)$ is finitely annihilated. Also, $P \subseteq \ell_R(N) \subseteq ass(U) = P$ whence $P = \ell_R(N)$. It follows that there is a $1-1$ map $R/P \to N^{(n)}$ for some n. Since any non-zero uniform submodule of R/P must imbed into N, N must be a nonsingular R/P-module. Therefore, U is tame.

For the second statement, let $0 \neq N \subseteq M$ be a submodule. Then $NR = m_1 R + \cdots + m_n R$ for some $m_1, \cdots, m_n \in N$. It is easy to see that $\ell_R(N) = \ell_R(NR) = \ell_R(m_1, \cdots, m_n)$. The result follows from the first statement.

If M is tame, then M contains a finitely generated submodule $Y \neq 0$ such that $PY = 0$ and Y is nonsingular over R/P where $P = ass(M)$. Thus, as a R/P-module, Y is finitely annihilated. The conclusion follows from $\ell_{R/P}(Y) = \ell_R(Y)/P$.

For the converse, suppose that $0 \neq N \subseteq M$ is a finitely annihilated submodule. Then there is a $1-1$ map $R/\ell_R(N) \to N^{(k)}$ for some k, and so any non-zero uniform submodule of $R/\ell_R(N)$ imbeds into M. Since R is Noetherian, $R/\ell_R(N)$ is a tame left R-module by the second result. Therefore, M is tame.

For a module M, let $\mathcal{T}(M) = \{K \subseteq M \mid K$ is a submodule of M and M/K is tame $\}$ partially ordered by inclusion. From [12], the *classical Krull dimension* of M, denoted $Cldim(M)$ is defined to be the deviation of $\mathcal{T}(M)$ as described in [15; 6.1.2]. Thus, $Cldim(M) = -1$ if and only if no factor M/K is tame for any submodule $K \subset M$. It follows that $Cldim(M) = 0$ provided $Cldim(M) \neq -1$ and every chain $M = K_0 \supseteq K_1 \supseteq \cdots$ in $\mathcal{T}(M)$ is eventually constant. In general, $Cldim(M) = \alpha$ provided $Cldim(M) \not< \alpha$ and for any descending chain $K_0 \supseteq K_1 \supseteq \cdots$ of members of $\mathcal{T}(M)$, $Cldim(K_i/K_{i+1}) < \alpha$ for all but finitely many i.

In case M is finitely generated, $Cldim(M)$ always exists by [17; 3.1.4]. It is important to note that if R satisfies the left second layer condition, then by [12; 2.9], $Cldim(R/I)$ coincides with the usual value of classical Krull dimension for any ideal $I \subseteq R$. The following includes some results from [12] that we will use throughout the rest of this paper.

Proposition 2.3. Let R be a left Noetherian ring that satisfies the left second layer condition and let M be a finitely generated module with $N \subseteq M$ a submodule.

1) $Cldim(M) = sup\{Cldim(N), Cldim(M/N)\}$

2) If M is tame, then $Cldim(M) = Cldim(R/\ell_R(M))$.

The following hold in case R is Noetherian and satisfies the second layer condition on both sides.

3) If M is a Noetherian $R-R$ bimodule, then $Cldim(_RM) = Cldim(M_R)$.

4) If M is tame and $N \leq_e M$, then $Cldim(N) = Cldim(M)$.

5) If $P \in Spec(R)$ and $0 \neq F + L/L \subset E/L$ is a finitely generated submodule where $E = E(R/P)$ and $L = ann_E(P)$, then $Cldim(F + L/L) = Cldim(R/P)$.

Proof. Statements (1), (2) and (3) are Theorem 2.4, Corollary 2.5 and Proposition 2.10, respectively, in [12].

(4) For an ordinal β, let $\mathcal{A} = \{D \subseteq R \mid D$ is a left ideal and $Cldim(R/D) < \beta\}$. Using (1), it is easy to verify that $\mathcal{A}$ is an idempotent filter and hence defines a torsion theory τ on left R-modules given by $\tau(M) = \Sigma\{Rx \subset M \mid Cldim(Rx) < \beta\}$. As in the proof of [12; 3.7], $\tau(M)$ is closed in M for any finitely generated tame module M. Thus, there cannot exist a submodule $N \leq_e M$ with $Cldim(N) < Cldim(M)$.

(5) Let $G = F + L/L$. If $Q \in Ass(G)$, then $P \rightsquigarrow Q$ by 2.1. By [10; 8.2.8], $Cldim(R/Q) = Cldim(R/P)$. Furthermore, $Q = \ell_R(V) = ass(V)$ for some uniform submodule $0 \neq V \subseteq G$. Since R satisfies the second layer condition, G is tame and so by (2) above, $Cldim(V) = Cldim(R/Q) = Cldim(R/P)$. The result now follows from [7; 11.6].

A useful device that we will need in the next section is the notion of a module that is critical with respect to classical Krull dimension. A non-zero module M is called α-*critical* for an ordinal $\alpha > -1$ provided $Cldim(M) = \alpha$ and for all $0 \neq K \in \mathcal{T}(M)$, $Cldim(M/K) < \alpha$. A module M is called α-*homogeneous* provided $Cldim(F) = \alpha$ for all finitely generated submodules

$0 \neq F \subseteq M$. Note that unlike the corresponding definitions in [12], M is not required to be tame. The following demonstrates that there is an ample supply of critical modules.

Proposition 2.4. Let R be a left Noetherian ring that satisfies the left second layer condition.

1) If a module M contains no non-zero submodule with classical Krull dimension -1, then M contains a critical submodule.

2) Every submodule of an α-homogeneous, α-critical module is α-critical.

Proof. (1) Assume that M contains no critical submodules. It follows from the hypothesis that we may choose a non-zero submodule $N_1 \subseteq M$ with minimal classical Krull dimension $\alpha > -1$. Clearly, N_1 is α-homogeneous. Since N_1 is not α-critical, there exists $N_2 \in \mathcal{T}(N_1)$ such that $Cldim(N_1/N_2) = \alpha$. Since N_2 is not α-critical, there exists $N_3 \in \mathcal{T}(N_2)$ with $Cldim(N_2/N_3) = \alpha$. Furthermore, since $N_2/N_3 \subset N_1/N_3$ is a tame submodule and N_1/N_2 is tame, $N_3 \in \mathcal{T}(N_1)$ by [12; 2.2]. Thus, continuing this construction yields an infinite descending chain in $\mathcal{T}(N_1)$ where all of the corresponding factors have classical Krull dimension α. This contradicts $Cldim(N_1) = \alpha$. Therefore, M contains a critical submodule.

(2) Let M be α-critical and α-homogeneous, and let $0 \neq N \subseteq M$ be a submodule. If $T \in \mathcal{T}(N)$ and $N/T \leq_e M/T$, then M/T is tame and hence $Cldim(N/T) \leq Cldim(M/T) < \alpha$. If N/T is not essential in M/T, then $N/T \oplus Y/T \leq_e M/T$ for some closed submodule $Y/T \neq 0$. In this case, $N/T \cong N + Y/Y \leq_e M/Y$ whence M/Y is tame. Thus, $Cldim(N/T) \leq Cldim(M/Y) < \alpha$. Therefore, since $Cldim(N) = \alpha$, N is α-critical.

3. NORMALIZING EXTENSIONS

Throughout, R will be a Noetherian ring that satisfies the second layer condition, S will be a Noetherian normalizing extension of R, and $1 \in X \subseteq S$ is a set of normal elements of S that generates S as a R-module. The goal of this section is to examine the R-module structure of certain tame injective S-modules associated with a link closed subset Ω of $Spec(S)$. In the next section, we will use the machinery developed here to determine when Ω satisfies the second layer condition.

Since we are interested in results that apply to centralizing extensions as well as certain twisted polynomial rings, it seems reasonable to work within a sufficiently general abstract setting that includes these rings as special cases. A convenient means of defining this setting is provided by statements *1-5* below. Our motivation for doing so is supplied by Proposition 3.1. In the following, $\Omega \neq \phi$ is a link closed subset of $Spec(R)$.

1) For every $Q \in Spec(R)$ and $x \in X$, $xQ = Q'x$ for some $Q' \in Spec(R)$.

2) For fixed $Q \in Spec(R)$, the set $\Gamma(Q) = \{Q' \in Spec(R) : xQ = Q'x$ for some $x \in X\}$ satisfies $Cldim(R/Q') = Cldim(R/Q)$ for all $Q' \in \Gamma(Q)$.

3) If $P \in \Omega, Q \in Spec(R)$ and $P \cap R \subseteq Q'$ for some $Q' \in \Gamma(Q)$, then $P \cap R \subseteq Q$.

4) For every $P \in \Omega$ and $x \in X, x(P \cap R) = (P \cap R)x$.

5) If either $P' = Q$ or $P' \rightsquigarrow Q$ where P' is minimal over $P \cap R, P \in \Omega$, then there exists a non-empty finite subset $\Gamma' \subseteq \Gamma(Q)$ such that $J = \cap \Gamma'$ satisfies $Jx \subseteq xQ$ for all $x \in X$.

In the following, R has a locally finite link graph provided the set $\{Q \in Spec(R) : P \rightsquigarrow Q\}$ is finite for all $P \in Spec(R)$.

Proposition 3.1. Statements *1-5* are true in the following cases.

1) S is a centralizing extension and $\Omega = Spec(S)$.

2) $S = R[y, y^{-1}; \sigma]$ and $\Omega = Spec(S)$ where R has a locally finite link graph.

3) $S = R[y; \sigma]$ and $\Omega = \{P \in Spec(S) : y \notin P\}$ where R has a locally finite link graph.

In particular, *1-5* are true in cases (2) and (3) if R is either a FBN ring or a polynormal ring.

Proof. (1) Clearly, *1, 2* and *4* are true. Since $\Gamma(Q) = \{Q\}$ for all $Q \in Spec(R)$, *3* and *5* are true.

(2), (3) For $S = R[y; \sigma], X = \{y^n : n = 0, 1, 2, \cdots\}$ while for $S = R[y, y^{-1}; \sigma], X = \{y^n : n \in \mathbf{Z}\}$. Since y is a normal element, $Sy = yS$ satisfies the left and right AR-property by [15; 4.1.10]. Thus, if $P \rightsquigarrow Q$, then $y \in P$ if and only if $y \in Q$ by [7; 11.16]. Thus, in (3), Ω is link closed.

For either ring and any $x \in X, Q \in Spec(R), xQ = Q'x$ where $Q' = \sigma^m(Q)$ for some m. Thus, *1* and *2* are true. From [15; 10.6.4], $P \cap R$ is σ-stable, i.e., $\sigma(P \cap R) = P \cap R$. Thus, if $P \cap R \subseteq Q' \in Spec(R)$ where $xQ = Q'x$ then $Q' = \sigma^m(Q) \supseteq P \cap R$ whence $Q \supseteq \sigma^{-m}(P \cap R) = P \cap R$ proving *3*. Statement *4* is an easy consequence of stability and the definition of multiplication in either ring.

Let $P \in \Omega$. By [15; 10.6.12], $P \cap R = P_1 \cap \cdots \cap P_n$ where each P_i is minimal over $P \cap R, \sigma(P_i) = P_{i+1}$ for $i = 1, \cdots n-1$ and $\sigma(P_n) = P_1$. It then follows from the relation $y^m P_i = \sigma^m(P_i) y^m$ that $\Gamma(P_i)$ is finite. Also, if $P_i \rightsquigarrow Q, \Gamma(Q) = \{\sigma^m(Q) : m \in N\}$ where N is the set of whole numbers or integers depending on which extension is being considered. In any case, since $\sigma^m(P_i) \rightsquigarrow \sigma^m(Q)$ and R has a locally finite link graph, the set $\Gamma(Q)$ is finite. Clearly $J = \cap \Gamma(P_i)$ or $J = \cap \Gamma(Q)$ satisfies the conclusion of *5*. The last statement follows from [7; 14.22] and [14; 8] together with [7; 11.18].

For the remainder of this section, it will be assumed that R, S and Ω satisfy conditions *1-5* above. When it is necessary to refer to any one of these statements, we will use the appropriate italic numeral. We begin with some preparatory results aimed at examining the structure of the S-injective hulls $E_S(S/P)$ where $P \in \Omega$.

Lemma 3.2. Let $_SM \neq 0$ be an S-module and let $0 \neq F \subseteq {_R}M$ be a finitely generated R-submodule. For all $x \in X$, $T/ann_F(x) \subseteq F/ann_F(x)$ is an R-submodule such that F/T is tame iff xF/xT is tame. In particular, xF is tame iff $F/ann_F(x)$ is tame.

Proof. First, note that since each $x \in X$ is normal $ann_F(x)$ is an R-submodule of F.

Suppose that $_RF/T$ is tame. Choose $0 \neq U/xT \leq xF/xT$. Then $U = xV$ where V is the R-submodule $V = \{z \in F : xz \in F\}$. Since $xT \subset U, T \subset V$. From 2.1, $\ell_R(V/T) = \ell_R(v_1 + T, \cdots, v_n + T)$ for some $v_1, \cdots, v_n \in V$. We claim that $\ell_R(U/xT) = \ell_R(xv_1 + xT, \cdots, xv_n + xT)$. Let $r \in R$ with $rxv_i \in xT$ for all i. For some $r' \in R, rx = xr'$. Then for all $i, xr' v_i \in xT$ whence $xr' v_i = xt_i$ for some $t_i \in T$. Thus, $xr' v_i - xt_i = x(r' v_i - t_i) = 0$ and $r' v_i - t_i \in ann_F(x) \subseteq T$. It follows that $r' v_i \in T$ for all i and hence $r'V \subseteq T$. Now, $rU = rxV = xr' V \subseteq xT$. Therefore, $r \in \ell_R(U/xT)$. Since the other inclusion is trivial, our claim is proved.

As a consequence of the result established in the last paragraph, every non-zero submodule of xF/xT is finitely annihilated. By 2.2, xF/xT is a tame R-module.

The proof of the reverse implication is essentially a repeat of the proof the forward implication.

The last statement follows from the first with $T = ann_F(x)$.

Corollary 3.3. Let $M \neq 0$ be an S-module and let $0 \neq F \subseteq {_R}M$ be a finitely generated R-module. For all $x \in X$, the map $\mathcal{T}(F/ann_F(x)) \to \mathcal{T}(xF)$ given by $T/ann_F(x) \to xT$ is an order preserving bijection. In particular, $Cldim(F/ann_F(x)) = Cldim(xF)$.

Proof. The first statement is an easy consequence of 3.2. The last statement follows from the definition of classical Krull dimension.

Proposition 3.4. Let M be a finitely generated tame uniform S-submodule with $P = ass(_SM) \in \Omega$ and let $\beta = Cldim(R/P \cap R)$. Then

1) β is the smallest ordinal that occurs as the classical Krull dimension of a non-zero finitely generated R-submodule of M.

2) $A = \Sigma\{F \subseteq {_R}M \mid {_R}F$ is finitely generated with $Cldim(F) = \beta\}$ is a $P \cap R$-tame β-homogeneous R-submodule of M and is also an S-submodule of M.

3) $ann_M(P \cap R) \leq_e A$ as R-modules.

Proof. We will first show that A is tame. Let α be the smallest ordinal that occurs as the classical Krull dimension of a nonzero finitely generated R-submodule of M. Let $A = \Sigma\{F \subseteq {}_RM : {}_RF \text{ is f.g. and } Cldim(F) = \alpha\}$. If $F \subseteq {}_RM$ is f.g. with $Cldim(F) = \alpha$, then for all $x \in X$, xF is a R-submodule of M and $Cldim(F/ann_F(x)) \leq \alpha$. By 3.3, $Cldim(xF) = Cldim(F/ann_F(x)) \leq \alpha$ and so $Cldim(xF) = \alpha$ whenever $xF \neq 0$. Thus, $xF \subseteq A$. It follows that A is a left S-submodule of M and hence, ${}_SA$ is tame. By 2.2, A contains a nonzero finitely annihilated S-submodule Y. Now, $\ell_S(Y) = \ell_S(y_1, \cdots, y_k)$ for some $y_1, \cdots, y_k \in Y$. Therefore, $\ell_R(Y) = \ell_S(Y) \cap R = \ell_S(y_1, \cdots, y_k) \cap R = \ell_R(y_1, \cdots, y_k)$. Thus, there is a $1-1$ map $R/\ell_R(Y) \to Y^{(k)}$. By 2.2, $R/\ell_R(Y)$ is a tame left R-module. Consequently, $Y \subseteq A$ contains a nonzero tame R-submodule. This also shows that $\alpha > -1$.

Assume that A also contains a nonzero wild R-submodule. Then there exist closed R-submodules $V, W \subseteq A$ such that $V \cap W = 0$, $V \oplus W \leq_e A$ where V is tame, W is wild and both are nonzero. Note that $V \cong V \oplus W/W \leq_e A/W$ and so A/W is tame. Since ${}_RA$ is α-homogeneous, $A \supseteq {}_RZ$ an α-critical R-submodule by 2.4(1). Assume that Z is wild. If $Z \cap W = 0$, then there is $1-1$ map $Z \to A/W$ which is impossible. Thus, $Z \cap W \neq 0$, $Z/Z \cap W \cong Z + W/W \subseteq A/W$ is tame and hence, $Cldim(Z+W/W) < \alpha$. Now, if $Z+W/W \neq 0$, then the intersection $(V \oplus W/W) \cap (Z+W/W) \neq 0$, whence $V \cong V \oplus W/W$ contains a nonzero R-submodule with classical Krull dimension less than α. Therefore, it must be the case that $Z \subseteq W$.

Choose $x \in X$. Since $xZ \subseteq A$, $Cldim(xZ) = \alpha$ whenever $xZ \neq 0$. In this case, as the following shows, xZ is a wild α-critical module: First, if $0 \neq xY \subseteq xZ$ is a tame submodule, then $Y/ann_Y(x) \neq 0$ and by 3.2, $Y/ann_Y(x)$ is tame. Further, by 3.3, $\alpha = Cldim(xY) = Cldim(Y/ann_Y(x))$. However, by 2.4(2), Y is α-critical and so $Cldim(Y/ann_Y(x)) < \alpha$ which is impossible. Therefore, xZ is wild. Second, let $0 \neq xT \subseteq xZ$ be a submodule such that xZ/xT is tame. By 3.2, the module $Z/(T + ann_Z(x))$ is tame. As in 3.3, 'multiplication by x' defines an order preserving bijection $\mathcal{T}(Z/T + ann_Z(x)) \to \mathcal{T}(xZ/xT)$. Thus, $Cldim(xZ/xT) = Cldim(Z/(T + ann_z(x))) < \alpha$. Finally, since $Cldim(xZ) = \alpha$, xZ is α-critical.

From above, $xZ \subseteq W$ for all $x \in X$. Thus, the S-module $SZ \subseteq W$. Since SZ is a tame S-module, a repeat of the argument in the first paragraph with SZ in place of A yields that SZ contains a nonzero tame R-submodule. Clearly, this contradicts the wildness of ${}_RW$. Therefore, $W = 0$ and ${}_RA$ is tame.

(1) Let $L = ann_M(P)$ and let $N = A \cap L$. Since ${}_SM$ is uniform, N is a nonzero S-submodule of M. Further, $P \subseteq \ell_S(N) \subseteq ass({}_SM) = P$. Thus, $P = \ell_S(N)$ and since N is tame, N is a torsion-free S/P-module. Therefore, N is finitely annihilated and there is a $1-1$ S-homomorphism $S/P \to N^{(n)}$ for some n. Composing with $R/P \cap R \cong R + P/P \subseteq {}_RS/P$ yields a $1-1$ R-

homomorphism $R/P \cap R \to N^{(n)}$. According to [7; 7.27], $P \cap R = \bigcap_{i=1}^{n} P_i$ where $P_i \in Spec(R)$ for all i and as rings $R/P_i \cong R/P_j$ for all i,j. Using 2.3 (1), it's easy to see that $\beta = Cldim(R/P \cap R) = Cldim(R/P_i)$ for all i and that $Cldim(U) = \beta$ for all left ideals $0 \neq U \subseteq R/P \cap R$. Choose U to be uniform. Since there is a $1-1$ R-homomorphism $R/P \cap R \to N^{(n)}$, there exists a $1-1$ R-homomorphism $U \to N$. Since $N \subseteq A, \alpha = Cldim(N) = \beta$.

(2) and (3) From above $_RA$ is tame and $L \subseteq ann_M(P \cap R) \subseteq A$. Note that 4 insures that $ann_M(P \cap R)$ is a S-submodule of M. Choose a finitely generated uniform submodule $0 \neq U \subseteq A$ with $Q = ass(_RU) = \ell_R(U)$. Since $L \leq_e {_S}M, SU \cap ann_M(P \cap R) \neq 0$. Let $0 \neq z \in SU \cap ann_M(P \cap R)$. Then $z = x_1 u_1 + \cdots + x_n u_n$ where $x_i \in X$ and $u_i \in U$ for all $i = 1, \cdots, n$. By 1, there exists $Q_i \in Spec(R)$ such that $Q_i x_i = x_i Q$ and hence, $Q_i x_i u_i = x_i Q u_i = 0$ for all i. Let $H = Q_1 \cap \cdots \cap Q_n$. Then $HRz = Hz = 0$ and so $H \subseteq \ell_R(Rz) \subseteq \cap Ass(Rz)$. Therefore, there exists $Q' \in Ass(Rz)$ such that $Q_i \subseteq Q'$ for some i.

Now, $Rz \subseteq ann_M(P \cap R) \subseteq A$ and Rz tame implies that $Cldim(R/Q') = \beta$ and $Q' \supseteq P \cap R$ is minimal over $P \cap R$. Also, by 2, $Cldim(R/Q) = Cldim(R/Q_i) = \beta$. Therefore, $Q_i = Q' \supseteq P \cap R$ is minimal over $P \cap R$. By 3, $P \cap R \subseteq Q$. Thus, $U \subseteq ann_M(Q) \subseteq ann_M(P \cap R)$. It now follows that $ann_M(P \cap R) \leq_e A$. Therefore, (2) and (3) are true.

Corollary 3.5. Let $P \in \Omega$ and let $B = \Sigma\{F \subseteq {_R}E_S(S/P) \mid {_R}F \text{ is finitely generated and } Cldim(F) \leq \beta\}$ where $\beta = Cldim(R/P \cap R)$.

1) B is the unique largest $P \cap R$-tame β-homogeneous R-submodule of $E_S(S/P)$.

2) B is a closed R-submodule of $E_S(S/P)$.

3) B is an essential S-submodule of $E_S(S/P)$.

4) If $E_S(S/P)$ is injective as a R-module, then $E_S(S/P) = B \oplus C$ for some submodule C.

Proof. (1) is a clear consequence of 3.4. For (2), suppose that $B \leq_e B' \subseteq {_R}E_S(S/P)$. Choose $0 \neq F \subseteq B'$ a finitely generated submodule. Then $F \cap B \leq_e F$ and hence F is tame. By 2.3 (4), $Cldim(F) = Cldim(F \cap B) = \beta$ and so $F \subseteq B$. Therefore $B' = B$ and B is closed. For (3), B is a S-submodule of $E_S(S/P)$ by 3.4(2). Since $E_S(S/P) \cong W^{(n)}$ for some uniform injective S-module W, $B \leq_e {_S}E_S(S/P)$ by 3.4(1). Finally, for (4), recall that a closed R-submodule of an injective R-module is a direct summand.

Note that if S is a free R-module with the generating set X as a basis, then it follows from Baer's criterion for injectivity that injective S-modules are injective as R-modules. Thus, the last statement of 3.5 is true in case S is either of $R[y; \sigma], R[y, y^{-1}; \sigma]$.

If Ω fails to satisfy the left second layer condition, then according to 2.1(3), there exists $P \in \Omega$ and a finitely generated S-submodule $M \subseteq E =$

$E_S(S/P)$ such that $PM \neq 0$, $_SM/ann_M(P)$ is uniform and $Q = \ell_S(M) = \ell_S(M/ann_M(P)) = ass(M/ann_m(P)) \subset P$. We will call any $P \in Spec(S)$ *left restricted* provided that any $Q \in Spec(S)$ that arises in this fashion satisfies $Q \cap R \subset P \cap R$ and $P \cap R/Q \cap R \leq_e {_R}R/Q \cap R$. A subset Λ of $Spec(S)$ will be called *left restricted* if all $P \in \Lambda$ are left restricted and the same terminology applies to S if $Spec(S)$ is left restricted. *Right restricted* is defined similarly.

In case S is a centralizing extension, $P \cap R \in Spec(R)$ for all $P \in Spec(S)$ and so the second part of the definition of restricted is a consequence of the first part. It then follows from [4; A(a)] that $Spec(S)$ is restricted for S satisfying condition (†) in that paper. The chief advantage of this condition is brought out in [4; C] where it is shown that such a centralizing extension of a FBN ring or polynormal Noetherian ring satisfies the second layer condition.

In order to characterize restricted prime ideals of S, we need to establish a relationship between linked primes in S and their restrictions in R. In the following, B_P will denote the module B of 3.5 and β_P will denote the ordinal $Cldim(R/P \cap R)$.

Corollary 3.6. *If $P, Q \in \Omega$ with $P \rightsquigarrow Q$, then $\beta_P = \beta_Q$.*

Proof. Since $P \rightsquigarrow Q$, there exists an ideal $J \subseteq S$ with $PQ \subseteq J \subset P \cap Q$ such that $P \cap Q/J$ is a torsion-free left S/P-module and a torsion-free right S/Q-module. Let $M = P \cap Q/J$ and note that M is a Noetherian $S-S$ bimodule and also for any $0 \neq y \in M$, RyR is a Noetherian $R-R$ bimodule.

As a left S/P-module, there is a $1-1$ map $M \to M_1 \oplus \cdots \oplus M_n$ where each M_i is a finitely generated uniform torsion-free left S/P-module. Then M_i is tame and $PM_i = 0$ for each i. By 3.5(1), (3), every finitely generated left R-submodule of M_i has classical Krull dimension equal to β_P for each i. It follows that the same is true for every finitely generated left R-submodule of M. The right hand version of the same argument yields that every finitely generated right R-submodule of M has classical Krull dimension equal to β_Q. Since RyR is a Noetherian bimodule for any $0 \neq y \in M$ and since $Cldim()$ is symmetric on Noetherian bimodules by 2.3(3), $\beta_P = \beta_Q$.

Proposition 3.7. *Let $P \in \Omega$, $E = E_S(S/P)$ and $L = ann_E(P)$. Let H/L and K/L be closed S-submodules of E/L and B_P/L, respectively, such that $B_P/L \oplus H/L \leq_e {_S}E/L$ and $ann_E(P \cap R)/L \oplus K/L \leq_e {_S}B_P/L$.*

1) *If $Q \in Ass(_SH/L)$, then $Q \cap R \subset P \cap R$ and $P \cap R/Q \cap R \leq_e R/Q \cap R$.*
2) *If $Q \in Ass(_SK/L)$, then $P \rightsquigarrow Q$. Furthermore, $Q = \ell_S(M + L/L)$ where $M \leq {_S}K$ satisfies the hypothesis of Jategaonkar's Main Lemma and $P \rightsquigarrow Q$ via $\ell_S(M)$.*

Proof. Let $Q = \ell_S(M + L/L) = ass(M + L/L)$ where $M + L/L \neq 0$ is a finitely generated, uniform S-submodule of H/L or K/L (for the proof of

(1) or (2), respectively). By [10; p. 155], we may assume, in either case, that M satisfies the hypothesis of Jategaonkar's Main Lemma [10; 6.1.3]. By that result, either $Q = \ell_S(M) \subset P$ or $P \leadsto Q$ via $\ell_S(M)$.

(1) Let $M = SF$ for some finitely generated R-submodule F with $PF \neq 0$. If $P \leadsto Q$ via $\ell_S(M)$, then from 3.6, $\beta_Q = \beta_P$. Let $\beta = \beta_Q = \beta_P$. Then $Cldim(F/F \cap L) \leq \beta$ whence $Cldim(F) = \sup\{Cldim(F \cap L), Cldim(F/F \cap L)\} \leq \beta$. Thus, $F \subseteq B_P$ and so $M \subseteq B_P$ which is a contradiction. It follows that $Q = \ell_S(M) \subset P$. If $Q \cap R = P \cap R$, then $(P \cap R)(F + L/L) = 0$ and the above argument yields the same contradiction. Therefore, $Q \cap R \subset P \cap R$.

Let $Q_1, \cdots, Q_n$ be the distinct primes of R minimal over $Q \cap R$. Since $Q \cap R = \bigcap_{i=1}^n Q_i \subseteq \ell_R(F + L/L) \subseteq \cap Ass(F + L/L)$, there exists $D \in Ass(F + L/L)$ such that $Q_j \subseteq D$. Clearly, $Cldim(R/D) > \beta$. Thus, $Cldim(R/Q_j) > \beta$. Using $R/Q_i \cong R/Q_j$ (as rings) for all i and $\beta_P = \beta$, it is easy to see that $P \cap R/Q \cap R \leq_e R/Q \cap R$.

(2) Suppose that $Q = \ell_S(M) \subset P$. Again, let $M = SF$ where $F \leq_R M$ is finitely generated and $PF \neq 0$. If $Q \cap R = P \cap R$, then $(P \cap R)F = 0$ and $F \subseteq ann_M(P \cap R) \cap K = ann_M(P)$ which contradicts $PF \neq 0$. Therefore, $Q \cap R \subset P \cap R$.

Assume that $Cldim(R/Q \cap R) = \beta_P$. Then all of the prime ideals minimal over $P \cap R$ are minimal over $Q \cap R$. Thus, we can write $Q \cap R = (P \cap R) \cap \bigcap_{i=1}^n Q_i$ where $\{Q_1, \cdots, Q_n\}$ is a non-empty collection of primes minimal over $Q \cap R$, none of which contain $P \cap R$. Since all factors of R by primes minimal over $Q \cap R$ are isomorphic as rings, $\beta_P = Cldim(R/Q_i)$ for all i.

Now, $(\bigcap_{i=1}^n Q_i)(P \cap R)F \subseteq (Q \cap R)F = 0$. If $(P \cap R)^2 F = 0$, then $0 \neq (P \cap R)F \subseteq ann_M(P \cap R)$ and $P \cap R \subseteq \ell_R((P \cap R)F)$. Then, since $(P \cap R)F$ is $P \cap R$-tame, all prime ideals minimal over $\ell_R((P \cap R)F)$ are minimal over $P \cap R$. But $\bigcap_{i=1}^n Q_i \subseteq \ell_R((P \cap R)F)$ together with $Cldim(R/Q_i) = \beta_P$ for all i implies that for some $Q_j, \ell_R((P \cap R)F) \subseteq Q_j$. However, this contradicts the definition of the Q_i. Therefore, $(P \cap R)^2 F \neq 0$. In particular, $P(P \cap R)F \neq 0$. By our standing hypothesis on M, $\ell_S((P \cap R)F) = \ell_S(M) = Q$. Thus, $\bigcap_{i=1}^n Q_i \subseteq \ell_R((P \cap R)F) = \ell_S((P \cap R)F) \cap R = Q \cap R$ which is a contradiction. Therefore, there exists at least one prime minimal over $P \cap R$, say H, that is not minimal over $Q \cap R$. Then $H \supset H' \supset Q \cap R$ where H' is minimal over $Q \cap R$. Thus, $Cldim(R/H) = \beta_P < Cldim(R/H') = Cldim(R/Q \cap R)$.

Let $D \in Ass(_R F + L/L)$, say $D = \ell_R(G + L/L)$ where $G \leq _R F$. Then $D = \ell_R(G/ann_G(P \cap R))$ since $G \cap L = ann_G(P \cap R)$. From 2.3(5), $Cldim(R/D) = \beta_P$. Since $F \subseteq B_P$ and B_P is $P \cap R$-tame, $P' \leadsto D$ for some prime P' minimal over $P \cap R$. By 5, there exists a finite subset $\Gamma' \subseteq \Gamma(D)$ such that $J = \cap \Gamma'$ satisfies $Jx \subseteq xD$ for all $x \in X$. Then $(P \cap R)JxG \subseteq (P \cap R)xDG \subseteq (P \cap R)xL \subseteq (P \cap R)L = 0$. It follows that $(P \cap R)JSG = 0$ whence $J \subseteq \ell_S(SG + L/L) \cap R = Q \cap R$. Therefore, $Cldim(R/J) \geq Cldim(R/Q \cap R)$. Also, by 2, $\beta_P = Cldim(R/D) =$

$Cldim(R/J)$. Thus, $\beta_P \geq Cldim(R/Q \cap R)$ which is a contradiction. Consequently, our original assumption fails and $P \rightsquigarrow Q$ via $\ell_S(M)$.

Corollary 3.8 Let $P \in \Omega, E = E_S(S/P)$ and $L = ann_E(P)$. If $L = ann_E(P \cap R)$, then P is left restricted. In particular, if $P = S(P \cap R)$, then P is left restricted.

Proof. Let $Q = \ell_S(M + L/L) = ass(M + L/L) = \ell_S(M) \subset P$ where $M \leq {}_S E$ is finitely generated and satisfies the hypothesis of [10; 6.1.3]. Also, let H/L and K/L be the submodules of E/L as defined in 3.7. Since both H/L and K/L are closed S-submodules of E/L and $L = ann_E(P \cap R)$, we may assume that $M + L/L$ is contained in one of H/L or K/L. By 3.7(2) and [10; 6.1.3], $Q = \ell_S(M) \subset P$ cannot occur if $M + L/L \subseteq K/L$. Therefore, $M + L/L \subseteq H/L$ and the result follows from 3.7(1).

Corollary 3.9. Ω is left restricted iff for all $P \in \Omega$ for which $P \supset S(P \cap R), P/S(P \cap R)$ satisfies the left second layer condition in $S/S(P \cap R)$.

Proof. Let "′" denote images in $S' = S/S(P \cap R)$ and let $V = E_{S'}(S'/P')$. Then we may take $V = ann_E(P \cap R) \supset L = ann_E(P) = ann_V(P')$ where $E = E_S(S/P)$.

($\Rightarrow$) Let $Q' = \ell_{S'}(Y/ann_Y(P'))$ where $Y \leq_S V$ satisfies the hypothesis of [10; 6.1.3]. Then either $Q' = \ell_{S'}(Y) \subset P'$ or $P' \rightsquigarrow Q'$. In the first case, $Q = \ell_S(Y) \subset P$. Then since P is restricted, $Q \cap R \subset P \cap R$. But $(P \cap R)Y = 0$ whence $P \cap R \subseteq Q \cap R$ which is impossible. Therefore, the second alternative holds.

($\Leftarrow$) By 3.8, we need only consider the case $P \supset S(P \cap R)$. Suppose that $Q = \ell_S(M + L/L) = ass(M + L/L) = \ell_S(M) \subset P$ where $M \leq {}_S E$ is finitely generated and $PM \neq 0$. As in the proof of 3.7, we can assume that $M + L/L$ is contained in one of $ann_E(P \cap R)/L, K/L$ or H/L and that the hypothesis of [9; 6.1.3] is satisfied. In the first case, $Q' = \ell_{S'}(M) \subset P'$ which contradicts the second layer condition for P'. The second case contradicts the conclusion of 3.7(2). For the third case, the result follows from 3.7(1).

The proof of [4] that certain central extensions are restricted depends on Ore sets in R being Ore sets in the extension and this, in turn, follows from the relation $rx = xr$ for all $x \in X$. Since this will fail for $R[y; \sigma]$, we cannot apply the same proof to this ring. However, using 3.9, we can show that certain prime ideals in $R[y; \sigma]$ inherit the restricted property from $R[y, y^{-1}; \sigma]$.

Theorem 3.10. If either $S = R[y, y^{-1}; \sigma]$ and $\Omega = Spec(S)$ or $S = R[y; \sigma]$ and $\Omega = \{P \in Spec(S) : y \notin P\}$, then Ω is restricted.

Proof. For the first choice of S and Ω, let $Q \subset P$ be a bad pair of primes of S arising from [10; 6.1.3]. From [1; 7.2], $Q \cap R \subset P \cap R$. To show that $P \cap R/Q \cap R \leq_e R/Q \cap R$, it suffices to show that $P \cap R$ is not contained in any prime ideal that is minimal over $Q \cap R$. This follows easily from [1; 3.4] and the comment [1; p. 86].

Let $S = R[y;\sigma] \subset T = R[y, y^{-1}; \sigma]$. By 3.8 we can assume that $P \supset (P \cap R)S$. By 3.9, it suffices to show that $P/(P \cap R)S$ satisfies the second layer condition. It follows from [15; 10.6.4 (iii), 6.3 (iii)], that we can assume that $P \cap R = 0$. Now, let $E = E_S(S/P)$ and $Q = \ell_S(M + L/L) = ass(M + L/L) = \ell_S(M) \subset P$ where $M \leq {}_SE$ is finitely generated, $PM \neq 0$ and M satisfies the hypothesis of [10; 6.1.3]. Let $E' = E_T(T/TP)$. Note that $TP \in Spec(T)$ by [15; 10.6.4 (iv)].

We claim that E' is an injective left S-module. Let $0 \neq I \leq {}_SS$ and let $m : I \to E'$ be a S-homomorphism. Let J be the union of the chain $I \subset y^{-1}I \subset y^{-2}I \subset \cdots$. Then J is a left ideal of T. It is easy to see that $m^* : J \to E'$ defined by $m^*(y^{-n}s) = y^{-n}m(s)$ is a well defined T-homomorphism with $m^* \mid I = m$. By injectivity of E', there exists $e \in E'$ such that m^* is right multiplication by e. Restricting to I yields that m is right multiplication by e. Therefore, by Baer's criterion, E' is a injective S-module.

Since there is a $1 - 1$ S-homomorphism $S/P \to T/TP \subseteq E'$ and ${}_SE'$ is injective, there exists a $1 - 1$ S-homorphism $E \to E'$. Therefore, we can assume that $M \leq {}_SE \leq {}_SE'$. In this case, $TM \leq {}_TE'$. Let $A = \ell_T(TM)$. By [15; 10.6.4(ii)], Q is σ-stable and so $QT = TQ \subseteq A$. Also, $(A \cap S)M = 0$ implies that $A \cap S \subseteq Q$. By [15; 10.6.3(i)], $A = (A \cap S)T \subseteq QT$. Thus, $QT = A = \ell_T(TM)$. Since $Q \subset P, QT \subset PT$. Otherwise, by [15; 10.6.4(iii)], $Q = QT \cap S = PT \cap S = P$ which is a contradiction. It follows that the pair of primes (QT, PT) is an undesirable pair in the sense of Bell [1; p. 104].

Again, by [15; 6.3(iii)], $QT \cap R \subseteq QT \cap S = Q$ and hence $Q \cap R \subseteq QT \cap R \subseteq Q \cap R$. Thus, $QT \cap R = Q \cap R$. Similarly, $P \cap R = 0 = PT \cap R \supseteq Q \cap R$ whence $QT \cap R = PT \cap R$. However, this contradicts the conclusion of [1; 7.2]. Therefore $P \rightsquigarrow Q$ via $\ell_S(M)$. Since this alternative always occurs, P satisfies the left second layer condition. The right second layer condition follows by the symmetric argument.

4. The Second Layer Condition

We will assume throughout that R is a Noetherian ring that satisfies the second layer condition and that R, S and Ω satisfy conditions *1-5* listed at the beginning of Section 3.

Theorem 4.1. *The following are equivalent for Ω left restricted.*

1) Ω satisfies the left second layer condition.

2) For all $P \in \Omega, E_S(S/P)$ is β_P-homogeneous.

3) For all $P \in \Omega, ann_E(P \cap R) \leq_e {}_RE$ where $E = E_S(S/P)$. In particular, as a R-module, $E_S(S/P)$ is $P \cap R$-tame.

4) For all $P \in \Omega, B_P/L \leq_e {}_SE/L$ where $E = E_S(S/P)$ and $L = ann_E(P)$.

Proof. (1) ⇒ (2) Let $_SM \subseteq E_S(S/P)$ be a finitely generated S-submodule. Then M has an affiliated series $0 = M_0 \subset M_1 \subset \cdots \subset M_n = M$ with corresponding primes $P = P_1, P_2, \cdots, P_n$. According to [7; 11.6], each P_i is in the left link closure of $Ass(_SM) = \{P\}$ and each M_i/M_{i-1} is a torsion-free S/P_i-module. In particular, $P_1, \cdots, P_n \in \Omega$. By 3.6, $\beta_P = Cldim(R/P_i \cap R)$ for all i. Thus, from 3.4(3), the R-modules M_i/M_{i-1} are β_P-homogeneous. It follows from 2.3(1) that M is a β_P-homogeneous R-module.

(2) ⇔ (3) The forward implication is any easy consequence of 3.5. For the reverse implication, note that for any finitely generated R-submodule $F \subseteq ann_E(P \cap R), Cldim(F) \leq \beta_P$. Thus, by 3.5, $ann_E(P \cap R) \subseteq B_P$ and hence $B_P \leq_e {}_RE$. Therefore, by 2.3(4), $B_P = E$ and statement 2 is true.

(2) ⇔ (3) ⇒ (4) Clear.

(4) ⇒ (1) Let $Q \in Ass(_SE/L)$, say $Q = \ell_S(M/L) = ass(M/L)$ where M/L is uniform and satisfies the hypothesis of [10; 6.1.3]. Since $M/L \cap B_P/L \neq 0$, we may assume that $M/L \subseteq B_P/L$. Similarly, we may assume that M/L is contained in one of $ann_E(P \cap R)/L$ or K/L where K/L is as defined in 3.7. Now, if $Q = \ell_S(M) \subset P$, then $Q \cap R \subset P \cap R$. Thus, if $M \subseteq ann_E(P \cap R)$, then $P \cap R \subseteq Q \cap R$ which is impossible. Therefore, $M/L \subseteq K/L$. But then the conclusion of 3.7(2) contradicts $Q = \ell_S(M) \subset P$. Thus, $P \rightsquigarrow Q$. Therefore, Ω satisfies the second layer condition.

The following is an immediate consequence of 4.1.

Corollary 4.2. If Ω is restricted and every prime ideal of R is maximal, then Ω satisfies the second layer condition.

The following technical lemma allows us to apply 4.1 to some special cases. The proof utilizes Byun's technique for proving [4; B]. Here, an ideal I of R is called *stable* provided $IS = SI$.

Lemma 4.3. Suppose that Ω is left restricted. Further, suppose that for every $Q \in Spec(S)$ and $P \in \Omega$ with $Q \subset P$ and $P \cap R/Q \cap R \leq_e R/Q \cap R$, there exists $c \in P \cap R \cap \mathcal{C}(Q \cap R)$ such that $R(c + Q \cap R) \cap (c + Q \cap R)R$ contains a nonzero ideal $I/Q \cap R$ with I stable. Then Ω satisfies the left second layer condition.

Proof. By 4.1, it suffices to show that for every $P \in \Omega, B_P/L \leq_e {}_SE/L$ where $E = E_S(S/P)$. Suppose that $B_P/L \oplus H/L \leq_e {}_SE/L$ where $_SH/L \neq 0$. Choose $Q \in Ass(_SH/L)$. By 3.7(1), $Q \cap R \subset P \cap R$ and $P \cap R/Q \cap R \leq_e R/Q \cap R$. Let I and c be as stated in the hypothesis. Without loss of generality, we may assume that $Q \cap R = 0$ and, in this case, $0 \neq I \subseteq Rc \cap cR \subseteq P \cap R$.

Let $Q = \ell_S(M + L/L) = ass(M + L/L)$ where $0 \neq M + L/L \leq {}_SH/L$ is finitely generated and uniform. Since $I \neq 0$ is stable and not contained in Q, IM is a nonzero S-submodule of M. Therefore, $IM \cap L \neq 0$. Then $I \subseteq cR$ implies that there exists $m \in M$ such that $0 \neq cm \in L$. Since $c \in P \cap R, m \notin L$. Thus, $I \subseteq Rc \subseteq \ell_R(m + L)$, whence $I \subseteq \ell_R(Rm + L/L)$

and $Rm + L/L \neq 0$. Now, I stable yields that $IS \subseteq \ell_S(Sm + L/L) = Q$. However, it follows that $I \subseteq Q \cap R = 0$ which is a contradiction. Therefore, $B_P/L \leq_e {}_S E/L$.

4.4 Corollary. If Ω is restricted and R is a principal ideal ring, then Ω satisfies the second layer condition.

Proof. Let $Q \subset P \in \Omega$ with $Q \in Spec(S)$ and $P \cap R/Q \cap R \leq_E R/Q \cap R$. Since $R/Q \cap R$ is a principle ideal ring, $P \cap R/Q \cap R$ is generated by a normal element $n + Q \cap R$ by [1; 6.3]. Further, by 4, $P \cap R$ is stable. The left second layer condition now follows from 4.3 with the obvious choice for I and c. Similarly, Ω satisfies the right second layer condition.

If P is a prime ideal of $R[y; \sigma]$ with $y \notin P$ or P is any prime ideal of $R[y, y^{-1}; \sigma]$, then according to [15; 10.6.4 (ii), (iii)], $P \cap R$ is σ-stable. In this case, σ induces an automorphism $R/P \cap R \to R/P \cap R$ which will be denoted by σ_P.

4.5 Corollary. Let $S = R[y; \sigma]$ with $\Omega = \{P \in Spec(S) : y \notin P\}$ or let $S = R[y, y^{-1}; \sigma]$ with $\Omega = Spec(S)$ where, in either case, one of the following holds.

a) R is a FBN ring and for all $P \in \Omega, \sigma_P$ has finite order.

b) R is a polycentral ring and for all $P \in \Omega, \sigma_P$ restricted to the center of $R/P \cap R$ has finite order.

Then Ω satisfies the second layer condition.

Proof. For (a) or (b) it suffices to show that the hypothesis of 4.3 is satisfied (a similar argument works on the right). Thus, let $Q \subset P \in \Omega$ with $Q \in Spec(S)$ and $P \cap R/Q \cap R \leq_e R/Q \cap R$. For either extension, $Q \in \Omega$ and $Q \cap R$ is σ-stable. Also, in either case, $R/Q \cap R$ is FBN or polycentral. Thus, we may assume that $Q \cap R = 0$.

(a) Since R is semiprime, $(P \cap R) \cap \mathcal{C}$ is nonempty where $\mathcal{C}$ is the set of regular elements of R. Choose $c \in (P \cap R) \cap \mathcal{C}$. Since $Rc \leq_e {}_R R, 0 \neq \ell_R(R/Rc) \leq_e {}_R R$ is an ideal contained in Rc by [7; 8F]. Since $\ell_R(R/Rc)$ is essential in ${}_R R$, it contains a regular element and hence is also essential in R_R. Similarly, cR contains a nonzero ideal essential in both ${}_R R$ and R_R. It follows that $Rc \cap cR \supseteq J$, J an ideal with $J \leq_e {}_R R$ and $J \leq_e R_R$. Then, since σ_Q has finite order, some intersection, say $I = J \cap \sigma_Q(J) \cap \cdots \cap \sigma_Q^m(J) \subseteq Rc \cap cR$ is stable and nonzero.

(b) Let $Q \cap R = 0 = Q_1 \cap \cdots \cap Q_n$. For each i, define $J_i = \bigcap_{j \neq i} Q_j$. Then each $J_i \neq 0$ is a closed semiprime ideal of R such that $J_i \oplus Q_i \leq_e R$. Since R is a polycentral ring, there exists for each $i, 0 \neq z_i \in J_i \cap Center(R)$. The ideal $Rz_i + Q_i/Q_i \leq_e R/Q_i$ and hence $z_i \in \mathcal{C}(Q_i)$. Let $c = z_1 + \cdots + z_n$. If $rc = 0$, then $rz_i = 0$ for each i whence $r \in \bigcap_{i=1}^n Q_i = 0$. Thus, c is a central regular element. Let $J = Rc = cR$ and repeat the argument in part (a) above.

If $\sigma = 1$, then for R FBN or polycentral, both (a) and (b) of 4.5 are satisfied. In this case, 4.5(a) is included in [4; C]. If R is commutative, $S = R[y; \sigma]$ and every $P \in \Omega$ has $P \supset S(P \cap R)$, then each σ_P must have finite order by [8; 4.3]. For any FBN ring A and finite group G, the group ring AG is FBN by [13; 4.9] and the automorphism σ induced by any $g \in G$ has finite order. For other examples, see [5].

In some instances, any meaningful discussion of the second layer condition should revolve around $Spec(S) - \Omega$. For example, let $R = k[x]$ where k is an algebraically closed field, and let $\sigma \in Aut(R)$ be defined by $\sigma(x) = x + 1$. Then $S = R[y; \sigma]$ is a prime ring and, by [15; 4.3.8 (ii)], $y \in P$ for all $0 \neq P \in Spec(S)$. Also, by [10; A.3.6], S satisfies the second layer condition. Using a result of Musson [16], we generalize this result by showing that $Spec(S) - \Omega$ inherits the second layer condition from R.

4.6 Theorem. Let $S = R[y; \sigma]$. Then $\Lambda = \{P \in Spec(S) : y \in P\}$ satisfies the second layer condition.

Proof. For all $P \in \Lambda, P = P \cap R + Sy$ and, both as rings and R-modules, $S/P \cong R/P \cap R$. In particular, $P \cap R \in Spec(R)$. Note that $y(S/P) = 0$. Make $E_R(S/P)$ into a left S-module by defining $ye = 0$ for all $e \in E_R(S/P)$. Let $A = E_R(S/P)$. Clearly, as S-modules, $S/P \leq_e A$. Since $A \leq_e E_S(A)$ (as S-modules), $E_S(A) \cong E_S(S/P)$.

According to [16; 1.2], $E_S(A)$ may be described in the following way: for each $n \geq 0$, let AY_n be an isomorphic copy of the abelian group A, i.e., $A \cong AY_n$ under $a \to aY_n$ with addition in AY_n defined in the obvious way. Define each AY_n to be an R-module under a new action " $\cdot$ " given by $r \cdot aY_n = \sigma^n(r)aY_n$ for all $r \in R$ and $a \in A$, where the product $\sigma^n(r)a$ comes from the old action of R on A. Note that for all $r \in R$ and $a \in A, r \cdot aY_0 = raY_0$. Thus, under $\cdot$, $A \cong AY_0$ as R-modules. Let $A^+ = \bigoplus_{n=0}^{\infty} AY_n$ be the corresponding R-module with operations defined coordinate-wise. Make A^+ into a left S-module by defining $yaY_0 = 0$ for all $a \in A$ and for $n \geq 1, yaY_n = aY_{n-1}$ for all $a \in A$. By [16; 1.2], $A^+ \cong E_S(A)$ as S-modules.

Consider an R-module summand AY_n of A^+. By definition of $_RAY_n$, we may view AY_n as the abelian group $E_R(S/P)$ with the R-module action $\cdot$ given by $r \cdot e = \sigma^n(r)e$ for all $r \in R$ and $e \in E_R(S/P)$. Choose $0 \neq e \in E_R(S/P)$. Then there exists $t \in R$ such that $0 \neq te \in S/P$. Since σ^n is an automorphism, $\sigma^n(t') = t$ for some $t' \in R$. Thus, $te = \sigma^n(t')e = t' \cdot e$ and hence $0 \neq t' \cdot e \in S/P$. Therefore, S/P is an essential R-submodule (under $\cdot$) of $E_R(S/P)$. Furthermore, under this action, the annihilator in R of S/P is the prime ideal $\sigma^{-n}(P \cap R)$. Using $\mathcal{C}\,(\sigma^{-n}(P \cap R)) = \sigma^{-n}(\mathcal{C}\,(P \cap R))$ it is easy to check that S/P is a torsion-free $R/\sigma^{-n}(P \cap R)$-module. Therefore, under $\cdot$ as defined in the last paragraph, AY_n is a $\sigma^{-n}(P \cap R)$-tame R-module. From 2.3(4) every R-submodule (under $\cdot$) $0 \neq F \subseteq AY_n$ has $Cldim(F) = Cldim(R/\sigma^{-n}(P \cap R)) = Cldim(R/P \cap R)$. Let $\beta = Cldim(R/P \cap R)$. It follows that, as an R-module, $E_S(S/P)$ is tame and β-homogeneous.

Now, if $P \in \Lambda$ fails to satisfy the second layer condition, then by 2.1(3), there exists a finitely generated S-submodule $0 \neq M \subseteq E_S(S/P)$ such that $L = ann_M(P) \subset M$ and $Q = \ell_S(M) = \ell_S(M/L)$ is a prime ideal with $Q \subset P$. Since $Sy \subseteq P, SyL = 0$. Since y is a normal element, Sy satisfies the left AR-property by [15; 4.1.10]. Then $L \leq_e M$ yields that $(Sy)^m M = Sy^m M = 0$ for some m. Thus, $Sy \subseteq Q$ and hence, $yM = 0$ and also $Q = Q \cap R + Sy$. It follows from the first conclusion that M must be finitely generated as a R-module. It follows from the second conclusion that $Q \cap R \subset P \cap R$. Therefore, $\ell_R(M) = R \cap \ell_S(M) = Q \cap R \subset P \cap R$ whence $Cldim(R/\ell_R(M)) > \beta$. However, from the end of the last paragraph and 2.3(2), $Cldim(R/\ell_R(M)) = Cldim(M) = \beta$ which is a contradiction. Therefore, Λ satisfies the second layer condition.

4.7 Corollary. Let $S = R[y; \sigma]$. If every prime ideal of R is maximal or R is a principal ideal ring or if one of (a) or (b) in 4.5 is true, then S satisfies the second layer condition.

Proof. This follows from 4.2, 4.4, 4.5 and 4.6.

5. CENTRALIZING EXTENSIONS

As before, R will always denote a Noetherian ring that satisfies the second layer condition. However, it will be assumed here that S is a centralizing extension of R. Our goal is to use the results of Section 4 to examine S in case R is AR-separated. Note that for any ideal $I \subseteq R, IS = SI$ is an ideal of S. Also, from [18; 2.12.39], $P \cap R \in Spec(R)$ for all $P \in Spec(S)$. The following is then an easy consequence of 4.3. This result is also a corollary of [4; C].

5.1 Proposition. If S is restricted, then S satisfies the second layer condition in case R is either an FBN ring or a polynormal ring.

According to [14; 8], a polynormal Noetherian ring R is an example of an AR-separated ring, i.e., for all pairs of prime ideals $Q \subset P$, there exists an ideal $Q \subset I \subseteq P$ such that the ideal $I/Q \subseteq R/Q$ satisfies the AR-property. By [7; 11.14], AR-separated rings satisfy the strong second layer condition (see [7; p. 183]).

While we have been unable to determine if 5.1 remains true in case R is AR-separated, the following result shows that S must satisfy a diluted version of AR-separation if S satisfies the strong second layer condition.

5.2 Theorem. Let R be an AR-separated ring and let S be restricted. If S satisfies the strong second layer condition, then for all pairs $Q \subset P$ in $Spec(S)$ with $Q \cap R \subset P \cap R$, there exists an ideal $Q \subset J \subseteq P$ such that the ideal $J/Q \subseteq S/Q$ satisfies the AR-property.

Proof. Let $Q \subset P$ be as described in the hypothesis. Since R is AR-separated, there exists an ideal I with $Q \cap R \subset I \subseteq P \cap R$ such that $I/Q \cap R \subseteq R/Q \cap R$ satisfies the AR-property. Note that $SI + Q/Q \neq 0$

is an ideal of S/Q and S/Q satisfies the strong second layer condition. We claim that $SI + Q/Q$ satisfies the AR-property in S/Q.

Let $K \leq_e {}_S H$ where H is a finitely generated S/Q-module and $(SI)K = 0$. By [7; 11.4], H is annihilated by a finite product of prime ideals from the right link closure of $Ass({}_{S/Q}H)$. Let $P'/Q \in Ass({}_{S/Q}H) = Ass({}_{S/Q}K)$ and suppose that $P'/Q \rightsquigarrow Q'/Q$. Then $Q' = \ell_S(M/L)$ where $0 \neq {}_S M \subseteq E_S(S/P')$ is finitely generated and $QM = 0$. Let $M = SF$ where $0 \neq {}_R F \subseteq E_S(S/P')$ is finitely generated.

Now, $(Q \cap R)F = 0$. Also, $SI \subseteq P'$ whence $I \subseteq P' \cap R$. By 4.1, $E_S(S/P')$ is $P' \cap R$-tame and so $L = ann_F(P' \cap R) \leq_e F$. Since $I/Q \cap R \subseteq R/Q \cap R$ satisfies the AR-property and $IL = 0$, $I^n F = 0$ for some n. Thus, $(SI^n)M = SI^n SF = SI^n F = 0$ and so $SI^n = (SI)^n \subseteq Q'$. Therefore, $SI + Q/Q \subseteq Q'/Q$.

Note that since $SI \subseteq Q'$, a repeat of the above argument yields that $SI + Q/Q \subseteq Q''/Q$ for $Q'/Q \rightsquigarrow Q''/Q$. It follows that $SI + Q/Q$ is contained in the finitely many prime ideals in the left link closure of $Ass({}_{S/Q}H)$ whose product annihilates H. Therefore, $(SI)^m H = 0$ for some m. The right-hand version follows by symmetry. By [7; 11.11], $SI + Q/Q \subseteq S/Q$ satisfies the AR-property.

Acknowledgements

Support for this work was provided by a Fall 1993 sabbatical granted by the University of Wisconsin Centers Grants Committee. The author would like to thank Allen D. Bell, Ian M. Musson, Mark L. Teply and the Department of Mathematics at the University of Wisconsin-Milwaukee for their support and hospitality during his Fall semester visit.

6. References

[1] A.D. Bell, *Localization and ideal theory in Noetherian strongly group-graded rings*, J. Algebra **105** (1987), 76-115.

[2] K.A. Brown, *Module extensions over Noetherian rings*, J. Algebra **69** (1981), 247-260.

[3] K.A. Brown and R.B. Warfield, *The influence of ideal structure on representation theory*, J. Algebra **116** (1988), 294-315.

[4] L.H. Byun, *The second layer condition for certain centralizing extensions of FBN rings and polynormal rings*, Comm. in Algebra **21** (1993), 2175-2184.

[5] R.F. Damiano and J. Shapiro, *Twisted polynomial rings satisfying a polynomial identity*, J. Algebra **92** (1985), 116-127.

[6] K.R. Goodearl, *Linked injectives and Ore localizations*, J. London Math. Soc. **37** (1988), 404-420.

[7] K.R. Goodearl and R.B. Warfield, *An Introduction to Noncommutative Rings*, London Math. Soc. Student Text **16**, Cambridge University Press, Cambridge, 1989.

[8] R.S. Irving, *Prime ideals of Ore extensions over commutative rings*, J. Algebra **56** (1979), 315-342.

[9] A.V. Jategaonkar, *Solvable Lie algebras, polycyclic-by-finite groups and bimodule Krull dimension*, Comm. in Algebra **10** (1982), 19-69.

[10] A.V. Jategaonkar, *Localization in Noetherian Rings*, London Math. Soc. Lecture Note Series **98**, Cambridge University Press, London/New York, 1985.

[11] K.A. Kosler, *Module extensions and the second layer condition*, Comm. in Algebra **20** (1992), 69-91.

[12] K.A. Kosler, Classical Krull dimension and the second layer condition, in "Ring Theory, Proceedings of the Biennial Ohio State-Denison Conference 1992", World Scientific Publishing Co., Singapore/New Jersey, 1993.

[13] E.S. Letzter, *Prime ideals in finite extensions of Noetherian rings*, J. Algebra **135** (1990), 412-439.

[14] J.C. McConnell, *Localization in enveloping rings*, J. London Math. Soc. **43** (1968), 421-428.

[15] J.C. McConnell and J.C. Robson, *Noncommutative Noetherian Rings*, John Wiley and Sons, Chichester, 1987.

[16] I.M. Musson, *Conditions for a module to be injective and some applications to Hopf-algebra duality*, Proc. of the American Math. Soc. **123** (1995), 693-702.

[17] C. Nastasescu and F. van Oystaeyen, *Dimensions of Ring Theory*, D. Reidel Publishing, Dordrecht, 1987.

[18] L.H. Rowen, *Ring Theory*, Vol. I, Academic Press, Inc., San Diego, 1988.

CURRENT ADDRESS: University of Wisconsin at Waukesha, 1550 University Drive, Waukesha, WI 53188-2799.

GENERATORS OF SUBGROUPS OF FINITE INDEX IN $GL_m(\mathbb{Z}G)$

GREGORY T. LEE AND SUDARSHAN K. SEHGAL

ABSTRACT. Let G be a finite group, and $\mathbb{Z}G$ its integral group ring. We provide a set of generators of a subgroup of finite index in the general linear group, $GL_m(\mathbb{Z}G)$, provided $m \geq 3$. We also provide partial results in the case $m = 2$.

1. INTRODUCTION

Let $\mathbb{Z}G$ be the integral group ring of a finite group G. Let $GL_m(\mathbb{Z}G)$ be the group of invertible $m \times m$ matrices over $\mathbb{Z}G$. The purpose of this paper is to give explicitly a finite set of generators of a subgroup of finite index in $GL_m(\mathbb{Z}G)$, if $m \geq 3$. Partial results are given if $m = 2$. The case $m = 1$ has been extensively dealt with by Ritter-Sehgal and Jespers-Leal (see [6], [7] and [2], [3]).

To state our main result, we need to introduce some units of $M_m(\mathbb{Z}G)$. The Bass cyclic units, $\mathcal{B}_1$ of $\mathbb{Z}G$ are well-known and also defined in the next section. We let $\mathcal{B}_1 I_m$ denote the subgroup of $GL_m(\mathbb{Z}G)$ generated by the matrices bI_m, for all $b \in \mathcal{B}_1$. Let $E_{p,q}$ be the matrix with a 1 in the (p,q) position and zeroes elsewhere. Then we define $\mathcal{E}$ to be the subgroup of $GL_m(\mathbb{Z}G)$ generated by all the matrices $I_m + gE_{p,q}$, $p \neq q$, $g \in G$. Our main result is

Theorem 1. *Let $GL_m(\mathbb{Z}G)$ be the group of invertible $m \times m$ matrices over $\mathbb{Z}G$, the integral group ring of a finite group. If $m \geq 3$, then $\langle \mathcal{B}_1 I_m, \mathcal{E} \rangle$ is a subgroup of finite index in $GL_m(\mathbb{Z}G)$.*

1991 *Mathematics Subject Classification.* Primary 16U60; Secondary 16S34, 20H25.
This work was supported by NSERC of Canada.

2. Notation and Needed Results

We will use the following notation throughout. Let G be a finite group, and e_i, $1 \leq i \leq t$, the primitive central idempotents of $\mathbb{Q}G$. Let $\theta_i : \mathbb{Q}Ge_i \to M_{n_i}(D_i)$ be an isomorphism, where n_i is a natural number, and D_i is a division ring which is finite-dimensional over its centre F_i, an algebraic number field. (Here, $M_{n_i}(D_i)$ denotes the ring of $n_i \times n_i$ matrices over D_i.) Let O_i be the ring of integers of F_i. Define $\pi_i : \mathbb{Q}G \to M_{n_i}(D_i)$ via $\pi_i(\eta) = \theta_i(\eta e_i)$. Let Λ_i be a $\mathbb{Z}$-order (or, simply, order) in $M_{n_i}(D_i)$ containing $\pi_i(\mathbb{Z}G)$. Let $\mathcal{O}_i$ be an order in D_i which contains O_i (such as any maximal order). Then $M_{n_i}(\mathcal{O}_i)$ is another order in $M_{n_i}(D_i)$.

Furthermore, $M_m(\Lambda_i)$ and $M_{mn_i}(\mathcal{O}_i)$ are both orders in $M_{mn_i}(D_i)$. We have
$$M_m(\mathbb{Q}G) = M_m(\bigoplus \mathbb{Q}Ge_i) \cong M_m(\bigoplus M_{n_i}(D_i)) \cong \bigoplus M_{mn_i}(D_i).$$
Let $\tau : M_m(\mathbb{Q}G) \to \bigoplus M_{mn_i}(D_i)$ be this isomorphism.

For any ring, R, with identity, and any natural number n, we let $GL_n(R)$ denote the group of invertible $n \times n$ matrices over R. If D is a division ring which is finite-dimensional over the rationals, and if R is any unital subring thereof, let $SL_n(R)$ be the subgroup of $GL_n(R)$ consisting of those matrices which have reduced norm one. We let $E_n(R)$ denote the subgroup of $GL_n(R)$ generated by the elementary matrices. (It is easy to see that $E_n(R) \leq SL_n(R)$.) If W is any ideal of R, then we say that an elementary matrix is W-elementary if the nonzero entry off the diagonal is in W. Let $E_n(W)$ denote the subgroup of $E_n(R)$ which is generated by these matrices, and let $\tilde{E}_n(W)$ denote its normal closure in $E_n(R)$.

We will use the Congruence Subgroup Theorem of Bass, Milnor, and Serre. We state it here, and refer the reader to [8, Theorem 19.32].

Lemma 2.1. *Let W be any nonzero ideal in $\mathcal{O}_i$. Then, for any natural number $n \geq 3$, $|SL_n(\mathcal{O}_i) : \tilde{E}_n(W)| < \infty$.*

In fact, we need to strengthen this result. The following lemma is due to Vaserstein [9]. Let us note that if i, j and k are pairwise distinct integers, and a, b are in our ring, then the commutator $[I + aE_{i,j}, I + bE_{j,k}]$ is equal to $I + abE_{i,k}$.

Lemma 2.2. *Let W be any nonzero ideal in $\mathcal{O}_i$. Then, for any natural number $n \geq 3$, $\tilde{E}_n(W^{2^{4n-2}}) \leq E_n(W)$. Hence $|SL_n(\mathcal{O}_i) : E_n(W)| < \infty$.*

Proof. Let Q be any ideal in $\mathcal{O}_i$. First, suppose

(*) $$B = \begin{pmatrix} A & 0 \\ 0 & 1 \end{pmatrix} \in GL_n(\mathcal{O}_i)$$

where $A \in GL_{n-1}(\mathcal{O}_i)$. We claim that $B^{-1}E_n(Q^2)B \subseteq E_n(Q)$. Take $C = I_n + q_1q_2E_{i,j}$, with $q_1, q_2 \in Q$, $1 \leq i \neq j \leq n$. This is an arbitrary generator of $E_n(Q^2)$. If $i = n$, then $B^{-1}CB = \prod_{k=1}^{n-1}(I_n + q_1q_2a_kE_{n,k})$, where a_k is the (j, k) entry of A. Thus, $B^{-1}CB \in E_n(Q^2)$. Similarly, if $j = n$, then $B^{-1}CB \in E_n(Q^2)$. Otherwise, i, j and n are pairwise distinct, and $C = [I_n + q_1E_{i,n}, I_n + q_2E_{n,j}]$, which means that

$$B^{-1}CB = [B^{-1}(I_n + q_1E_{i,n})B, B^{-1}(I_n + q_2E_{n,j})B].$$

Repeating the above argument with Q in place of Q^2, we see that each term of this commutator is in $E_n(Q)$, so $B^{-1}CB \in E_n(Q)$.

The only restriction on B was that the nth row and column resemble those of I_n. But the nth row and column are not distinguished, and since $n \geq 3$, any elementary matrix has at least one j such that the jth row and column resemble those of I_n. It follows immediately that if B is an elementary matrix, then $B^{-1}CB \in E_n(Q)$.

Now, the proof of [1, Theorem 4.2] shows us that for any $B \in GL_n(\mathcal{O}_i)$ (and therefore, in particular, for any $B \in E_n(\mathcal{O}_i)$), there exists $X \in E_n(\mathcal{O}_i)$ such that X is the product of at most $3n - 2$ elementary matrices, and

$$XB = \begin{pmatrix} A & 0 \\ \gamma & 1 \end{pmatrix}$$

with $A \in GL_{n-1}(\mathcal{O}_i)$, $\gamma \in (\mathcal{O}_i)_{1 \times n-1}$. Let $\gamma A^{-1} = (\gamma_1 \cdots \gamma_{n-1})$, and for each j, $1 \leq j < n$, write $X_j = I_n - \gamma_j E_{n,j}$. Then

$$\prod_{j=1}^{n-1} X_j = \begin{pmatrix} I_{n-1} & 0 \\ -\gamma A^{-1} & 1 \end{pmatrix}$$

and

$$(\prod_{j=1}^{n-1} X_j)XB = \begin{pmatrix} A & 0 \\ 0 & 1 \end{pmatrix}.$$

Therefore,

$$B = X^{-1}(\prod_{j=1}^{n-1} X_j^{-1})\begin{pmatrix} A & 0 \\ 0 & 1 \end{pmatrix}$$

which is a product of at most $4n - 2$ matrices, each of which is elementary or of the form (*) above. Thus, iterating our argument, if $B \in E_n(\mathcal{O}_i)$ and $C \in E_n(W^{2^{4n-2}})$, then $B^{-1}CB \in E_n(W)$. Hence, $\tilde{E}_n(W^{2^{4n-2}}) \leq E_n(W)$, and Lemma 2.1 completes the proof. □

We will need to use the following results. For part (1), see [8, Lemma 4.6]. The other parts are easy. We denote by $\mathcal{U}(R)$ the group of units of a ring R.

Lemma 2.3. *Let A be a finite-dimensional algebra over the rationals. Then*
(1) *if $\Lambda_1 \subseteq \Lambda_2$ are orders in A, then $|\mathcal{U}(\Lambda_2) : \mathcal{U}(\Lambda_1)| < \infty$;*
(2) *with Λ_1 and Λ_2 as in (1), $|\Lambda_2 : \Lambda_1| < \infty$;*
(3) *the centre of an order in A is an order in the centre of A;*
(4) *if Λ_1 and Λ_2 are any two orders in A, and H is a subgroup of finite index in $\mathcal{U}(\Lambda_1)$, then $|\mathcal{U}(\Lambda_2) : H \cap \mathcal{U}(\Lambda_2)| < \infty$. Thus, if C is a group containing a subgroup of finite index in Λ_1, then it also contains a subgroup of finite index in Λ_2.*

Finally, let us introduce the Bass cyclic units. For any $x \in G$, write
$$\hat{x} = 1 + x + \cdots + x^{|x|-1}.$$
Let φ be the Euler function. Then, for each $1 < i < |x|$ with $(i, |x|) = 1$, the elements of the form
$$(1 + x + \cdots + x^{i-1})^{\varphi(|G|)} + \frac{1 - i^{\varphi(|G|)}}{|x|}\hat{x}$$
are called the Bass cyclic units of $\mathbb{Z}G$. (It is easily shown that these are units of $\mathbb{Z}G$. See [8, p.33].) Let $\mathcal{B}_1$ denote the subgroup of $\mathcal{U}(\mathbb{Z}G)$ which is generated by these units, for all $x \in G$, and all such i. Bass and Milnor showed that these units generate a subgroup of finite index in $\mathcal{U}(\mathbb{Z}G)$ when G is abelian, and they provide a useful reduction in the nonabelian case.

We will denote by $\mathcal{B}_1 I_m$ the subgroup of $GL_m(\mathbb{Z}G)$ which is generated by the matrices of the form bI_m, for $b \in \mathcal{B}_1$.

3. A Reduction

Our reduction is precisely the higher-dimensional analogue of [7, Lemma 3.2].

Lemma 3.1. *Let C be a subgroup of $GL_m(\mathbb{Z}G)$ such that for each i, C contains a subgroup C_i satisfying $\tau(C_i) = 1$ in every component except for the ith, and in the ith component, $\tau(C_i)$ contains $E_{mn_i}(q\mathcal{O}_i)$, for some natural number q. Then $\langle C, \mathcal{B}_1 I_m \rangle$ is of finite index in $GL_m(\mathbb{Z}G)$, if $m \geq 3$.*

Proof. We shall first show that $\langle C, \mathcal{B}_1 I_m \rangle$ contains a subgroup of finite index in the centre, $Z(GL_m(\mathbb{Z}G))$. Take $z \in Z(\mathcal{U}(\mathbb{Z}G))$. In the proof of [7, Lemma 3.2], it is shown that there exist a natural number l and $b \in \mathcal{B}_1$ such that for each i, the reduced norm $nr(\pi_i(z^l b^{-1})) = 1$. Thus,

$$nr\begin{pmatrix} \pi_i(z^l b^{-1}) & & \\ & \ddots & \\ & & \pi_i(z^l b^{-1}) \end{pmatrix} = 1$$

or, in other words, this matrix is in $SL_{mn_i}(D_i)$. Clearly, $\pi_i(z^l b^{-1}) \in \pi_i(\mathbb{Z}G) \subseteq \Lambda_i$, and the same holds for its inverse. Hence, $\pi_i(z^l b^{-1}) \in \mathcal{U}(\Lambda_i)$. Since $M_{n_i}(\mathcal{O}_i)$ is another order in $M_{n_i}(D_i)$, Lemma 2.3 gives

$$|\mathcal{U}(\Lambda_i) : \mathcal{U}(\Lambda_i \cap M_{n_i}(\mathcal{O}_i))| = r_i < \infty.$$

Letting $r = \prod r_i$, we have $\pi_i(z^l b^{-1})^r \in SL_{n_i}(\mathcal{O}_i)$. Thus,

$$\begin{pmatrix} \pi_i(z^{lr} b^{-r}) & & \\ & \ddots & \\ & & \pi_i(z^{lr} b^{-r}) \end{pmatrix} \in SL_{mn_i}(\mathcal{O}_i).$$

By the Congruence Subgroup Theorem, $|SL_{mn_i}(\mathcal{O}_i) : E_{mn_i}(q\mathcal{O}_i)| = k_i < \infty$. Letting $k = \prod k_i$, we obtain

$$\begin{pmatrix} \pi_i(z^{klr} b^{-kr}) & & \\ & \ddots & \\ & & \pi_i(z^{klr} b^{-kr}) \end{pmatrix} \in E_{mn_i}(q\mathcal{O}_i).$$

By assumption,

$$(1, \ldots, 1, \begin{pmatrix} \pi_i(z^{klr} b^{-kr}) & & \\ & \ddots & \\ & & \pi_i(z^{klr} b^{-kr}) \end{pmatrix}, 1, \ldots, 1) \in \tau(C)$$

for each i. Multiplying these together for the various components, we discover that

$$\tau \begin{pmatrix} z^{klr} b^{-kr} & & \\ & \ddots & \\ & & z^{klr} b^{-kr} \end{pmatrix} \in \tau(C).$$

Since τ is an isomorphism,

$$\begin{pmatrix} z^{klr} b^{-kr} & & \\ & \ddots & \\ & & z^{klr} b^{-kr} \end{pmatrix} \in C.$$

Therefore,

$$\begin{pmatrix} z & & \\ & \ddots & \\ & & z \end{pmatrix}^{klr} \in \langle C, \mathcal{B}_1 I_m \rangle.$$

Since the elements of $Z(GL_m(\mathbb{Z}G))$ are precisely the multiples of the identity matrix by elements of $Z(\mathcal{U}(\mathbb{Z}G))$, we conclude that

$$Z(GL_m(\mathbb{Z}G))/(Z(GL_m(\mathbb{Z}G)) \cap \langle \mathcal{B}_1 I_m, C \rangle)$$

is a torsion group. But $Z(\mathcal{U}(\mathbb{Z}G))$ is a finitely generated abelian group, and therefore so is $Z(GL_m(\mathbb{Z}G))$. Thus,

$$|Z(GL_m(\mathbb{Z}G))/(Z(GL_m(\mathbb{Z}G)) \cap \langle \mathcal{B}_1 I_m, C \rangle)| < \infty$$

as claimed.

It follows that $\tau(\langle \mathcal{B}_1 I_m, C \rangle)$ contains a subgroup of finite index in the group $\tau(Z(GL_m(\mathbb{Z}G)))$. Therefore by Lemma 2.3, for each i, $\tau(\langle \mathcal{B}_1 I_m, C \rangle)$ contains

$$(1, \ldots, 1, K_i, 1, \ldots, 1)$$

where K_i is of finite index in $Z(GL_{mn_i}(\mathcal{O}_i))$. By our assumption, $\tau(C)$ contains

$$(1, \ldots, 1, E_{mn_i}(q\mathcal{O}_i), 1, \ldots, 1)$$

for each i. Since $mn_i \geq 3$, Lemma 2.2 allows us to infer that $\tau(C)$ contains a subgroup of finite index in

$$(1, \ldots, 1, SL_{mn_i}(\mathcal{O}_i), 1, \ldots, 1).$$

From the proof of [8, Theorem 21.6], we deduce that $\tau(\langle \mathcal{B}_1 I_m, C \rangle)$ contains a subgroup of finite index in

$$(1, \ldots, 1, GL_{mn_i}(\mathcal{O}_i), 1, \ldots, 1)$$

for each i. Hence, we have a subgroup of finite index in $\prod GL_{mn_i}(\mathcal{O}_i)$.

Since $\tau(\mathbb{Z}G)$ is an order in $\bigoplus M_{mn_i}(D_i)$, it follows immediately from Lemma 2.3 that $\tau(\langle \mathcal{B}_1 I_m, C \rangle)$ is of finite index in $\tau(GL_m(\mathbb{Z}G))$. We conclude that $\langle \mathcal{B}_1 I_m, C \rangle$ is of finite index in $GL_m(\mathbb{Z}G)$. □

4. Proof of Theorem 1

Before we prove the main result, let us introduce some notations for various types of matrix units. Let $E_{p,q}$ denote the (p,q) matrix unit in $M_m(\mathbb{Z}G)$. (Clearly, $I_m + \alpha E_{p,q} \in GL_m(\mathbb{Z}G)$ for any $\alpha \in \mathbb{Z}G$, provided $p \neq q$, since its inverse is $I_m - \alpha E_{p,q}$.) Denote by $E_{p,q}^*$ the (p,q) matrix unit in $M_{n_i}(D_i)$. Regard a matrix in $M_{mn_i}(D_i)$ as an $m \times m$ grid of $n_i \times n_i$ matrices. Then, denote by $E'_{p,q}$ the matrix which contains the $n_i \times n_i$ identity matrix in its (p,q) block, and zeroes elsewhere. Finally, ignoring this block structure, let $E''_{p,q}$ be the (p,q) matrix unit in $M_{mn_i}(D_i)$. Now, we have the

Proof of Theorem 1. Fix $p \neq q$. For $g, h \in G$, and $u, v \in \mathbb{Z}$, we have

$$(I_m + gE_{p,q})^u (I_m + hE_{p,q})^v = I_m + (ug + vh)E_{p,q}$$

and therefore, $\langle \mathcal{B}_1 I_m, \mathcal{E} \rangle$ contains $I_m + \alpha E_{p,q}$ for all $\alpha \in \mathbb{Z}G$. Since $e_i \in \mathbb{Q}G$, we may choose a natural number k_i such that $k_i e_i \in \mathbb{Z}G$. Certainly, then, $\langle \mathcal{B}_1 I_m, \mathcal{E} \rangle$ contains $I_m + k_i \alpha e_i E_{p,q}$, for all $\alpha \in \mathbb{Z}G$. Now, the jth component of $\tau(I_m + k_i \alpha e_i E_{p,q})$ is

$$I_{mn_j} + \pi_j(k_i \alpha e_i) E'_{p,q} = I_{mn_j} + \theta_j(k_i \alpha e_i e_j) E'_{p,q}.$$

If $i \neq j$, then $e_i e_j = 0$, so this is the identity matrix. If $i = j$, then this is $I_{mn_i} + k_i \pi_i(\alpha) E'_{p,q}$.

Let $\mathcal{E}_{i,p,q}$ be the subgroup of $\mathcal{E}$ consisting of the elements of the form $I_m + k_i \alpha e_i E_{p,q}$, for all $\alpha \in \mathbb{Z}G$. We have just seen that

$$\tau(\mathcal{E}_{i,p,q}) = (1, \ldots, 1, I_{mn_i} + k_i \pi_i(\mathbb{Z}G) E'_{p,q}, 1, \ldots, 1).$$

Since $\pi_i(\mathbb{Z}G)$ and $M_{n_i}(\mathcal{O}_i)$ are orders in $M_{n_i}(D_i)$, there is a natural number t_i such that $t_i M_{n_i}(\mathcal{O}_i) \subseteq \pi_i(\mathbb{Z}G)$. Thus, for any $A \in M_{n_i}(\mathcal{O}_i)$,

$$(1, \ldots, 1, I_{mn_i} + k_i t_i A E'_{p,q}, 1, \ldots, 1) \in \tau(\mathcal{E}_{i,p,q}).$$

In particular, if we take $\omega \in \mathcal{O}_i$, and $1 \leq r, s \leq n_i$, then letting $A = \omega E^*_{r,s}$, we have

$$(1, \ldots, 1, I_{mn_i} + k_i t_i \omega E^*_{r,s} E'_{p,q}, 1, \ldots, 1) \in \tau(\mathcal{E}_{i,p,q}).$$

Now, the matrices of the form $E^*_{r,s} E'_{p,q}$ are matrix units in $M_{mn_i}(D_i)$. If we regard these matrices as $m \times m$ grids of $n_i \times n_i$ matrices, then the 1 cannot go in the (p,p) block, for any p. Let $\mathcal{E}_i$ be the subgroup of $\mathcal{E}$ generated by the $\mathcal{E}_{i,p,q}$, for all $p \neq q$. Then, $\tau(\mathcal{E}_i)$ contains

$$(1, \ldots, 1, I_{mn_i} + k_i t_i \omega E''_{r,s}, 1, \ldots, 1)$$

for all $\omega \in \mathcal{O}_i$, and all pairs (r,s) which do not correspond to a (p,p) block.

However, the pairs (r,s), with $r \neq s$ which do correspond to a (p,p) block will come to us for free. Indeed, suppose (r,s) is such a pair. Then there exists an integer w, with $1 \leq w \leq m$, such that $(w-1)n_i + 1 \leq r, s \leq wn_i$. Since $m > 1$, we may choose y with $1 \leq y \leq mn_i$, such that y does not fall between $(w-1)n_i + 1$ and wn_i. Then, we have seen that

$$(1, \ldots, 1, I_{mn_i} + k_i t_i \omega E''_{r,y}, 1, \ldots, 1)$$

and

$$(1, \ldots, 1, I_{mn_i} + k_i t_i E''_{y,s}, 1, \ldots, 1)$$

are in $\tau(\mathcal{E}_i)$. Therefore, so is their commutator, namely

$$(1, \ldots, 1, I_{mn_i} + k_i^2 t_i^2 \omega E''_{r,s}, 1, \ldots, 1).$$

Taking $q = \prod k_i^2 t_i^2$, we conclude that $\tau(\mathcal{E}_i)$ contains

$$(1, \ldots, 1, E_{mn_i}(q\mathcal{O}_i), 1, \ldots, 1)$$

for all i. We apply Lemma 3.1 to complete the proof. $\square$

5. The Case $m = 2$

The proof of Theorem 1 breaks down if $m = 2$, as Lemma 2.2 does not hold in this case. However, if D_i is commutative, and so equals F_i, and $\mathcal{O}_i = O_i$, then the lemma holds provided that F_i is not $\mathbb{Q}$ or an imaginary quadratic extension of the rationals. Notice that $\mathbb{Q}$ is always a simple component of $\mathbb{Q}G$ due to the existence of the augmentation map, which implies that $M_2(\mathbb{Q})$ is a component of $M_2(\mathbb{Q}G)$. Luckily, we can allow one exceptional component, as shown in [8, Lemma 22.10], provided the projection to this component is also of finite index. In fact, we have the following result.

Theorem 2. *Suppose that G is a finite group satisfying*
  a) $\mathbb{Q}G$ does not have a noncommutative division ring as a simple component; and,
  b) the factor commutator group G/G' has no element of order 2 or 3.
Then $\langle \mathcal{B}_1 I_2, \mathcal{E} \rangle$ is a subgroup of finite index in $GL_2(\mathbb{Z}G)$.

Proof. The assumptions guarantee that the simple components of $\mathbb{Q}G$ will include just one copy of $\mathbb{Q}$ and no imaginary quadratic extensions of $\mathbb{Q}$. Indeed, if $\mathbb{Q}Ge_i$ is such a component, then Ge_i is cyclic of order 1, 2, 3, 4 or 6. If Ge_i is trivial, then $e_i = (1/|G|)\sum_{g \in G} g$. Otherwise, G has the cyclic group of order 2 or 3 as a homomorphic image, contradicting b).

In view of the above comments, we only have to notice that if
$$M_2(\mathbb{Q}G) \cong M_2(\mathbb{Q}) \oplus \cdots$$
then
$$\tau \begin{pmatrix} 1 & 1 \\ 0 & 1 \end{pmatrix} = \left(\begin{pmatrix} 1 & 1 \\ 0 & 1 \end{pmatrix}, \cdots \right)$$
and
$$\tau \begin{pmatrix} 1 & 0 \\ 1 & 1 \end{pmatrix} = \left(\begin{pmatrix} 1 & 0 \\ 1 & 1 \end{pmatrix}, \cdots \right).$$
Moreover, we know that the elementary matrices generate $SL_2(\mathbb{Z})$ (see [8, Lemma 19.4]). □

We can also go around the difficulties with the $m = 2$ case by enlarging the base field. Let $F = \mathbb{Q}(\epsilon)$, where ϵ is a primitive n^{th} root of unity, be a splitting field of G. (For example, let the exponent of G divide n.) Suppose further that $n \neq 1, 2, 3, 4, 6$. Let $O = \mathbb{Z}[\epsilon]$.

Let $\mathcal{B}_1^O$ be the Bass cyclic units constructed by using $\epsilon^i x$, $x \in G$ (instead of x), as explained on page 141 of [8]. Let $\mathcal{E}^O$ be the group generated by the elementary matrices $I_m + \epsilon^i g E_{p,q}$, $p \neq q$, $1 \leq i \leq n$, $g \in G$. We have

Theorem 3. *Suppose $F = \mathbb{Q}(\epsilon)$ is a splitting field of G, where ϵ is a primitive n^{th} root of unity. Let $O = \mathbb{Z}[\epsilon]$. Suppose that $n \neq 1,2,3,4,6$, if $m = 2$. Then $\langle \mathcal{B}_1^O I_m, \mathcal{E}^O \rangle$ is of finite index in $GL_m(OG)$, for any $m \geq 2$.*

Remark. Theorem 1 was contained in the Master's thesis of the first named author [5]. One can also deduce Theorem 1 from the results in section 4 of Jespers-Wang [4], which deals more generally with the problem of constructing units in $\mathbb{Z}S$, where S is a semigroup. We have included our proof here because it is simple and direct, and also necessary if one is to prove Theorems 2 and 3.

In addition, we have included the proof of Lemma 2.2 because there is a gap in the proof which is used throughout the literature, and it has been fixed here (cf. [2], [7], or [8, Theorem 19.33]).

References

1. H. Bass, *K-Theory and Stable Algebra*, Publ. Math. Inst. Hautes Études Sci. **22** (1964), 5–60.
2. E. Jespers, G. Leal, *Generators of Large Subgroups of the Unit Group of Integral Group Rings*, Manuscripta Math. **78** (1993), 303–315.
3. E. Jespers, G. Leal, *Degree 1 and 2 Representations of Nilpotent Groups and Applications to Units of Group rings*, Manuscripta Math. **86** (1995), 303–315.
4. E. Jespers, D. Wang, *Units of Integral Semigroup Rings*, J. Algebra **181** (1996), 395–413.
5. G.T. Lee, *Generators of Large Subgroups of General Linear Groups over Group Rings*, M.Sc. Thesis, University of Alberta, 1995.
6. J. Ritter, S.K. Sehgal, *Construction of Units in Integral Group Rings of Finite Nilpotent Groups*, Bull. Amer. Math. Soc. **20** (1989), 165–168.
7. J. Ritter, S.K. Sehgal, *Construction of Units in Integral Group Rings of Finite Nilpotent Groups*, Trans. Amer. Math. Soc. **324** (1991), 603–621.
8. S.K. Sehgal, *Units in Integral Group Rings*, Longman, New York, 1993.
9. L.N. Vaserstein, *The Structure of Classical Arithmetic Groups of Rank Greater Than One*, Math. USSR Sbornik **20** (1973), 465–492.

DEPARTMENT OF MATHEMATICAL SCIENCES, UNIVERSITY OF ALBERTA, EDMONTON, ALBERTA, CANADA T6G 2G1

Dedicated to Professor Bruno J. Müller on his 60th birthday

WEAK RELATIVE INJECTIVE M-SUBGENERATED MODULES

Saroj Malik and N. Vanaja

ABSTRACT. We study weak relative injective and relative tight modules in the category $\sigma[M]$, where M is a right R-module. Many of the known results in the category of right R-modules are extended to $\sigma[M]$ without assuming either M is projective or finitely generated. Conditions are given for a A-tight module to be weakly A-injective in $\sigma[M]$. Modules for which every submodule is weakly injective (tight) in $\sigma[M]$ are characterized. Modules M for which every module in $\sigma[M]$ is weakly injective and for which weakly injective modules are closed under direct sums are studied.

Introduction

Weakly injective modules have been widely studied by Al-Huzli, Jain, López-Permouth, Rizvi, Yousif and several others in [1, 2, 6, 10]. In this paper we study the weakly injective modules in $\sigma[M]$, the full subcategory of mod-R, consisting of all the submodules of M-generated modules, where M is any right R-module. The motivation for this paper came from [6, 10, 11]. Amongst several results proved some are new and some are generalisation from mod-R to $\sigma[M]$.

Let M be an R-module and $A \in \sigma[M]$. Section 2 contains some basic properties about A-tight and weakly A-injective modules in $\sigma[M]$. We consider the conditions under which a A-tight module in $\sigma[M]$ is weakly A-injective in $\sigma[M]$. Suppose every factor module of A which is embeddable in $\widehat{N}$, the injective hull of N in $\sigma[M]$, is finite dimensional. Then N is A-tight if and only if N is weakly A-injective in $\sigma[M]$. We show that a locally finite dimensional module N (i.e. every finitely generated submodule of N has finite Goldie dimension) is tight in $\sigma[M]$ if and only if it is weakly injective in $\sigma[M]$.

In Section 3 we consider SWI (every submodule is weakly injective) and ST (every submodule is tight) modules in $\sigma[M]$. We prove that SWI (ST) modules are closed under taking essential extensions. A finitely generated module N is an SWI module in $\sigma[M]$ if and only if every essential submodule

---

1991 *Mathematics Subject Classification*. Primary 16A52, 16A53; Secondary 16A33.
The first author was supported by NBHM and the Department of Mathematics, University of Mumbai.

Typeset by $\mathcal{A}_{\mathcal{M}}\mathcal{S}$-TEX

of N is weakly injective in $\sigma[M]$, if and only if N is weakly injective and every essential submodule of N is weakly N-injective. If M is an SWI module in $\sigma[M]$, then every non-M-singular module is weakly injective in $\sigma[M]$ and if M is also projective in $\sigma[M]$, then M is non-M-singular.

In Section 4 we generalise the Theorem in [1] (also Theorem 2.6 in Jain and López-Permouth [6]) to $\sigma[M]$, where we do not assume either M is projective or finitely generated. In the final section we show that if every subfactor of a module $N \in \sigma[M]$ is tight in $\sigma[M]$, then every subfactor of M is weakly injective in $\sigma[M]$ and characterise the weakly semi-simple modules (i.e. the modules M for which every module in $\sigma[M]$ is weakly injective in $\sigma[M]$).

1. Preliminaries

For basic notions we refer to [5]. Throughout this paper all rings are associative rings with identity and all modules are right unital modules. R denotes a ring and M denotes an R-module. The category $\sigma[M]$ is the full subcategory of R-modules which are submodules of M-generated modules and for any $N \in \sigma[M]$, $\widehat{N}$ denotes the injective hull of N.

Suppose N and $A \in \sigma[M]$. We say N is **weakly A-injective in** $\sigma[M]$, if for any homomorphism $\phi : A \to \widehat{N}$, there exists a submodule X of $\widehat{N}$ such that $\phi(A) \subseteq X \simeq N$. N is called **weakly injective in** $\sigma[M]$ if N is weakly A-injective for all finitely generated modules A in $\sigma[M]$.

Suppose N is weakly A-injective $\sigma[M]$. If every A/K embeds in $\widehat{N}$, then every A/K embeds in N. A module with this property is called A-**tight** in $\sigma[M]$. If N is A-tight for all finitely generated modules A in $\sigma[M]$ then N is called **tight in** $\sigma[M]$.

If $M = R$, then all these definitions coincide with the usual ones. It is easy to check that if a module N is injective in $\sigma[M]$, then it is weakly injective in $\sigma[M]$ and a weakly injective module in $\sigma[M]$ need not be weakly injective in $\sigma[R]$, i.e. it need not be a weakly injective R-module. For, if M is a simple non-injective R-module, then M is injective in $\sigma[M]$ and hence weakly injective in $\sigma[M]$; but M is not weakly injective in the usual sense.

We call a module $N \in \sigma[M]$ an **SWI (ST)** module if every submodule of N is weakly injective (tight) in $\sigma[M]$. We say a module is an **l.f.d.** (**q.f.d.**) module if every finitely generated submodule (factor module) of it has finite Goldie dimension. We say $\sigma[M]$ is **l.f.d.** if every module in $\sigma[M]$ is an l.f.d. module.

The following abbreviations for various terms are used in this paper: **WI** for weakly injective, **f.g.** for finitely generated, **e.f.g.** for essentially finitely generated, and **f.d.** for finite Goldie dimensional.

2. Weakly A-injective and A-tight modules in $\sigma[M]$.

Let M be an R-module. We prove below some basic results regarding weakly A-injective and A-tight modules in $\sigma[M]$. Many of the results are proved in the case when $M = R$ in [2] and [6].

Proposition 2.1. *Let $N \in \sigma[M]$ and $L, K \in \sigma[N]$. If L is weakly K-injective (K-tight) in $\sigma[M]$ then L is weakly K-injective (K-tight) in $\sigma[N]$. Hence, if L is weakly injective (tight) in $\sigma[M]$ then L is weakly injective (tight) in $\sigma[N]$.*

Proof. We have $L \subseteq \overline{L} \subseteq \widehat{L}$, where $\overline{L}$ and $\widehat{L}$ denote the injective hulls of L in $\sigma[N]$ and in $\sigma[M]$ respectively. Suppose $X \simeq L$ and $X \subseteq \widehat{L}$. The isomorphism $f : L \to X$ can be extended to an enodomorphism of $\widehat{L}$ and since $\overline{L}$ is fully invariant in $\widehat{L}$, we must have $X \subseteq \overline{L}$. Now the lemma is clear. □

The next two results are generalisation of Lemmas 1.8 and 1.9 in [8].

Lemma 2.2. *Let $L, K \in \sigma[M]$. If K is L-injective and L is N-tight in $\sigma[M]$ for all cyclic modules $N \in \sigma[M]$, then K is $\widehat{L}$-injective.*

Proof. If $x \in \widehat{L}$, then xR is embeddable in L and hence K is xR-injective. Thus K is $\widehat{L}$-injective. □

Proposition 2.3. *Suppose a module $L \in \sigma[M]$ is N-tight in $\sigma[M]$, for all cyclic modules N in $\sigma[M]$. If L is self-injective then L is injective in $\sigma[M]$.*

Proof. Take $K = L$ in Lemma 2.2. □

Corollary 2.4. *A self-injective R-module is an injective R-module if and only if it is R-tight.*

The following is a generalisation of Remark 2.4 in [11] and can be easily proved using the definitions.

Proposition 2.5. *Let $0 \to A \to B \to C \to 0$ be an exact sequence of modules in $\sigma[M]$. If $N \in \sigma[M]$ is weakly B-injective (B-tight) in $\sigma[M]$, then N is weakly A-injective (A-tight) and weakly C-injective (C-tight) in $\sigma[M]$.*

The converse of 2.5 is not true. For example the $\mathbb{Z}$-module $\mathbb{Z}/p\mathbb{Z}$ is weakly $\mathbb{Z}/p\mathbb{Z}$-injective but is not weakly $\mathbb{Z}/p^2\mathbb{Z}$-injective.

Any finitely generated module in $\sigma[M]$ is a subfactor of M^n for some $n \in \mathbb{N}$. Hence we have,

Corollary 2.6. *Let $N \in \sigma[M]$. If N is weakly M^n-injective (M^n-tight) in $\sigma[M]$ for all $n \in \mathbb{N}$, then N is WI (tight) in $\sigma[M]$. The converse is true if M is f.g.*

Corollary 2.7. *Suppose $N \in \sigma[M]$ is M^n-cyclic. If N is weakly M^{n+1}-injective (M^{n+1}-tight) then N is WI (tight) in $\sigma[M]$.*

Proof. Let N be weakly M^{n+1}-injective in $\sigma[M]$ and $f : A \to \widehat{N}$ be an epimorphism, where $A = M_1 \oplus M_2 \oplus \cdots \oplus M_k$, each $M_i = M$ and $k \in \mathbb{N}$. Put $X_0 = M$. By hypothesis $X_0 + f(M_1) \subseteq X_1(\simeq N) \subseteq \widehat{N}$. By induction we can choose X_i, for $1 \leq i \leq k$, such that $X_{i-1} + f(M_i) \subseteq X_i(\simeq N) \subseteq \sigma[N]$. Then $f(A) \subseteq X_k \simeq N \subseteq \widehat{N}$. So N is weakly M^k-injective in $\sigma[M]$ for all $k \in \mathbb{N}$. By 2.6 N is WI in $\sigma[M]$. Similarly the Corollary can be proved when N is M^{n+1}-tight in $\sigma[M]$. □

Corollary 2.8. *[Lemma 1.12, 8] A cyclic R-module is WI if and only if it is weakly R^2-injective.*

Proposition 2.9. *Let $K \in \sigma[M]$ and L be essential in K. Then*
(1) *if L is weakly A-injective (A-tight) in $\sigma[M]$, then so is K;*
(2) *if L is WI (tight) in $\sigma[M]$, then so is K.*

Proof. Straight forward. □

We prove below that the converse of Proposition 2.9 is true when L is weakly K-injective (K-tight) in $\sigma[M]$.

Proposition 2.10. *Let $L, K \in \sigma[M]$ and $\widehat{L} = \widehat{K}$. Then*
(1) *if $\sigma[M]$, L is weakly K-injective (K-tight) in $\sigma[M]$ and K is weakly A-injective (A-tight) in $\sigma[M]$, then L is weakly A-injective (A-tight) in $\sigma[M]$;*
(2) *if L is weakly K-injective (K-tight) in $\sigma[M]$ and K is WI (tight) in $\sigma[M]$, then L is WI (tight) in $\sigma[M]$.*

Proof. If L is K-tight and $\widehat{L} = \widehat{K}$, then K can be embedded in L. Hence there exists a submodule X of $\widehat{K}$ containing K and isomorphic to L. The proposition follows from 2.9. □

Next we get some sufficient conditions for an A-tight module N in $\sigma[M]$ to be weakly A-injective in $\sigma[M]$.

Lemma 2.11. *Let a module $N \in \sigma[M]$. Suppose L is a f.d. submodule of $\widehat{N}$. If L is embeddable in N, then there exists a submodule K of $\widehat{N}$ such that $L \subseteq K \simeq N$.*

Proof. Let $f : L \to N \subseteq \widehat{N}$ be a monomorphism. The map $f^{-1} : f(L) \to L$ can be extended to an isomorphism $g : \widehat{f(L)} \to \widehat{L}$. We have $\widehat{N} = \widehat{f(L)} \oplus T = \widehat{L} \oplus T'$. As $\widehat{L}$ is self-injective and f.d., $T \simeq T'$ by remark after Theorem 1.29 in [**12**]. Thus the isomorphism g can be extended to an isomorphism $h : \widehat{N} \to \widehat{N}$. Let $K = h(N)$. Then $N \simeq K \supseteq hf(L) = L$. □

Proposition 2.12. *Let a module $N \in \sigma[M]$ be A-tight in $\sigma[M]$. If either A is a q.f.d. module or N is f.d., then N is weakly A-injective in $\sigma[M]$.*

Theorem 2.13. *Let $N \in \sigma[M]$ be an l.f.d. module and A be a f.g. module in $\sigma[M]$. N is A-tight in $\sigma[M]$ if and only if it is weakly A-injective in $\sigma[M]$.*

Proof. Let N be A-tight in $\sigma[M]$. Suppose $\phi : A \to L \subseteq \widehat{N}$ be an epimorphism. As N is A-tight, L is embeddable in N. Since L is f.d. there exists a submodule K of $\widehat{N}$ such that $L \subseteq K \simeq N$ (2.11). Consequently N is weakly A-injective in $\sigma[M]$. The other part of the Theorem is obvious. □

Corollary 2.14. *An l.f.d. tight module in $\sigma[M]$ is WI in $\sigma[M]$. In particular, any f.d. (uniform) tight module in $\sigma[M]$ is WI in $\sigma[M]$.*

Corollary 2.15. *[Proposition 2.2, 7] Let N be an R-module such that the injective hull $E(N)$ of N is a sum of indecomposables. Then N is tight if and only if it is WI.*

If M is a locally noetherian module, then every injective module in $\sigma[M]$ is a direct sum of indecomposable modules [27.4, **14**]. Hence we get,

Corollary 2.16. *Let M be a locally noetherian module. A module $N \in \sigma[M]$ is tight in $\sigma[M]$ if and only if it is WI in $\sigma[M]$.*

Lemma 2.17. *Let N be an injective module in $\sigma[M]$. If every cyclic submodule is f.d., then so is every f.g. submodule of N.*

Proof. Let $C = A+B$, where A and B are cyclic submodules of N. Suppose $\widehat{C} = \widehat{A} \oplus K$ and $p: \widehat{C} \to K$ is the projection map along $\widehat{A}$; then C is essential in $\widehat{A} \oplus p(B)$ and $p(B)$ is cyclic. $\widehat{A} \oplus p(B)$ is f.d. and hence C is f.d. By finite induction we get the Lemma. □

Corollary 2.18. *Let M be an R-module. If every cyclic module in $\sigma[M]$ is f.d., then every tight module in $\sigma[M]$ is WI in $\sigma[M]$.*

Taking $M = R$ we get the following result proved by López-Permouth [Theorem 3.1, **9**].

Corollary 2.19. *[Theorem 3.1, **9**] Let R be a right q.f.d. ring. Then every tight right R-module is WI.*

Next we consider direct sum of weakly A-injective modules in $\sigma[M]$.

Proposition 2.20. *Let N_i, $i \in I$, be a family of weakly A-injective (A-tight) modules in $\sigma[M]$. If $\widehat{\oplus_{i \in I} N_i} = \oplus_{i \in I} \widehat{N_i}$, then $\oplus_{i \in I} N_i$ is weakly A-injective (A-tight) in $\sigma[M]$. In particular any finite direct sum of weakly A-injective (A-tight) modules is weakly A-injective (A-tight).*

Proof. Let each N_i be a weakly A-injective module in $\sigma[M]$ and X be a factor module of A contained in $L := \widehat{\oplus_{i \in I} N_i} = \oplus_{i \in I} \widehat{N_i}$. Suppose $f_i: L \to \widehat{N_i}$ be the projection map, for all $i \in I$. For each i, there exists $X_i \simeq N_i$ such that $f_i(X) \subseteq X_i \subseteq \widehat{N_i}$. Now $X \subseteq \oplus_{i \in I} f_i(X) \subseteq \oplus_{i \in I} X_i \simeq \oplus_{i \in I} N_i$. Hence $\oplus_{i \in I} N_i$ is weakly A-injective. The proof is similar for the case of tight modules. □

3. ST AND SWI MODULES

We first study ST modules. We recall that a module N in $\sigma[M]$ is called **compressible** if it embeds in each of its essential submodules. We give some connections between tightness and compressibility of modules and prove that every submodule of a module $N \in \sigma[M]$ is A-tight in $\sigma[M]$ if and only if every submodule of $\widehat{N}$ is A-tight in $\sigma[M]$.

Lemma 3.1. *Suppose M is an R-module and $N \in \sigma[M]$. If every essential submodule of N is N-tight in $\sigma[M]$. Then N is compressible.*

Proof. Let L be essential in N. Then $\widehat{L} = \widehat{N}$ and L is N-tight. Hence N is embeddable in L and so N is compressible. □

Theorem 3.2. *Let M be an R-module and $N \in \sigma[M]$. Then the following are equivalent:*
  (a) *every submodule of N is A-tight;*
  (b) *every quotient A/B of A which embeds in $\widehat{N}$ is compressible;*
  (c) *every submodule of $\widehat{N}$ is A-tight.*

Proof. $(a) \Rightarrow (b)$ Suppose the quotient A/B of A embeds in $\widehat{N}$. Since N is A-tight, A/B is embeddable in N. Therefore every submodule of A/B is A-tight and hence is A/B-tight (2.5). By Lemma 3.1, A/B is compressible.

$(b) \Rightarrow (c)$ Let L be a submodule of $\widehat{N}$ and A/B be a quotient of A which embeds in $\widehat{L}$. As $\widehat{L} \subseteq \widehat{N}$, A/B is compressible. $L \cap A/B$ is essential in A/B. Therefore A/B is embeddable in $L \cap A/B$ and hence in L. Thus L is A-tight.

$(c) \Rightarrow (a)$ is obvious. □

Corollary 3.3. *Every module in $\sigma[M]$ is A-tight in $\sigma[M]$ if and only if every quotient A/B of A is compressible.*

By taking $M = R$ in 3.2 and 3.3 we get Lemma 3.1 and Theorem 3.3 in López-Permouth and Rizvi [11].

We next give conditions for a module N in $\sigma[M]$ to be ST in $\sigma[M]$ which may be compared with [Proposition 3.7, 10]. We show that any module N in $\sigma[M]$ is ST in $\sigma[M]$ if and only if $\widehat{N}$ is so.

Theorem 3.4. *Let N be a module in $\sigma[M]$. Then the following are equivalent:*
  (a) *N is ST in $\sigma[M]$;*
  (b) *every f.g. submodule of $\widehat{N}$ is ST in $\sigma[M]$.*
  (c) *every f.g. submodule of $\widehat{N}$ is compressible.*
  (d) *$\widehat{N}$ is ST in $\sigma[M]$.*

Proof. $(a) \Rightarrow (b)$ Any f.g. submodule of $\widehat{N}$ is embeddable in N, as N is tight in $\sigma[M]$.

$(b) \Rightarrow (c)$ follows from 3.1.

$(c) \Rightarrow (d)$ follows from 3.2 and $(d) \Rightarrow (a)$ is obvious. □

In the previous section we saw that an l.f.d. module is tight in $\sigma[M]$ if and only if it is WI in $\sigma[M]$. We show below that an SWI module is indeed l.f.d.

Proposition 3.5. *Let N be an e.f.g. module in $\sigma[M]$ such that every essential submodule of N is weakly N-injective in $\sigma[M]$. Then N is f.d.*

Proof. Let L be a f.g. essential submodule of N. Then any essential submodule of L is weakly N-injective in $\sigma[M]$ and hence is weakly L-injective in $\sigma[M]$ (2.5).

Let $A = \oplus_{i \in I} L_i$ be essential in L. Then A is weakly L-injective and $\widehat{A} = \widehat{L}$. Hence $L \subseteq X(\simeq A) \subseteq \widehat{A} = \widehat{L}$. As L is f.g. and essential in X, we get that $|I|$ is finite. Thus L and so N is f.d. □

If a module N is generated by n elements, any weakly R^n-injective R-module is weakly N-injective. Hence taking $M = R$ in 3.5 we get Lemma 3.2 in [**10**].

Corollary 3.6. *[Lemma 3.2, 10] Let N be a f.g. R-module generated by n elements. If every essential submodule of N is weakly R^n-injective, then N has finite Goldie dimension.*

Corollary 3.7. *[Lemma 3.2, 6] Let R be a ring such that each right ideal is weakly R-injective. Then R has finite uniform (Goldie) dimension.*

Corollary 3.8. *Let N be a f.g. module in $\sigma[M]$. If every essential submodule in N is WI in $\sigma[M]$, then N is f.d.*

By 3.8 any SWI module in $\sigma[M]$ is l.f.d. We prove that an l.f.d. module $N \in \sigma[M]$ is an SWI module if and only if every essential submodule of N is WI in $\sigma[M]$ and hence if a module $N \in M$ satisfies the hypothesis of 3.8, then it is an SWI module in $\sigma[M]$. We consider the following condition $(*)$ for a module N, which is satisfied by l.f.d. modules.

$(*)$ **Every non-zero (cyclic) submodule of N contains a uniform submodule.**

We show that an ST module in $\sigma[M]$ satisfies $(*)$ if and only if it is an SWI module in $\sigma[M]$.

Proposition 3.9. *Let $N \in \sigma[M]$ satisfy $(*)$.*
(1) If every essential submodule of N is L-tight in $\sigma[M]$, then every uniform and hence every f.d. submodule of N is weakly L-injective in $\sigma[M]$.
(2) If N is f.g. and every essential submodule of N is tight, then N is f.d. and SWI in $\sigma[M]$.

Proof. (1) We first show that if $\oplus_{i \in I} U_i$ essential in N, where each U_i is a uniform module, then each U_i is weakly L-injective in $\sigma[M]$. Fix an $i \in I$. Let $J = \{j \in I | \widehat{U_j} \simeq \widehat{U_i}\}$. For each $j \in J$ we can choose a non-zero submodule V_j of U_j such that V_j can be embedded in U_i. Consider $B = \oplus_{k \in I} V_k$, where $V_k = U_k$, for $k \in I \setminus J$. Then B is essential in N and hence is L-tight. Let $f : L \to \widehat{U_i}$ be a non-zero homomorphism. There exists an embedding $g : f(L) \to B$. Since $f(L)$ is uniform $gf(L)$ is isomorphic to a submodule V_k [7.5, 13], for some $k \in I$. By the choice of J, it is easy to see that this $k \in J$. Hence $f(L)$ is embeddable in U_i and hence U_i is L-tight.

Let U be a uniform submodule of N. By hypothesis there exists $A = \oplus_{i \in I} U_i$ essential in N, where each U_i is a uniform module. Now $A \cap U$ is a uniform submodule of A and hence is isomorphic to a submodule V_{i_0} of U_{i_0} for some $i_0 \in I$ [7.5,13]. Then $\oplus_{i \in I_0} U_i \oplus V_{i_0}$, where $I_0 = I \setminus \{i_0\}$, is essential in N and hence V_{i_0} is weakly L-injective. By 2.9 U is weakly L-injective.

(2) In view of (1) it is enough to prove that N is f.d. There exists $A := \oplus_{i \in I} U_i$ essential in N, where each U_i is uniform and cyclic. Now N is compressible by 3.1 and hence A contains a copy of N. Since N is f.g., N can be embedded in a finite direct sum of the U_i's and hence N is f.d. □

Proposition 3.10. *Let N be a module in $\sigma[M]$ such that $\widehat{N}$ is a direct sum of indecomposable modules.*
(1) If every cyclic uniform submodule of N is A-tight in $\sigma[M]$, then every submodule of $\widehat{N}$ is weakly A-injective in $\sigma[M]$.
(2) If every essential submodule of N is A-tight in $\sigma[M]$, then every submodule of $\widehat{N}$ is weakly A-injective in $\sigma[M]$.

Proof. (1) Suppose $L \subseteq \widehat{N}$. Then there exists $K = \oplus_{i \in I} U_i$ essential in L, where each U_i is a cyclic uniform submodule of N. Since $\widehat{N}$ is a direct sum of indecomposable modules, any local summand of $\widehat{N}$ is a summand of $\widehat{N}$ by Mohamed and Müller [Theorem 2.25 and Proposition 2.24,12]. Thus $\widehat{K} = \oplus_{i \in I} \widehat{U_i}$. Each U_i is A-tight and hence weakly A-injective in $\sigma[M]$ (2.12). Now K is weakly A-injective in $\sigma[M]$ (2.20) and hence L is weakly A-injective in $\sigma[M]$ (2.9).

(2) The module N satisfies (∗). Every uniform submodule of N is A-injective $\sigma[M]$ (3.9). By (1) every submodule of $\widehat{N}$ is weakly A-injective in $\sigma[M]$. □

Corollary 3.11. *Let M be a locally noetherian module and $N \in \sigma[M]$. If every essential (cyclic uniform) submodule of N is A-tight (tight) in $\sigma[M]$, then every submodule of $\widehat{N}$ is weakly A-injective (WI) in $\sigma[M]$.*

Theorem 3.12. *Let N be a module in $\sigma[M]$. Then the following are equivalent:*
  (a) *N is an SWI module in $\sigma[M]$;*
  (b) *every f.g. submodule of $\widehat{N}$ is an SWI module in $\sigma[M]$;*
  (c) *every f.g. (cyclic) submodule of $\widehat{N}$ is f.d. and every f.g. submodule of $\widehat{N}$ is compressible;*
  (d) *$\widehat{N}$ is an SWI module in $\sigma[M]$;*
  (e) *N is ST in $\sigma[M]$ and satisfies (∗);*
  (f) *N is ST in $\sigma[M]$ and is l.f.d.*
  If further N is f.g., then the above are equivalent to the following;
  (g) *N is WI in $\sigma[M]$ and every essential submodule of N is weakly N-injective;*
  (h) *every essential submodule of N is WI in $\sigma[M]$;*
  (i) *N satisfies (∗) and every essential submodule of N is tight in $\sigma[M]$.*

Proof. (a) ⇒ (b) follows from the fact that any f.g. submodule of $\widehat{N}$ is embeddable in N.

(b) ⇒ (c) follows from 3.1 and 3.8.

(c) ⇒ (d) From 3.4 it follows that $\widehat{N}$ is ST and from 2.14 we get that $\widehat{N}$ is an SWI module.

(d) ⇒ (e), (e) ⇒ (f) and (f) ⇒ (a) follow from 3.8, 3.9(2) and 2.14 respectively.

(a) ⇒ (g) is obvious. (g) ⇒ (h), (h) ⇒ (i) and (i) ⇒ (a) follow from 2.10(2), 3.8 and 3.9(2) respectively. □

Corollary 3.13. *Let N be an e.f.g. module in $\sigma[M]$. N is SWI in $\sigma[M]$ if and only if every essential submodule of N is WI in $\sigma[M]$.*

Proof. Let T be a f.g. essential submodule of N. Suppose every essential submodule of N is WI in $\sigma[M]$. Then every essential submodule of T is WI in $\sigma[M]$ and hence by 3.12 T is an SWI module in $\sigma[M]$. Again by 3.6 SWI modules are closed under essential extensions and hence N is an SWI module. The converse is trivial. □

By taking $M = R$ in the Theorem 3.12 we get the equivalence of conditions (3), (4) and (5) of [Theorem 3.9, **10**].

Corollary 3.14. *[Theorem 3.9, **10**] The following conditions are equivalent for a ring R:*
   (a) *every right ideal of R is WI;*
   (b) *every essential right ideal of R is WI;*
   (c) *every essential right ideal of R is weakly R-injective and R_R is WI.*

Let M and N be R-modules. N is called **singular in** $\sigma[M]$ or ***M*-singular** if $N \simeq L/K$, for an $L \in \sigma[M]$ and K essential in L. For any $N \in \sigma[M]$, the largest M-singular submodule is denoted by $Z_M(N)$. If $Z_M(N) = 0$, then N is called **non-singular in** $\sigma[M]$ or **non-*M*-singular**. The following Corollary can be easily deduced from 3.12.

Corollary 3.15. *Let M be an R-module. Then*
   (1) *every non-M-singular module is tight in $\sigma[M]$ if and only if every f.g. non-M-singular module is compressible;*
   (2) *every non-M-singular module is WI in $\sigma[M]$ if and only if every f.g. non-M-singular module in $\sigma[M]$ is f.d. and compressible;*
   (3) *every uniform module in $\sigma[M]$ is tight (WI) in $\sigma[M]$ if and only if every f.g. uniform module is compressible.*

Proposition 3.16. *If a module M is an SWI module in $\sigma[M]$, then every non-M-singular module in $\sigma[M]$ is WI in $\sigma[M]$. The converse is true if $Z_M(M) = 0$.*

Proof. By 3.8 M is l.f.d. We claim that any non-M-singular module is l.f.d. and contains a uniform submodule isomorphic to a submodule of M.

Let A be a f.g. non-zero non-M-singular module. It is enough to prove that A is f.d. and A contains a uniform submodule isomorphic to a submodule of M. As $\widehat{A}$ is a homomorphic image of a direct sum of copies of M, we can find a module T such that $A \subseteq T \subseteq \widehat{A}$ and an epimorphism $f : L \to T$, where L is a finite (external) direct sum of cyclic submodules of M. As M is l.f.d. and T is non-M-singular, T is f.d. and hence A is f.d. Since T is non-M-singular and f is an epimorphism, $f \neq 0$ on at least one uniform submodule U of L and $f|_U$ is an isomorphism. As A is essential in T, A contains a uniform submodule isomorphic to a submodule of M.

By 3.10 it is enough to show that every cyclic uniform non-M-singular module U is WI in $\sigma[M]$. This follows from 2.9 as U contains a uniform submodule isomorphic to a submodule of M and M is an SWI module. □

Theorem 3.17. *Let M be a f.g. R-module such that $Z_M(M) = 0$ and M is WI in $\sigma[M]$. Let $S = \text{End}_R(M)$. The following are equivalent:*

(a) *M is an SWI module in $\sigma[M]$;*

(b) *M is f.d. compressible and $\text{Hom}_R(M, N) \neq 0$, for all non-zero $N \subseteq M$;*

(c) *M is f.d., S is semiprime and $\text{Hom}_R(M, N) \neq 0$, for all non-zero $N \subseteq M$;*

(d) *$\text{End}_R(\widehat{M})$ is right semisimple and is the classical right quotient of S, and $\text{Hom}_R(M, N) \neq 0$, for all non-zero $N \subseteq M$.*

Proof. $(a) \Rightarrow (b)$ By 3.1 and 3.8 M is l.f.d. and compressible. Let $0 \neq L \subseteq M$. Let K be a complement of L in M. Then $L \oplus K$ is essential in M and hence is weakly M-injective. Therefore $L \oplus K$ contains an essential submodule M' isomorphic to M. If $p : L \oplus K \to L$ is the projection along K, then its restriction to M' is non-zero as L is non-zero non-M-singular.

$(b) \Rightarrow (a)$ It is enough to show that every essential submodule of M is tight in $\sigma[M]$ (3.10). Suppose N is an essential submodule of M and X is a f.g. submodule of $\widehat{N}$. Then $\widehat{M} = \widehat{N}$ and M is WI implies that X can be embedded in M. Since M is compressible it contains a submodule isomorphic to M and thus N contains a submodule isomorphic to X. Thus N is tight in $\sigma[M]$.

It may be noted that polyform modules M are precisely the modules M for which $Z_M(M) = 0$. From [5.19, 5] we get the equivalence of (b), (c) and (d). □

Next result shows that if M is a projective SWI module in $\sigma[M]$ then M is non-M-singular. In fact it is enough to assume that the M-singular submodule of N is tight in $\sigma[M]$.

Theorem 3.18. *Let N be a projective module in $\sigma[M]$. If $Z_M(N)$ is tight in $\sigma[M]$, then $Z_M(N) = 0$.*

Proof. Let $A := Z_M(N)$ and B be a complement of A in N. Then $\widehat{N} = \widehat{A} \oplus \widehat{B}$. As A is tight in $\sigma[M]$, it is easy to see that $\widehat{A}$ is M-singular. Let $f : N \to \widehat{A}$ be the restriction of the projection map $p : \widehat{N} \to \widehat{A}$ along $\widehat{B}$. By [4.5(2), 5] $\text{Ke} f$ is essential in N. Since $\widehat{B}$ is non-M-singular, N is non-M-singular and $A = 0$. □

Theorem 3.19. *Let N be a projective module in $\sigma[M]$. If every essential submodule of N is N-tight in $\sigma[M]$, then $Z_M(N) = 0$.*

Proof. Let $A := Z_M(N)$ and B be a complement of A in N. Then $A \oplus B$ is essential in N and hence is tight in $\sigma[M]$. So N is embeddable in $A \oplus B$. Consider $f : N \to A$, induced by the projection map of $A \oplus B$ onto A along B. Then $\operatorname{Ke} f$ is essential in N [4.5(2), 5]. Since $\operatorname{Ke} f$ is non-M-singular N is non-M-singular and $A = 0$. □

Corollary 3.20. *[Lemma 3.1, 10] An arbitrary ring R is right non-singular if either the singular right ideal of R or every essential submodule of R is weakly R-injective.*

A ring R is semiprime Goldie if and only if R is right non-singular, R_R is f.d. and every f.g. non-singular R-module is compressible. The following is a generalisation of [Theorem 3.9, 10] and the referee has pointed out that this result has also been proved by Y.Zhou in a forthcoming paper [Corollary 3.6, 15].

Proposition 3.21. *Let M be a projective module in $\sigma[M]$. Then the following are equivalent:*
  (a) *M is SWI in $\sigma[M]$;.*
  (b) *$Z_M(M) = 0$ and every non-M-singular module in $\sigma[M]$ is WI in $\sigma[M]$;*
  (c) *M is l.f.d., $Z_M(M) = 0$ and every f.g. submodule of $\widehat{M}$ is compressible;*
  (d) *$\widehat{M}$ is SWI in $\sigma[M]$.*

If M is also f.g., then $S = \operatorname{End}_R(M)$ is semiprime and $\operatorname{End}_R(\widehat{M})$ is the classical right quotient ring of S.

Proof. $(a) \Rightarrow (b)$ follows from 3.18 and 3.16.
  $(b) \Rightarrow (c)$ follows from 3.4 and 3.8.
  $(c) \Rightarrow (d)$ By 3.4 $\widehat{M}$ is an ST module. Since M is l.f.d., M is an SWI module (2.14).
  $(d) \Rightarrow (a)$ is obvious.
  The rest of the proof of the Theorem follows from 3.17. □

We end this section by proving that for any R-module M, every SWI module in $\sigma[M]$ is semiprime. We recall that a submodule N of M is called a **prime submodule** of M, if whenever $AI \subseteq N$, where $A \subseteq M$ and I is a right ideal of R, either $A \subseteq N$ or $MI = 0$. A module M is called

prime if 0 is a prime submodule of M. A submodule N of M is called a **semiprime submodule of** M if whenever $mI^2 \subseteq N$, where $m \in M$ and I is a right ideal of R, $mI \subseteq N$. A module M is called **semiprime** if the trivial submodule 0 is semiprime in M. Any prime submodule of a module M is a semiprime submodule and an irreducible semiprime module is prime.

Lemma 3.16. *Let N be a module satisfying (*) such that every cyclic submodule is compressible. Then N is semiprime.*

Proof. We can find $L = \oplus_{i \in I} U_i$ essential in N, where each U_i is cyclic uniform and compressible. It is easy to check that a uniform compressible module is prime. Since direct sum of semiprime modules is semiprime L is semiprime.

In order to show that N is semiprime it is enough to show that every cyclic submodule of N is semiprime. Let A be a cyclic submodule of N. Then $A \cap L$ is essential in A and since A is compressible, $A \cap L$ contains a copy of A and hence L contains a copy of A. As a submodule of a semiprime module is a semiprime module, A is semiprime. □

Using 3.8, 3.1 and 3.16 we get the following.

Corollary 3.17. *Let N be an SWI module in $\sigma[M]$. Then N is semiprime and l.f.d.*

4. Direct sum of weak injective modules in $\sigma[M]$

In this section we prove that for any module M, $\sigma[M]$ is l.f.d. if and only if direct sums of arbitrary collections of weakly injective (tight) modules in $\sigma[M]$ are weakly injective (tight) in $\sigma[M]$. It has been proved in [1] that every cyclic R-module is f.d. if and only if every direct sum of weakly (tight) injective right R-modules is weakly injective (tight). Our result contains this theorem as a special case (take $M = R$). We do not make any finiteness or projectivity assumption on the module M. We recall that for a module M, every factor module is f.d. if and only if every factor module of M has finite dimensional socle [4].

It has been pointed out by the referee that the equivalence of (a), (b), and (c) has also been proved by Brodski, Saleh, Thuyet and Wisbauer in [Proposition 10, **3**].

Theorem 4.1. Let M be an R-module. The following are equivalent:
- (a) $\sigma[M]$ is l.f.d.;
- (b) every direct sum of injectives in $\sigma[M]$ is WI in $\sigma[M]$;
- (c) every direct sum of weakly injectives in $\sigma[M]$ is WI in $\sigma[M]$;
- (d) every direct sum of tights in $\sigma[M]$ is tight in $\sigma[M]$;
- (e) every direct sum of weakly injectives in $\sigma[M]$ is A-tight for each cyclic module $A \in \sigma[M]$;
- (f) every direct sum of indecomposable injectives in $\sigma[M]$ is A-tight for each cyclic module $A \in \sigma[M]$;
- (g) every cyclic module in $\sigma[M]$ is finite dimensional;

Proof. (a) $\Rightarrow$ (b) Let $M_i \in \sigma[M]$, $i \in I$, be injective modules in $\sigma[M]$. We define $L := \oplus_{i \in I} M_i$. Let N be any finitely generated submodule of $\widehat{L}$. By hypothesis N is finite dimensional; hence it contains an essential submodule say $A = U_1 \oplus \cdots \oplus U_k$, where each U_i is uniform. Let $0 \neq x_i \in (L \cap U_i)$, for $1 \leq i \leq k$. Put $T = \oplus_{i=1}^k x_i R$. There exists a finite subset J of I such that $T \subseteq \oplus_{j \in J} M_j$. Also T is essential in N. Since $\oplus_{j \in J} M_j$ is injective in $\sigma[M]$, it contains an injective hull $\widehat{T}$ of T in $\sigma[M]$. Then $L = \widehat{T} \oplus F$, for some $F \in \sigma[M]$. As T is essential in N, $\widehat{T} \simeq \widehat{N}$ and $\widehat{N} \cap F = 0$. Then $N \subseteq (\widehat{N} \oplus F) \simeq L$. Thus L is WI in $\sigma[M]$.

(b) $\Rightarrow$ (c) Let $M_i \in \sigma[M]$, $i \in I$, be such that each M_i is WI in $\sigma[M]$. Let $L := \oplus_{i \in I} M_i$ and N be any f.g. submodule of $\widehat{L}$. The module $T := \oplus_{i \in I} \widehat{M_i}$ is WI in $\sigma[M]$ and $\widehat{T} = \widehat{L}$. Hence there exists a module X such that $N \subseteq X \simeq T \subseteq \widehat{L}$. As N is finitely generated there exists a finite subset J of I such that $N \subseteq \oplus_{i \in J} \widehat{M_i}$. Now $\oplus_{i \in J} M_i$ is WI in $\sigma[M]$, being a finite direct sum. Hence there exits a module Y such that $N \subseteq Y \simeq \oplus_{i \in J} \widehat{M_i} \subseteq L$. Thus L is WI in $\sigma[M]$.

(c) $\Rightarrow$ (d) can be proved along the same lines as the proof of (b) $\Rightarrow$ (c).

(d) $\Rightarrow$ (e) and (e) $\Rightarrow$ (f) are obvious.

(f) $\Rightarrow$ (g) By [4] it is enough to show that every cyclic module N in $\sigma[M]$ has finite dimensional socle. If $\text{Soc}\, N = 0$, we are done. Suppose $\text{Soc}\, N \neq 0$ and T is a complement of $\text{Soc}\, N$ in N. Then $\text{Soc}\, N \oplus T$ is essential in N/T. Hence $\text{Soc}\, N/T$ is essential in N/T and N/T is cyclic; so we may without loss of generality assume $\text{Soc}\, N$ is essential in N. Let $\text{Soc}\, N = \oplus_{i \in I} S_i$ and $L = \oplus_{i \in I} \widehat{S_i}$. Clearly L is N-tight by hypothesis and $\widehat{N} = \widehat{\text{Soc}\, N} = \widehat{L}$. Thus N embeds in $\widehat{L}$ and hence by tightness, N embeds in L. As N is finitely generated there exists a finite subset J of I such that N embeds in $\oplus_{i \in J} \widehat{S_i}$ and hence N is f.d.

(g) $\Rightarrow$ (a) follows from 2.17. $\square$

5. WEAKLY SEMISIMPLE MODULES

Following [6] we call a module M **weakly semisimple** if every module in $\widehat{M}$ is WI in $\sigma[M]$. Weakly semisimple rings have been studied by Jain and

López-Permouth in [6]. Zhou has characterised weakly semisimple modules in terms of compressible and tight modules in [16]. We give below some characterisations of weakly semisimple modules using direct summands and SWI modules.

Lemma 5.1. *Let N be a f.g. module such that every factor module of N is compressible. Then every factor module of N is f.d.*

Proof. By [4] it is enough to show that the socle of every factor module of N is f.d. Let L be a factor module of N such that $Soc\, L \neq 0$. There exists a submodule T of L such that $Soc\, L \oplus T$ is essential in L. Then $Soc\,(L/T)$ is essential in L/T. Since L/T is compressible, L/T is embeddable in $Soc\,(L/T)$. As L/T is f.g., L/T is f.d. It is easy to see that $Soc\, L$ is finitely generated. □

Using $(h) \Rightarrow (a)$ in Theorem 4.1, and 5.1 we get the following.

Corollary 5.2. *Let M be an R-module. If every cyclic module in $\sigma[M]$ is compressible, then $\sigma[M]$ is l.f.d.*

We recall that a **subfactor** of a module N is a factor module of a submodule of N or equivalently a submodule of a factor module of N.

Proposition 5.3. *Let N be a module in $\sigma[M]$. If every subfactor of N is tight in $\sigma[M]$, then every subfactor of N is WI in $\sigma[M]$. Hence if every module in $\sigma[M]$ is tight, then every module in $\sigma[M]$ is WI.*

Proof. It is enough to show that every subfactor of N is l.f.d. (2.14) and this follows from 3.4 and 5.1. □

Theorem 5.4. *Let M be an R-module. Then the following are equivalent:*
(a) *M is weakly semisimple;*
(b) *$\sigma[M]$ is l.f.d. and every (cyclic) uniform module is tight;*
(c) *$\sigma[M]$ is l.f.d. and every f.g. uniform module in $\sigma[M]$ is compressible;*
(d) *M is SWI in $\sigma[M]$ and every M-singular module in $\sigma[M]$ is WI (tight);*
(e) *every M-singular and every non-M-singular module in $\sigma[M]$ is WI (tight) in $\sigma[M]$;*
(f) *direct summand of every tight module in $\sigma[M]$ is tight in $\sigma[M]$;*

(g) direct summand of every WI module in $\sigma[M]$ is WI in $\sigma[M]$.

Proof. $(a) \Rightarrow (b)$ follows from 3.8.

$(b) \Rightarrow (a)$ By 4.1 direct sum of WI modules in $\sigma[M]$ is WI. Any cyclic uniform tight module is WI in $\sigma[M]$ and an essential extension of a WI module is WI in $\sigma[M]$ (3.12). As every module in $\sigma[M]$ contains an essential submodule which is a direct sum of uniform modules we get (a).

The equivalence of (c) and (b) follows from 3.4.

$(a) \Rightarrow (d)$ is obvious and $(d) \Rightarrow (e)$ follows from 3.16.

$(e) \Rightarrow (a)$ Let $N \in \sigma[M]$. Let L be a complement of $Z_M(N)$ in N. Then $Z_M(N) \oplus L$ is essential in N. By (d) both $Z_M(N)$ and L are WI (tight) in $\sigma[M]$. By 2.9 N is WI (tight). Now (a) follows from 5.3.

$(a) \Rightarrow (f)$ and $(a) \Rightarrow (g)$ are obvious.

$(f) \Rightarrow (a)$ Let N be any module in $\sigma[M]$. Consider $L := \widehat{N \oplus N^{\aleph}}$, where $\aleph$ is any infinite cardinal. Since $\widehat{L}$ is isomorphic to a submodule of L, L is tight in $\sigma[M]$ and hence N is tight in $\sigma[M]$. By 5.3 M is weakly semisimple.

$(g) \Rightarrow (a)$ Let K be any semisimple module in $\sigma[M]$ and $L := \widehat{K \oplus K^{\aleph}}$, where $\aleph$ represents an infinite cardinal greater than the cardinality of R and the number of summands in K. Then as in the proof of Proposition 2.1 [9], it can be shown that L is WI in $\sigma[M]$. Hence any semisimple module in $\sigma[M]$ is WI and hence is injective in $\sigma[M]$ (2.3). Hence M is locally noetherian [27.3, **14**]. Therefore any tight module in $\sigma[M]$ is WI in $\sigma[M]$. From the proof of $(f) \Rightarrow (a)$ we get that any module in $\sigma[M]$ is a direct summand of a tight module in $\sigma[M]$. Hence (a) follows. □

REFERENCES

1. A.H. Al-Huzali, S.K. Jain and S.R. López-Permouth, *Rings whose cyclics have finite Goldie dimension*, J. Algebra **153** (1992), 37-40.
2. A.H. Al-Huzali, S.K. Jain and S.R. López-Permouth, *On the weak relative injectivity of rings and modules*, Lecture notes in Math. **1448** (1990), 93-98.
3. G.M.Brodskii, M.Saleh, Le Van Thuyet and R.Wisbauer, *On weak injectivity of direct sum of modules*, to be published.
4. V.P. Camillo, *Modules whose quotients have finite Goldie dimension*, Pacific J. Math., **69** (1977), 337-338.
5. N.V. Dung, D.V. Huynh, P.F. Smith and R. Wisbauer, *"Extending modules"*, Pitman Research Notes in Mathematics series, Longman, Harlow, 1994.
6. S.K. Jain and S.R. López-Permouth, *A survey on theory of weakly injective modules*, Computational Algebra, Lecture notes in Pure and applied Math.**151**, Dekker, New York (1994), 205-232
7. S.K. Jain and S.R. López-Permouth, *Weakly injective modules over hereditary noetherian prime rings*, J. Australian Math. Soc., (series A), **58**(1995), 287-297.

8. S.K. Jain, S.R. López-Permouth and S.Singh, *On a class of QI rings*, Glasgow Math. J. **34** (1992), 75-81.
9. S.R. López-Permouth, *Rings characterized by their weakly injective modules*, Glasgow Math. J. **34** (1992), 349-353.
10. S.R. López-Permouth, S.Tariq Rizvi and M.F. Yousif, *Some characterisations of semiprime Goldie rings*, Glasgow Math. J. **35**(1993), 357-365.
11. S.R. López-Permouth, and S.Tariq Rizvi, *On certain classes of QI-rings*, Methods in Module Theory (Colorado Springs C.O 1991), Lecture notes in pure and applied Math. **140**, Dekker, New York, 1993, 227-235.
12. S.M. Mohamed and B.J. Müller, *"Continuous and discrete modules"*, London Math. Soc. Lecture Notes Series **147**, Cambridge, 1990.
13. N. Vanaja, *All finitely generated M-subgenerated modules are extending*, Comm. Algebra, **24**(1996), 543-578.
14. R. Wisbauer, *"Foundations of module and ring theory,"* Gordon and Breach, Reading, 1991.
15. Y.Zhou, *Weak injectivity and module classes*, to be published.
16. Y.Zhou, *Notes on weakly-semisimple rings*, Bull. Austr. Math. Soc., **53**(1996), 517-525.

DEPARTMENT OF MATHEMATICS, HINDU COLLEGE, NEW DELHI 110-007, INDIA.

DEPARTMENT OF MATHEMATICS, UNIVERSITY OF MUMBAI, MUMBAI 400-098, INDIA.
E-mail address: mathbu.ernet.in!vanaja

DIRECT PRODUCT AND POWER SERIES FORMATIONS OVER 2-PRIMAL RINGS

Greg Marks

ABSTRACT. We show that the direct product of an infinite set of 2-primal rings (or even rings satisfying (PS I)) need not be a 2-primal ring, and we develop some sufficient conditions on the rings for their direct product to be 2-primal. We also show that the ring of formal power series over a 2-primal ring (or even a ring satisfying (PS I)) need not be 2-primal.

1. Introduction

All rings under consideration will be associative, and will contain a unity element unless otherwise noted. Ideals not designated as "right" or "left" will be two-sided ideals.

A ring S is called *2-primal* if its prime (Baer lower nil) radical, $Nil_*(S)$, coincides with the set of nilpotent elements of S. The 2-primal property can be regarded as a generalization of commutativity, since in a commutative ring, the prime radical (which is the intersection of all the prime ideals of the ring) *always* coincides with the set of nilpotent elements of the ring. In general, the prime radical of a ring equals the set of strongly nilpotent elements of the ring (see [9, Ex. 10.17]); in what follows we will make use of this characterization without comment.

Some of the fundamental properties of 2-primal rings are developed in [1], [4], [10], and [11]. The terminology throughout these papers is somewhat diverse. The term "2-primal" was coined by Birkenmeier, Heatherly, and E. K. Lee in [1]. Hirano, in [4], gives the name "N-rings" to what we call 2-primal rings. Sun, in [11], defines as "weakly symmetric" those rings S for which there exists an ideal I inside $Nil_*(S)$ with the property that $abc \in I$ implies $bac \in I$. (If $I = 0$ has this property, then the ring is said to be "symmetric.") Sun shows ([11, p. 188]) that a ring is weakly symmetric if and only if it is what we are calling 2-primal. Shin, in [10], does not give any name to the class of rings satisfying the 2-primal condition, while proving various necessary and various sufficient conditions for a ring to belong to this class.

1991 *Mathematics Subject Classification*. Primary 16N40, 16N60, 16S99; Secondary 16P40, 16P60.

Typeset by $\mathcal{A}_{\mathcal{M}}\mathcal{S}$-TEX

Among the fundamental properties of 2-primal rings are the following. A ring S is 2-primal if and only if every minimal prime ideal $P \subseteq S$ is completely prime ([10, Prop. 1.11]), i.e., the ring S/P is not only prime, but a domain. Any subring (possibly without unity) of a 2-primal ring is 2-primal ([1, Prop. 2.2]). If S is 2-primal, then the ring of upper triangular n by n matrices over S is 2-primal for any $n \in \mathbb{N}$ ([11, Prop. 2.6(2)] or [1, Prop. 2.5(1)]). If the ring R_i is 2-primal for every i in an arbitrary indexing set I, then the ring (possibly without unity) $S = \bigoplus_{i \in I} R_i$ is 2-primal ([1, Prop. 2.2]). If the ring R is 2-primal, and T is an arbitrary set of variables that commute with each other and with every element of R, then the polynomial ring $S = R[T]$ is 2-primal ([1, Prop. 2.6]).

Having proved the previous two facts, Birkenmeier, Heatherly, and E. K. Lee pose the following questions ([1, p. 373]): Is the direct product of an arbitrary set of 2-primal rings a 2-primal ring? Is the ring of formal power series over a 2-primal ring a 2-primal ring? In [2, Example 1.6], an example due to E. P. Armendariz resolves the first of these questions in the negative. In Armendariz's example, the 2-primal rings whose direct product is not 2-primal are all non-isomorphic. Below we see a further counterexample, in which an infinite direct product of copies of a single 2-primal ring is not 2-primal. The same 2-primal ring also resolves the second Birkenmeier-Heatherly-Lee question in the negative.

2. Two counterexamples

Example 1. We follow the constructions of [7, p. 233] and [9, Ex. 10.10B]. Let k be a field of characteristic zero, and put

$$S := k[x_1, x_2, x_3, \ldots] \Big/ \left(\{x_i x_j\}_{i \neq j}, \{x_i^{i+1}\}_{i \in \mathbb{N}} \right).$$

The nilradical,

$$N := Nil_*(S) = \sum_{i=1}^{\infty} x_i S,$$

is obviously not nilpotent. Put

$$R_0 := \begin{bmatrix} N & S \\ N & N \end{bmatrix},$$

and adjoin a unity element: $R := R_0 \oplus \mathbb{Z}$. As proved in [7], $Nil_*(R) = R_0$, which we can easily see as follows. Certainly R_0 is an ideal of R, and

$$R_0^2 = \begin{bmatrix} N & S \\ N & N \end{bmatrix}^2 = \begin{bmatrix} N & N \\ N^2 & N \end{bmatrix}$$
$$\subseteq \begin{bmatrix} N & N \\ N & N \end{bmatrix} = \mathbb{M}_2\left(Nil_*(S)\right) = Nil_*\left(\mathbb{M}_2(S)\right)$$

(see [9, Ex. 10.22]). So any element of R_0^2 is strongly nilpotent in $M_2(S)$; a fortiori, it is strongly nilpotent in the subring R of $M_2(S)$. Then $R_0^2 \subseteq Nil_*(R)$ implies $R_0 \subseteq Nil_*(R)$. Now suppose $\alpha = (\alpha_0, n) \in R$ is nilpotent. Then $n \in \mathbb{Z}$ is nilpotent; hence $\alpha \in R_0$. Thus,

$$\{\text{nilpotent elements of } R\} \subseteq R_0 \subseteq Nil_*(R);$$

hence, R is 2-primal.

Let $A := \prod_{n=1}^{\infty} R$, and let

$$x := \left(\begin{bmatrix} 0 & 1 \\ 0 & 0 \end{bmatrix}\right)_{n=1}^{\infty} \in A.$$

Obviously x is nilpotent of index 2. Define

$$r_1 := \left(\begin{bmatrix} 0 & 0 \\ x_i & 0 \end{bmatrix}\right)_{i=1}^{\infty} \in A, \quad r_2 := xr_1x, \quad r_{n+2} := r_{n+1}r_1r_{n+1} \quad (n \in \mathbb{N}).$$

We calculate

$$r_{n+1} = \left(\begin{bmatrix} 0 & x_i^{2^n-1} \\ 0 & 0 \end{bmatrix}\right)_{i=1}^{\infty} \quad (n \in \mathbb{N}).$$

So $r_2 \in xAx$, $r_{n+2} \in r_{n+1}Ar_{n+1}$ $(n \in \mathbb{N})$, and $r_m \neq 0$ for all m. Thus, x is not strongly nilpotent, and A is not 2-primal.

This example has a further use. Shin, in [10], defines the condition (PS I), which for a ring $\mathfrak{R}$ with unity, is that the ring $\mathfrak{R}/\text{ann}_r(a\mathfrak{R})$ be 2-primal for every $a \in \mathfrak{R}$. He shows ([10, Theorem 2.5(a)]) that if a direct product of a set of rings satisfies (PS I), then so must every ring in the set. Whether the converse holds is left open. (See the remark preceding Lemma 2.7 in [10].) This example shows the converse to be false, as follows.

Since (PS I) is a stronger condition than 2-primal, A does not satisfy (PS I).

Let $\alpha = (\alpha_0, n) \in R$ be nonzero, and suppose $\gamma = (\gamma_0, m) \in \text{ann}_r(\alpha R)$. We claim that $m = 0$. The nontrivial case is when $n = 0$; suppose for a contradiction that $m \neq 0$ in this case. Then

$$\alpha\gamma = (\alpha_0\gamma_0 + m\alpha_0, 0) = 0 \implies \alpha_0 \in \bigcap_{i=1}^{\infty} R_0^i,$$

by induction on i, since $\gamma_0 \in R_0$ and m is invertible in k. But

$$R_0^{2j} = \begin{bmatrix} N^j & N^j \\ N^{j+1} & N^j \end{bmatrix} \quad \text{and} \quad R_0^{2j+1} = \begin{bmatrix} N^{j+1} & N^j \\ N^{j+1} & N^{j+1} \end{bmatrix}$$

for every $j \in \mathbb{N}$. Therefore

$$\bigcap_{j=1}^{\infty} N^j = 0 \implies \bigcap_{i=1}^{\infty} R_0^i = 0 \implies \alpha_0 = 0 \implies \alpha = 0,$$

a contradiction, which proves the claim. Hence, for any nonzero $\alpha \in R$, $\text{ann}_r(\alpha R) \subsetneq R_0$. So if $\gamma + \text{ann}_r(\alpha R) \in R/\text{ann}_r(\alpha R)$ is nilpotent, then some power of γ lies in R_0, hence $\gamma \in R_0$. Then

$$\gamma \in R_0 = Nil_*(R) \implies \gamma + \text{ann}_r(\alpha R) \in Nil_*\left(R/\text{ann}_r(\alpha R)\right),$$

showing $R/\text{ann}_r(\alpha R)$ is 2-primal; hence, R satisfies (PS I).

Example 2. Define R as in Example 1, and let $B := R[[\lambda]]$ be the ring of formal power series over R. Put

$$c_i := \begin{bmatrix} 0 & 0 \\ x_i & 0 \end{bmatrix} \in R, \qquad d_i := \begin{bmatrix} 0 & x_i \\ 0 & 0 \end{bmatrix} \in R \qquad (i \in \mathbb{N});$$

and put

$$F(\lambda) := \sum_{i=1}^{\infty} c_i \lambda^i \in B, \qquad \Phi_1(\lambda) := \sum_{i=1}^{\infty} d_i \lambda^i \in B.$$

Since $d_i d_j = 0$ for all $i, j \in \mathbb{N}$, we see that $\Phi_1(\lambda)$ is nilpotent of index 2. Define

$$\Phi_{n+1}(\lambda) := \Phi_n(\lambda) F(\lambda) \Phi_n(\lambda) \qquad (n \in \mathbb{N}).$$

Noting that $c_i d_j = c_j d_i = 0$ if $i \neq j$, we calculate

$$\Phi_n(\lambda) = \sum_{i=1}^{\infty} \beta_{in} \lambda^{i(2^n-1)}, \qquad \text{where} \qquad \beta_{in} = \begin{bmatrix} 0 & x_i^{2^n-1} \\ 0 & 0 \end{bmatrix} \in R.$$

For any $n \in \mathbb{N}$, $\beta_{in} \neq 0$ for sufficiently large i. Thus, $\Phi_n(\lambda) \neq 0$ for all $n \in \mathbb{N}$, and $\Phi_{n+1}(\lambda) \in \Phi_n(\lambda) B \Phi_n(\lambda)$; so $\Phi_1(\lambda)$ is not strongly nilpotent. Hence B is not 2-primal.

Additional counterexamples to the two Birkenmeier-Heatherly-Lee questions have been produced by Huh, Kim, and Y. Lee in [6].

3. Affirmative results

We have used a somewhat unusual ring R in the above counterexamples; in fact, it was necessary for R to be somewhat unusual. Most of the constructions of 2-primal rings in [1] do not afford such counterexamples, as we will now see. First, let us observe that the question whether $S = \prod_{i \in I} R_i$ is 2-primal, for an arbitrary indexing set I and R_i belonging to some class of 2-primal rings, can be reduced to the case where I is countable: Given an element $\mathbf{r} = (r_i)_{i \in I} \in S$ that is nilpotent but not strongly nilpotent, we have a non-vanishing sequence $\mathbf{a}_0, \mathbf{a}_1, \mathbf{a}_2, \ldots$ in S, where $\mathbf{a}_0 = \mathbf{r}$ and $\mathbf{a}_n = \mathbf{a}_{n-1} \mathbf{s}_n \mathbf{a}_{n-1}$ (some $\mathbf{s}_n \in S$) for every $n \in \mathbb{N}$. If for each $n \in \mathbb{N}$, we pick some component $i_n \in I$ at which $\mathbf{a}_n$ is nonzero, then considering only these designated components of $\mathbf{r}$, the $\mathbf{a}_n$'s, and the $\mathbf{s}_n$'s, we see that $\mathbf{r}' := (r_{i_1}, r_{i_2}, r_{i_3}, \dots) \in \prod_{n=1}^{\infty} R_{i_n}$ is nilpotent but not strongly nilpotent. (Conversely, of course, $\prod_{i \in I} R_i$ 2-primal implies $\prod_{n=1}^{\infty} R_{i_n}$ 2-primal by [1, Prop. 2.2].)

Recall that a ring is called *right duo* (resp. *left duo*) if all of its right (resp. left) ideals are two-sided (cf. [9, Ex. 22.4A, Ex. 22.4B]). As defined by Shin in [10], a ring is said to satisfy (S I) if the right annihilator of any element of the ring is an ideal. The (S I) property is left-right symmetric ([10, Lemma 1.2]). Reduced rings are symmetric ([10, Lemma 1.1]), symmetric rings satisfy (S I) ([10, Prop. 1.4]), and rings satisfying (S I) satisfy (PS I) ([10, Prop. 1.6(a)]). Schematically, we have:

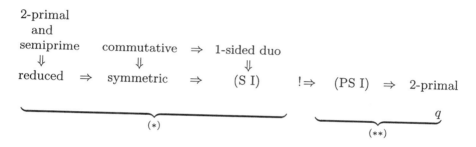

As proved in [10, Theorem 2.3(a)], a direct product of rings satisfying (S I) also satisfies (S I). Thus, if all rings in some set belong to one of the (∗) classes, then their direct product is 2-primal. If all rings in some set belong to one of the (∗∗) classes, then their direct product need not be 2-primal (as shown in Example 1).

Proposition 3. *Suppose that a ring R is 2-primal and $Nil_*(R)$ is a nilpotent ideal. Then $S = \prod_{i \in I} R$ is 2-primal.*

Proof. Put $\mathfrak{N} := Nil_*(R)$, and suppose $\mathfrak{N}^m = 0$. Suppose $\mathbf{r} = (r_i)_{i \in I} \in S$ is nilpotent, and consider an arbitrary sequence $\mathbf{a}_0, \mathbf{a}_1, \mathbf{a}_2, \dots$ in S for which $\mathbf{a}_0 = \mathbf{r}$ and $\mathbf{a}_n \in \mathbf{a}_{n-1} S \mathbf{a}_{n-1}$ for all $n \in \mathbb{N}$. Write $\mathbf{a}_k = (a_{ik})_{i \in I}$; then fix an element $i \in I$. We know r_i is nilpotent, hence $r_i \in \mathfrak{N}$. Then $a_{i0} = r_i$ and $a_{in} \in a_{i(n-1)} R a_{i(n-1)}$ imply that $a_{in} \in \mathfrak{N}^{2^n}$ for all $n \in \mathbb{N}$. Thus, for all $k \geq \log_2 m$, $a_{ik} = 0$. Since $i \in I$ is arbitrary here, we have $\mathbf{a}_k = \mathbf{0} \in S$ for all $k \geq \log_2 m$. Since all such sequences $\mathbf{a}_0, \mathbf{a}_1, \mathbf{a}_2, \dots$ are eventually zero, $\mathbf{r}$ is strongly nilpotent. $\square$

Remark. Proposition 3 was discovered independently—and strengthened—by Huh, Kim, and Y. Lee in [6].

Remark. As we see from its proof, Proposition 3 generalizes: if $\{R_i\}_{i \in I}$ is a set of 2-primal rings whose prime radicals are nilpotent of bounded index, then $\prod_{i \in I} R_i$ is 2-primal. If we omit the words "of bounded index" from this sentence, however, it is no longer true. For instance, Armendariz's example ([2, Example 1.6]), gives R_i's that are *finite* 2-primal rings whose direct product is not 2-primal.

Corollary 4. *Suppose a 2-primal ring R is one-sided Goldie* **or** *satisfies the ascending chain condition on both left and right annihilators. Then $S = \prod_{i \in I} R$ is 2-primal.*

Proof. Fisher, in [3, Cor. 1.7, Cor. 1.8], shows that under the hypotheses of the corollary, every nil ideal of R is nilpotent (recovering earlier theorems of C. Lanski, I. N. Herstein, and L. Small). So $Nil_*(R)$ is nilpotent, and Proposition 3 applies. □

Corollary 5. *Suppose A is a one-sided noetherian 2-primal ring, and R is the ring of upper triangular n by n matrices over A. Then $S = \prod_{i \in I} R$ is 2-primal.*

Proof. Recall that R is a 2-primal ring ([1, Prop. 2.5(1)]). Suppose that A is left noetherian. We show that R is left noetherian by induction on n, the case $n = 1$ being trivial. Define T to be the ring of upper triangular $n-1$ by $n-1$ matrices over A, and define the (A,T)-bimodule $M := A^{n-1}$. Then
$$R \cong \begin{bmatrix} A & M \\ 0 & T \end{bmatrix}.$$
Since A is left noetherian, T is left noetherian (by the inductive hypothesis), and M is a noetherian left A-module, it follows from [8, Theorem 1.22] that R is left noetherian. Then R is left Goldie, and Corollary 4 applies. The right noetherian case is analogous. □

Remark. Birkenmeier, Heatherly, and E. K. Lee pose the further question ([1, p. 373]): Is R 2-primal whenever $Nil_*(R)$ contains every nilpotent element of index two? This question was recently resolved (also in the negative) by Hirano, van Huynh, and Park in [5].

REFERENCES

1. Birkenmeier, G. F.; Heatherly, H. E.; Lee, E. K.; *Completely prime ideals and associated radicals;* in Jain, S. K.; Rizvi, S. T. (eds.); *Ring Theory: Proc. Biennial Ohio State–Denison Conference, May 1992,* World Scientific; Singapore and River Edge, N.J. (1993), 102–129, 373.

2. Birkenmeier, G. F.; Heatherly, H. E.; Lee, E. K.; *Completely prime ideals and radicals in near-rings;* in Fong, Y.; Bell, H. E.; Ke, W.-F.; Mason, G.; Pilz, G. (eds.); *Near-Rings and Near-Fields: Proc. Conference on Near-Rings and Near-Fields, Fredericton, New Brunswick, Canada, July 1993,* Kluwer; Dordrecht, Boston, London (1995), 63–73.
3. Fisher, J. W., *On the nilpotency of nil subrings,* Canad. J. Math **22** (1970), 1211–1216.
4. Hirano, Y., *Some studies on strongly π-regular rings,* Math. J. Okayama Univ. **20** (1978), 141–149.
5. Hirano, Y.; van Huynh, D.; Park, J. K.; *On rings whose prime radical contains all nilpotent elements of index two,* Arch. Math. **66** (1996), 360–365.
6. Huh, C.; Kim, H. K.; Lee, Y.; *Questions on 2-primal rings,* Department of Mathematics, Busan National University, Busan 609-735, South Korea (preprint).
7. Jacobson, N., *Structure of Rings,* second edition, AMS Colloq. Pub. 37; American Mathematical Society; Providence, R.I. (1964).
8. Lam, T. Y., *A First Course in Noncommutative Rings,* Graduate Texts in Mathematics 131, Springer-Verlag; New York, N.Y. (1991).
9. Lam, T. Y., *Exercises in Classical Ring Theory,* Problem Books in Mathematics, Springer-Verlag; New York, N.Y. (1995).
10. Shin, G., *Prime ideals and sheaf representations of a pseudo symmetric ring,* Trans. Amer. Math. Soc. **184** (1973), 43–60.
11. Sun, S.-H., *Noncommutative rings in which every prime ideal is contained in a unique maximal ideal,* J. Pure Appl. Algebra **76** (1991), 179–192.

DEPARTMENT OF MATHEMATICS, UNIVERSITY OF CALIFORNIA, BERKELEY, CALIFORNIA 94720

LOCALIZATION IN NOETHERIAN RINGS

MICHAEL MCCONNELL AND FRANCIS L. SANDOMIERSKI

Dedicated to Herrn Prof. Dr. Friederich Kasch on his 75th birthday.

ABSTRACT. When A is a left Noetherian ring with nilradical N, then there is a unitary subring B of A and Σ a left denominator set in B such that Q, the ring of left fractions of B with respect to Σ is left Artinian. Furthermore, for $P = Q \otimes_B A$, P is a flat right A-module of type FP such that M, a left A-module, is $C(N)$-torsion if and only if $P \otimes_A M = 0$. For the functors $T = P \otimes_A (\cdot) : A\text{-mod} \to Q\text{-mod}$ and $S = Hom_Q(P, \cdot) : Q\text{-mod} \to A\text{-mod}$, the natural transformation $1 \to ST$, $M \mapsto ST(M)$ is the localization of M in A-mod with respect to the torsion theory on A-mod corresponding to the multiplicative set $C(N)$.

When I is a semiprime ideal of a left Noetherian ring, then for each positive integer n, a ring Q_n is constructed as above for $N = I/I^n$ the nilradical of A/I^n and a sequence $Q_{n+1} \to Q_n$ of surjective ring homomorphisms with inverse limit Q a semiperfect ring.

1. PRELIMINARIES

In this presentation, unless otherwise noted, all rings are assumed to be unitary (i.e. have a multiplicative identity 1).

We refer the reader to Goldman [10] and Stenström [21] for torsion theories and localization in R-mod, the category of left R-modules for a ring R. Each hereditary torsion theory on R-mod determines a unique left exact radical t and is uniquely determined by t. Also, each hereditary torsion theory determines (and is uniquely determined by) a Gabriel topology [21] (or idempotent topologizing filter [10]) of left ideals of R.

If F_R is a flat module in mod-R, then $\Phi(F)$, the set of all left ideals L of R for which $FL = F$ is a Gabriel topology for R (see, e.g. Stenström [21] or Morita [15]).

If Σ is a multiplicative set in a ring R, then $\Phi(\Sigma)$, the set of all left ideals L of R for which $Lr^1 \cap \Sigma \neq \emptyset$ for all $r \in R$ is a Gabriel topology, where

---

Date: October 10, 1996.
1991 *Mathematics Subject Classification*. Primary 16A08, 16A33, 16A63.
Portions of this paper are from the first author's Ph.D dissertation at Kent State University.

$Lr^{-1} = \{x \in R : xr \in L\}$. We will denote by t_Σ the left exact radical on R-mod determined by Σ, where $t_\Sigma(M) = \{m \in M : Lm = 0 \text{ for some } L \in \Phi(\Sigma)\}$ for each M in R-mod. An R-module M is Σ-torsion provided $t_\Sigma(M) = M$ or, equivalently, that $M = \{m \in M : \sigma m = 0 \text{ for some } \sigma \in \Sigma\}$.

Definition 1.1. Let Σ_1, Σ_2 be multiplicative sets in a ring R, not necessarily unitary. Then Σ_1 is left equivalent to Σ_2, denoted $\Sigma_1 \lambda \Sigma_2$, provided that whenever M is a left R module, M is Σ_1-torsion if and only if M is Σ_2-torsion.

Lemma 1.2. *Let R be a ring, not necessarily unitary. Then the following statements are true:*

1. *The relation λ is an equivalence relation on the multiplicative sets in R.*
2. *If R is unitary and Σ_1, Σ_2 are multiplicative sets in R, then $\Sigma_1 \lambda \Sigma_2$ if and only if $\Phi(\Sigma_1) = \Phi(\Sigma_2)$*
3. *If Σ is a multiplicative set in R and $x \in R$ such that $\Sigma x \subseteq \Sigma$ ($x\Sigma \subseteq \Sigma$), then $\Sigma \lambda \Sigma x$ ($\Sigma \lambda x \Sigma$).*

Proof. It is easily verified that the first two statements hold.

We will show (3). Since $\Sigma x \subseteq \Sigma$ ($x\Sigma \subseteq \Sigma$), it is clear that Σx ($x\Sigma$) is a multiplicative set in R. Clearly, if M is a left R-module and M is Σx-torsion ($x\Sigma$-torsion), then M is Σ-torsion. Suppose M is Σ-torsion. Then for $m \in M$, $xm \in M$, so there is a $\sigma \in \Sigma$ such that $\sigma x m = 0$ ($\sigma m = 0$ and hence $x\sigma m = 0$). So M is Σx-torsion ($x\Sigma$-torsion) and (3) is shown. ∎

Since we will be dealing with rings which are not necessarily unitary, we will need to consider some ideas and results which are only considered for unitary rings in [6], [11], and [14].

The following definition is easily seen to be equivalent to the definition on page 143 of [11]. We state it for easy reference.

Definition 1.3. Let Σ be a multiplicative set in a ring R, not necessarily unitary, and $R \to S$ a ring homomorphism. Then $R \to S$ is a ring of left fractions of R with respect to Σ provided the following conditions are satisfied.
  (i) The map $R \to S$ maps the elements of Σ to units in S, and
  (ii) the kernel and cokernel of $R \to S$, regarded as left R-modules, are Σ-torsion. Denote the ring of fractions by $S = R_\Sigma$.

The following result will be needed in later considerations.

Theorem 1.4. *Let $S \to T$ be a ring of left fractions of S with respect to a multiplicative set Σ. Suppose R is a subring of S, Π a multiplicative set in R for which $\Pi \subseteq \Sigma$ and $\Pi \lambda \Sigma$. Then the following statements are equivalent.*

1. *The ring homomorphism $R \to T$ given by the restriction of $S \to T$ to R is a ring of left fractions of R with respect to Π.*
2. *The cokernel of the inclusion map $R \to S$, regarded as a left R-module, is Π-torsion.*

Proof. For $x \in S$, denote its image in T by $\bar{x}$. Suppose (1) holds and $s \in S$. By (1), the cokernel of $R \to T$ is Π-torsion, so there is a $\rho \in \Pi$ such that

$\bar{\rho}\bar{s} = \bar{r}$ for some $r \in R$. Thus $\rho s - r$ is in the kernel of $S \to T$, which is Σ torsion and hence Π-torsion, so theres is a $\mu \in \Pi$ such that $\mu(\rho s - r) = 0$, or $\mu\rho s = \mu r$ and (2) follows.

Conversely, suppose (2) holds. Since $\Pi \subseteq \Sigma$, $R \to T$ maps elements of Π to units in T. Since the kernel of $S \to T$ is Σ-torsion, it is Π-torsion and it follows that the kernel of $R \to T$ is Π-torsion. All that remains to be shown is that the cokernel of $R \to T$ is Π-torsion. Since the cokernel of $S \to T$ is Σ-torsion, it is Π-torsion. So if $t \in T$, there is $\rho \in \Pi$ such that $\bar{\rho}t = \bar{s}$ for some $s \in S$. By (2) there is a $\mu \in \Pi$ such that $\bar{\mu}\bar{\rho}t = \bar{r}$ for some $r \in R$, and so it follows that the cokernel of $R \to T$ is Π-torsion when viewed as a left R-module. ∎

The following theorem and its proof in the unitary case can be found in [21]. We will give a proof in the non-unitary case using the theorem for the unitary case.

Theorem 1.5. *Let Π be a multiplicative set in a ring R, not necessarily unitary. The following are equivalent:*
1. *There is a ring homomorphism $R \to T$ which is a ring of left fractions of R with respect to Π.*
2. *The set Π is a left denominator set in R.*

Proof. The implication (1) implies (2) is just as in the unitary case.

Suppose (2) holds. Embed R as a subring of a unitary ring S, where S is a ring isomorphic to $\mathbb{Z} \times R$, $\mathbb{Z}$ the ring of integers with addition in $\mathbb{Z} \times R$ componentwise and multiplication given by $(m, x)(n, y) = (mn, nx+my+xy)$. Clearly R is an ideal of S. We now show that Π is a left denominator set in S.

Choose $\rho_0 \in \Pi$. If $s \in S$ and $\rho \in \Pi$, then, as $\rho_0 s \in R$, there is an $r' \in R$ and $\rho' \in \Pi$ for which $r'\rho = \rho'\rho_0 s$. Thus Π is a left Öre set in S. If $s \in S$, $\rho \in \Pi$ and $s\rho = 0$, then $\rho_0 s \rho = 0$. Since $\rho_0 s \in R$, there is $\rho' \in \Pi$ such that $\rho'\rho_0 s = 0$ and Π is left reversible in S. Thus, by the theorem in the unitary case, there is a ring homomorphism $S \to T$ which is a left ring of fractions of S with respect to Π. Since $\rho_0 S \subseteq R$, the cokernel of the inclusion map $R \to S$ is Π-torsion. Thus, by 1.4, the restriction $R \to T$ of $S \to T$ to R is a ring of left fractions of R with respect to Π. ∎

Lemma 1.6. *Suppose Σ is a left denominator set in a subring R, not necessarily unitary, of a unitary ring A. Then*

$$B = \{b \in A : \sigma b \in R \text{ for some } \sigma \in \Sigma\}$$

is a unitary subring of A containing R and Σ is a left denominator set in B. Furthermore, A/B is Σ-torsion free as a left B-module.

Proof. Clearly, B/R is the Σ-torsion submodule of the left R-moudle A/R, so B is an R-submodule of A. If $b, c \in B$, then there is a $\sigma \in \Sigma$ such that $\sigma b \in R$, and hence $\sigma b c \in B$. Therefore $bc \in B$, and B is a subring of A. Clearly $R \subseteq B$, and $1 \in B$.

If $b \in B$ and $\sigma \in \Sigma$, then there is a $\rho \in \Sigma$ with $\rho b \in R$, so there is $\sigma' \in \Sigma$, $r \in R$ with $\sigma'\rho b = r'\sigma$. Thus Σ is a left Öre set in B. If $b \in B$, $\sigma \in \Sigma$ with $b\sigma = 0$, then $\rho b \in R$ for some $\rho \in \Sigma$, and $\rho b \sigma = 0$. Since Σ is left reversible in R, there is a $\sigma' \in \Sigma$ such that $\sigma'\rho b = 0$. Hence Σ is left reversible in B and the lemma is shown. ∎

Notation: For a ring Q and X in Q-mod, if X has finite length as a Q-module, then its length will be denoted by $\lambda(_Q X)$.

2. Flat Modules of Type FP

If P is a left Q-, right A-bimodule, then as in [15], we denote the functors

$$Hom_Q(P, \cdot) = S : Q\text{-mod} \to A\text{-mod},$$

$$P \otimes_A (\cdot) = T : A\text{-mod} \to Q\text{-mod},$$

and associated natural transformations $\alpha : 1_A \to ST$, $\beta : TS \to 1_Q$. Morita [15] called the module P_A of type FP precisely when $\beta : TS \to 1_Q$ is an equivalance. If, in addition, P_A is flat, then $\alpha : 1_A \to ST$ is the localization with respect to the torsion theory on A-Mod corresponding to $\Phi(P)$.

Definition 2.1. The context where Σ is a left denominator set in a subring B of a unitary ring A such that A/B is Σ-torsionfree as a left B module, $Q = B_\Sigma$ and $P = Q \otimes_B A$ will be denoted by $[\Sigma, B, A, P]$. In addition, $Q \to P$ denotes the map given by $x \mapsto x \otimes 1$.

Proposition 2.2. *Given a context $[\Sigma, B, A, P]$, the following hold:*
1. *For $a \in A$, $1 \otimes a \in Im(Q \to P)$ if and only if $a \in B$. Furthermore, $Pa \subseteq Im(Q \to P)$ if and only if $Aa \subseteq B$.*
2. *If X is in Q-mod and $x \in X$, then $x \in Im(\beta_X)$ if and only if $x = (\sigma^{-1} \otimes 1)g$ for some $\sigma \in \Sigma$ and $g \in S(X)$.*
3. *If $\sigma \in \Sigma$, then the composite map*

$$Q \to P \xrightarrow{\hat{\sigma}} P$$

is a monomorphism, where $\hat{\sigma}$ is given by right multiplication by σ.
4. *The module P_A is flat and $\Phi(P_A) = \Phi(\Sigma)$.*

Proof. Proof of 1: If $1 \otimes a \in Im(Q \to P)$, then $1 \otimes a = \sigma^{-1} b \otimes 1 = \sigma^{-1} \otimes b$ for some $\sigma \in \Sigma$, $b \in B$, and hence $1 \otimes \sigma a = 1 \otimes b$. Therefore $\sigma a - b \in Ker(A \to Q \otimes_B A)$, which is the Σ-torsion submodule of the left B module A. Thus there is a $\tau \in \Sigma$ such that $\tau\sigma a = \tau b \in B$, and so $a \in B$. The converse is clear, and the latter part of (1) is now evident.

Proof of 2: If $x \in Im(\beta_X)$ then $x = \sum p_i g_i$ for some finite set $p_i \in P$ and $g_i \in S(X)$. There is a $\sigma \in \Sigma$ and $a_i \in A$ such that $p_i = \sigma^{-1} \otimes a_i$ for each i. Thus we have

$$x = \sum p_i g_i = \sum (\sigma^{-1} \otimes a_i)g_i = (\sigma^1 \otimes 1)(\sum a_i g_i).$$

So for $g = \sum a_i g_i$, $x = (\sigma^{-1} \otimes 1)g$. The converse of (2) is clear.

Proof of 3: If $x = \tau^{-1}b \in Q$, where $\tau \in \Sigma$, $b \in B$, and $x \in Ker(Q \to P \xrightarrow{\hat{\sigma}} P)$, then $0 = x \otimes \sigma = \tau^{-1} \otimes b\sigma$, and so $1 \otimes b\sigma = 0$. Thus $b\sigma \in Ker(A \to Q \otimes_B A)$, so $b\sigma$ is in the Σ-torsion submodule of the left B-module A and therefore $\mu b\sigma = 0$ for some $\mu \in \Sigma$. As Σ is left reversible in B, $\rho \mu b = 0$ for some $\rho \in \Sigma$. thus it follows that $x = 0$.

Proof of 4: The functor $P \otimes_A (\cdot)$ is naturally equivalent to the composition of the functors $A \otimes_A (\cdot): A\text{-mod} \to B \text{ mod}$ and $Q \otimes_B (\cdot): B\text{-mod} \to Q\text{-mod}$, each of which is exact, so it follows that P is a flat right A-module. For M in A-mod it is clear that $P \otimes_A M = 0$ if and only if $Q \otimes_B M = 0$, and $Q \otimes_B M = 0$ if and only if $_B M$ is Σ-torsion. But $_B M$ is Σ-torsion if and only if $_A M$ is Σ-torsion, so (4) follows. ∎

We now characterize those contexts $[\Sigma, B, A, P]$ for which P is a right A-module of type FP with $Q = End(P_A)$ in accordance with Morita [15].

Theorem 2.3. *Given a context $[\Sigma, B, A, P]$, the following statements are equivalent:*

1. *The right A-module P is of type FP with $Q = End(P_A)$.*
2. *There is a $c \in \Sigma$ such that $Ac \subseteq B$.*

Proof. Suppose (1) holds and $C = End(_Q P)$. By (1), P is a generator in Q-mod. If we denote by $\bar{Q} = Im(Q \to P)$ a Q-submodule of P, by 2.2.2 there is a $g \in S(\bar{Q})$ and $\sigma \in \Sigma$ such that $(\sigma^1 \otimes 1)g = 1 \otimes 1$, or $(1 \otimes 1)g = \sigma \otimes 1$. Since P_A is flat and by (1), the map $P \otimes_A A \to P \otimes_A C$ induced by the ring homomorphism $A \to C$, where $a \mapsto \hat{a}$, viewed in A-mod is an isomorphism, it follows by 2.2.4 that $Coker(A \to C)$ is Σ-torsion in A-mod. Thus there is a $\tau \in \Sigma$ such that $\tau g = \hat{a}$ for some $a \in A$. Therefore

$$1 \otimes a = (1 \otimes 1)\hat{a} = (1 \otimes 1)\tau g = \tau(1 \otimes 1)g = \tau(\sigma \otimes 1) = 1 \otimes \tau\sigma$$

and therefore $1 \otimes (a - \tau\sigma) = 0$, so $a - \tau\sigma$ is an element of the Σ-torsion submodule of $_B A$, hence $\rho a = \rho\tau\sigma$ for some $\rho \in \Sigma$. Denote $c = \rho a = \rho\tau\sigma \in \Sigma$. We now have

$$Pc \subseteq Pa = P\hat{a} \subseteq Pg \subseteq \bar{Q} = Im(Q \to P),$$

so by 2.2.1 $Ac \subseteq B$, and (2) follows.

Conversely, suppose $c \in \Sigma$ with $Ac \subseteq B$. For $x \in Q$, $x \otimes 1 = (xc^{-1} \otimes 1)c \in Pc$, so $\bar{Q} \subseteq Pc$. Clearly $Pc \subseteq \bar{Q}$ by (2). Thus right multiplication by c yields a surjective homomorphism $P \to \bar{Q}$. Since $\bar{Q}$ is isomorphic to Q in Q-mod, it follows that P is a generator in Q-mod.

Let $C = End(_Q P)$ and $g \in C$. Now $(1 \otimes 1)g = \sigma^{-1} \otimes a$ for some $\sigma \in \Sigma$ and $a \in A$, so $(1 \otimes \sigma)g = 1 \otimes a$. Let $\tau = c\sigma$. Then for $x \in A$,

$$(1 \otimes x)\tau g = (1 \otimes x)c\sigma g = xc(1 \otimes \sigma)g = xc(1 \otimes a) = (1 \otimes x)ca,$$

so $\tau g = \hat{ca}$, from which it follows that $Coker(A \to C)$ is Σ-torsion in A-mod. Since P is flat in A-mod, the map $P \otimes_A A \to P \otimes_A C$ is surjective by 2.2.4 and therefore so is the map $P \to P \otimes_A C$ given by $p \mapsto p \otimes 1$. There is also the canonical map $P \otimes_A C \to P$. Clearly the composition $P \to P \otimes_A C \to P$

is the identity map on P, and it follows that $P \otimes_A C \to P$ is an isomorphism. Thus, by Theorem 1.1 in [15], statement (1) follows. ∎

3. LEFT NOETHERIAN RINGS

If I is an ideal of a ring R, then $C(I)$ denotes the set of all elements of R whose image by the canonical homomorphism $R \to R/I$ are regular elements (non-zero-divisors) in R/I.

Goldie [7, 8] characterized those rings R for which $C(0)$ is a left denominator set in R and the left ring of fractions with respect to $C(0)$ is a simple (semisimple) left Artinian ring. In particular, if R is a left Noetherian prime (semiprime) ring then $C(0)$ is a left denominator set and the left ring of fractions with respect to $C(0)$ is a simple (semisimple) left Artinian ring.

Thus when R is a left Noetherian ring and I is a semiprime ideal of R, the image of $C(I)$ in R/I is a left denominator set. We will denote by Q_I the corresponding left ring of fractions. The kernel of the canonical ring homomorphism $R \to Q_I$ is the ideal I and the image of an element $r \in R$ by this ring homomorphism will be denoted by $\bar{r}$.

For a left Noetherian semiprime ring R, Goldie [9] defined the reduced rank of M for finitely generated M in R-mod, denoted $\rho(M)$, as the length, $\lambda(Q \otimes_R M)$, of $Q \otimes_R M$ as a left Q-module, where $Q = Q_0$. Subsequently Chatters, Goldie, Hajarnavis and Lenagan [3] extended this concept to the reduced rank for a finitely generated module M in R-mod, where R is a left Noetherian ring. We will not go into details here, but will list the salient points about the reduced rank function.

If I is a semiprime ideal in a left Noetherian ring R, then for M a finitely generated module in R-mod with $I^n M = 0$ for some positive integer n, the reduced rank associated with I, denoted $\rho_I = \rho$, has the following properties:

1. $\rho(M)$ is a non-negative integer.
2. $\rho(M) = 0$ if and only if M is $C(I)$-torsion.
3. If $0 \to M_1 \to M \to M_2 \to 0$ is exact in R-mod, then

$$\rho(M) = \rho(M_1) + \rho(M_2).$$

Theorem 3.1. *Let A be a left Noetherian ring, N its nilradical, and ρ its associated reduced rank function. If $c \in C(N)$ such that $\rho(Ac)$ is minimal among all $\sigma \in C(N)$, then $\Sigma = C(N)c$ is a left denominator set in Ac.*

Proof. Suppose $d \in C(N)$ and $Ac \to Ac$ is the map given by $x \mapsto xdc$. Denote $K = Ker(Ac \to Ac)$ and $H = Coker(Ac \to Ac)$. Clearly $\rho(Ac) = \rho(Acdc) + \rho(K) = \rho(Acdc) + \rho(H)$. By the minimality of $\rho(Ac)$, $\rho(Ac) = \rho(Acdc)$. Therefore $\rho(K) = \rho(H) = 0$, so K and H are $C(N)$-torsion, hence Σ-torsion. The theorem follows. ∎

Theorem 3.2. *Let A be a left Noetherian ring and N its nilradical. Then there is a context $[\Sigma, B, A, P]$, with $\Sigma \subseteq C(N)$, for which the following hold:*

1. P_A *is a flat module of type FP with $End(P_A) = B_\Sigma = Q$ and $\Phi(P) = \Phi(C(N))$.*

2. The map $P \to Q_N$ given by $\sigma^{-1} \otimes a \mapsto \bar{\sigma}^{-1}\bar{a}$ is a surjective group homomorphism with kernel PN. Furthermore, the composite map $Q \to P \to Q_N$, where $Q \to P$ is given by $x \mapsto x \otimes 1$ for $x \in Q$, is a surjective ring homomorphism with kernel $J(Q)$ the Jacobson radical of Q.

3. If M is a finitely generated module in A-mod, then
$$\rho(M) = \lambda(_Q P \otimes_A M).$$

4. Q is a left Artinian ring and P is a finitely generated Q module.

Proof. We first exhibit the purported context $[\Sigma, B, A, P]$.

Select $c \in C(N)$ so that $\rho(Ac)$ is a least element of $\{\rho(Ad) : d \in C(N)\}$. By 3.1, $\Sigma = C(N)c$ is a left denominator set in Ac. Since $\rho((Ac + S)/S) = \rho(A/S) > 0$ and $(Ac + S)/S$ is a homomorphic image of Ac, $\rho(Ac) > 0$. By 1.6 and 2.1, for $B = \{b \in A : \sigma b \in Ac$ for some $\sigma \in \Sigma\}$, we have a context $[\Sigma, B, A, P]$. The choice of Σ shows that $\Sigma \subseteq C(N)$.

Proof of (1): Since $c \in \Sigma$ and $Ac \subseteq B$, by 2.4 P_A is a flat module of type FP with $Q = B_\Sigma = End(P_A)$. By 2.3, $\Phi(P) = \Phi(\Sigma)$, and (1) follows.

Proof of (2): It is easily verified that there is a group homomorphism $P \to Q_N$ given by $\sigma^{-1} \otimes a \mapsto \bar{\sigma}^{-1}\bar{a}$ for $\sigma \in \Sigma$, and $a \in A$.

For the natural ring homomorphism $B \to Q_N$ each $\sigma \in \Sigma$ is mapped to a unit in Q_N, so there is a unique ring homomorphism $Q \to Q_N$ such that $B \to Q \to Q_N = B \to Q_N$. For $x \in Q$ write $x = \sigma^{-1}b$ for some $\sigma \in \Sigma$, $b \in B$. Then the image of x under $Q \to P \to Q_N$ is $\bar{\sigma}^{-1}\bar{b}$. Thus $Q \to P \to Q_N = Q \to Q_N$.

In order to show that $Q \to Q_N$ is surjective, it suffices to show that $\bar{a} \in Im(Q \to Q_N)$ for each $a \in A$. Clearly the image of $(ac)c^1$ in Q_N is $\bar{a}$. The surjectivity of $Q \to Q_N$ implies the surjectivity of $P \to Q_N$.

Denote $K = Ker(Q \to Q_N)$. Since $Q \to Q_N$ is surjective and Q_N is a semisimple ring, $J(Q) \subseteq K$. On the other hand, if $x \in Q$, then $x = \sigma^{-1}b$ for some $\sigma \in \Sigma$, $b \in B$. Thus $x \in K$ if and only if $\bar{b} = 0$, if and only if $b \in B \cap N$. Thus $K = Q(B \cap N)$. Since $Q(B \cap N)$ is an ideal of Q, $(B \cap N)Q \subseteq Q(B \cap N)$, and $(B \cap N)$ is nilpotent, $K = Q(B \cap N)$ is nilpotent. Hence $K \subseteq J(Q)$.

All that remains to establish (2) is to show that $Ker(P \to Q_N) = PN$. This is clear since $\bar{\sigma}^{-1}\bar{a} = 0$ if and only if $a \in N$.

Proof of (3): Since $Q \to Q_N$ is a surjective ring homomorphism, Q_N may be regarded as a left Q-module and, for any finitely generated $\bar{A}$-module M, where $\bar{A} = A/N$,
$$\rho(M) = \lambda(_Q(Q_N) \otimes_{\bar{A}} M).$$

Now regarding Q_N as a left Q- right A-bimodule, the map $P \to Q_N$ is a Q-A bimodule epimorphism with kernel PN. Thus $P \to Q_N$ induces a Q-A bimodule isomophism $\bar{P} \to Q_N$ where $\bar{P} = P/PN$. As functors from A-mod to Q-mod, $_Q(Q_N) \otimes_A (\cdot)$ and $_Q\bar{P} \otimes_A (\cdot)$ are naturally equivelent and clearly for M a finitely generated module in A/N-mod,
$$\rho(M) = \lambda(_Q(Q_N) \otimes_A M) = \lambda(_Q\bar{P} \otimes_A M).$$

Since $\bar{P} \otimes_A M$ and $P \otimes_A M$ are isomorphic in Q mod,
$$\rho(M) = \lambda(_Q P \otimes_A M).$$

Suppose now that M is a finitely generated module in A-mod and $N^n M = 0$. We assume that for a finitely generated module K in A-mod with $N^m K = 0$ for $m < n$,
$$\rho(K) = \lambda(_Q P \otimes_A K).$$

From the exact sequence $0 \to M_1 \to M \to M_2 \to 0$ in A-mod with $M_1 = NM$ and $M_2 = M/NM$ we have
$$\rho(M) = \rho(M_1) + \rho(M_2) = \lambda(_Q P \otimes_A M_1) + \lambda(_Q P \otimes_A M_2).$$
by induction. Since $0 \to P \otimes_A M_1 \to P \otimes_A M \to P \otimes_A M_2 \to 0$ is exact in Q-mod, we have that
$$\rho(M) = \lambda(_Q P \otimes_A M_1) + \lambda(_Q P \otimes_A M_2) = \lambda(_Q P \otimes_A M).$$

Proof of (4): By (3), $\rho(A) = \lambda(_Q P \otimes A) = \lambda(_Q P)$ is finite and, as P is a generator in Q-mod, it follows that Q is left Artinian and P is a finitely generated Q-module. ∎

In 3.2, the subring B of A and therefore the ring of fractions Q required a choice $c \in C(N)$ such that $\rho(Ac)$ was minimal. We will show that, up to ring isomorphism, Q is independent of the choice of c. First, a definition.

Definition 3.3. Let A be a left Noetherian ring and N its nilradical. If $c \in C(N)$ so that $\rho(Ac)$ is minimal among $\{\rho(Ad) : d \in C(N)\}$, then
$$B^{(c)} = \{b \in A : \sigma b \in Ac \text{ for some } \sigma \in C(N)c\}$$
and $B^{(c)} \to Q^{(c)}$ is the ring of left fractions of $B^{(c)}$ with respect to $C(N)c$.

In the case of definition 3.3, $Ac \to B^{(c)}$ will always denote the inclusion map and $Ac \to Q^{(c)}$ the composite $Ac \to B^{(c)} \to Q^{(c)}$. Since the cokernel of $Ac \to B^{(c)}$ is $C(N)c$-torsion as an Ac-module by 1.4, $Ac \to Q^{(c)}$ is a ring of left fractions of Ac with respect to $C(N)c$.

Lemma 3.4. *Let A be a left Noetherian ring, N its nilradical and $c, d \in C(N)$. Then*

1. *If $\rho(Ac)$ is minimal, then $Adc \to Ac \to Q^{(c)}$ is a ring of left fractions of Adc with respect to $C(N)dc$.*
2. *If $\rho(Ac)$ is minimal, then $dAc \to Ac \to Q^{(c)}$ is a ring of left fractions of dAc with respect to $dC(N)c$.*
3. *If $\rho(Ac), \rho(Ad)$ are minimal and $e \in C(N)d$, then there are unique mutually inverse ring isomorphisms $f_e : Q^{(c)} \to Q^{(d)}$, respectively $g_e : Q^{(d)} \to Q^{(c)}$, such that $f_e(ex) = xe$, respectively $g_e(xe) = ex$ for $x \in Ac$.*

Proof. Proof of (1): By the minimality of $\rho(Ac)$, $\rho(Ac) = \rho(Adc)$, so the cokernel of $Adc \to Ac$ is $C(N)$-torsion as a left A-module. Since $C(N)dc$, $C(N)c$, and $C(N)$ are left equivalent it follows that the cokernel of $Adc \to Ac$ is $C(N)dc$-torsion as an Adc-module, so by 1.4, (1) follows.

Proof of (2): By 1.2, $C(N)c$ and $dC(N)c$ are left equivalent in Ac. Since the cokernel of $dAc \to Ac$ is $dC(N)c$-torsion, as $dc(Ac) \subseteq dAc$, (2) follows from 1.4.

Proof of (3): If $x, y \in Ac$ and $ex = ey$, then $exe = eye$ in $Q^{(d)}$ and therefore $xe = ye$ in $Q^{(d)}$ as $e \in C(N)d$ is a unit in $Q^{(d)}$. Thus we have a map $eAc \to Q^{(d)}$ sending $ex \mapsto xe$ which is clearly additive. For $x, y \in Ac$, $exey$ is mapped to $xeye$ so the map is a ring homomorphism which maps the elements of $eC(N)c$ into units is $Q^{(d)}$, since $C(N)ce \subseteq C(N)d$. By (2), $Q^{(c)}$ is a ring of left fractions of eAc with respect to $eC(N)c$, so there is a unique homomorphism $Q^{(c)} \to Q^{(d)}$ which maps $ex \mapsto xe$ for $x \in Ac$.

Now if $x, y \in Ac$ and $xe = ye$, then $exec = eyec$, so in $Q^{(c)}$, $ex = ey$ since ec is a unit in $Q^{(c)}$. Thus me have a map $Ace \to Q^{(c)}$, given by $xe \mapsto ex$, which is clearly additive and, for $x, y \in Ac$, $xeye$ is mapped to $exey$, so the map is a ring homomorphism which maps $C(N)ce$ into $eC(N)c$ which are units in $Q^{(c)}$. Since, by (1), $Q^{(d)}$ is a ring of left fractions of Ace with respect to $C(N)ce$, there is a unique ring homomorphism $Q^{(d)} \to Q^{(c)}$ which maps $xe \mapsto ex$ for $x \in Ac$.

It is clear that these ring homomorphisms are mutually inverse and (3) follows. ∎

Suppose A is a let Noetherian ring and I a semiprime ideal of A. If n is a positive integer, then we will denote $A_n = A/I^n$, $I_n = I/I^n$, and $A \to A_n$, $A_{n+1} \to A_n$ the canonical maps. Thus I_n is the nilradical of A_n and $C(I_n)$ is the image of $C(I)$ by $A \to A_n$.

By theorem 3.1, whenever $c_n \in C(I)$ such that $\rho(A_n c_n)$ is minimal, then $C(I_n)c_n$ is a left denominator set in $A_n c_n$. We adopt the following convention: if $A_n c_n \to Q_n$ is a ring of left fractions of $A_n c_n$ with respect to $C(I_n)c_n$, then for $x \in Ac_n$, the image of $x + I^n$ in Q_n will be denoted by x and if $x \in C(I)c_n$ the inverse of the image of $x + I^n$ in Q_n will be denoted by x^{-1}. Thus each element of Q_n is expressible in the form $\sigma^{-1} a$ where $a \in Ac_n$, $\sigma \in C(I)c_n$.

Furthermore, if $c \in C(I)$ is chosen so that $\rho(A_n c)$ is minimal and $d \in C(I)$, then $\rho(A_n dc) = \rho(A_n c)$ since $A_n dc \subseteq A_n c$ and $\rho(A_n cd) = \rho(A_n c)$ since $A_n cd$ is a homomorphic image of $A_n c$.

Definition 3.5. If A is a left Noetherian ring and I a semiprime ideal of A, then a sequence, denoted $(c) = \{c_n : n = 1, 2, \dots\} \subseteq C(I)$ is called a left Öre sequence for I provided:
(i) $C(I_n)c_n$ is a left denominator set in $A_n c_n$
(ii) $C(I)c_{n+1} \subseteq C(I)c_n$
for $n = 1, 2, \dots$

Notation: If (c) is a left Öre sequence for a semiprime ideal I of a left Noetherian ring A, then $A_n c_n \to Q_n^{(c)}$ will denote the ring of left fractions of $A_n c_n$ with respect to $C(I_n)c_n$.

Theorem 3.6. *Let I be a semiprime ideal of a left Noetherian ring A. Then the following hold:*

1. There is a left Öre sequence for I.
2. If (c) is a left Öre sequence for I, then there are unique surjective ring homomorphisms $Q_{n+1}^{(c)} \to Q_n^{(c)}$ for $n = 1, 2, \ldots$ giving commuting diagrams:

$$\begin{array}{ccc} A_{n+1}c_{n+1} & \to & A_n c_n \\ \downarrow & & \downarrow \\ Q_{n+1}^{(c)} & \to & Q_n^{(c)} \end{array}$$

where $A_{n+1}c_{n+1} \to A_n c_n$ is the map induced by $A_{n+1} \to A_n$.

3. If (c) and (d) are left Öre sequences for I, then there exist ring isomorphisms $Q_n^{(c)} \to Q_n^{(d)}$ such that the diagrams

$$\begin{array}{ccc} Q_{n+1}^{(c)} & \to & Q_n^{(c)} \\ \downarrow & & \downarrow \\ Q_{n+1}^{(d)} & \to & Q_n^{(d)} \end{array}$$

commute for all $n = 1, 2, \ldots$.

Proof. Proof of (1): Choose $d_n \in C(I)$ so that $\rho(A_n d_n)$ is minimal. Let $c_n = d_n \ldots d_1$. Then $\rho(A_n c_n)$ is minimal and $(c) = \{c_n : n = 1, 2, \ldots\}$ is a left Öre sequence for I.

Proof of (2): The composite $A_{n+1}c_{n+1} \to A_n c_n \to Q_n^{(c)}$ is a ring homomorphism which maps the elements of $C(I_{n+1})c_{n+1}$ to units in $Q_n^{(c)}$, since $C(I)c_{n+1} \subseteq C(I)c_n$, so there is a unique ring homomorphism $Q_{n+1}^{(c)} \to Q_n^{(c)}$ such that the aforementioned diagram commutes.

If $a \in A$, then $ac_n = ac_n c_{n+1} c_{n+1}^{-1}$ viewed in $Q_n^{(c)}$. Since $ac_n c_{n+1}$ and c_{n+1}^{-1} as elements of $Q_n^{(c)}$ are in the image of $Q_{n+1}^{(c)} \to Q_n^{(c)}$, it follows that this ring homomorphism is surjective.

Proof of (3): Denote $e_n = d_1 \cdots d_n \in C(I)d_n \subseteq C(I)$. By lemma 3.4, the map from $Q_n^{(c)}$ to $Q_n^{(d)}$ given by $e_n x \mapsto x e_n$ for $x \in Ac_n$ is an isomorphism. The desired map $Q_n^{(c)} \to Q_n^{(d)}$ given by $e_n x \mapsto e_n(xe_n)e_n^{-1}$ is clearly an isomorphism of rings.

Suppose $e_{n+1}x \in Q_{n+1}^{(c)}$ with $x \in Ac_{n+1}$. By the composite

$$Q_{n+1}^{(c)} \to Q_{n+1}^{(d)} \to Q_n^{(d)},$$

$$e_{n+1}x \mapsto e_{n+1}(xe_{n+1})e_{n+1}^1 = e_n d_{n+1} x e_n d_{n+1}(e_n d_{n+1})^{-1}.$$

Thus

$$e_{n+1}x \mapsto e_n d_{n+1} x e_n e_n^{-1} \in Q_n^{(d)}.$$

Now by

$$Q_{n+1}^{(c)} \to Q_n^{(c)} \to Q_n^{(d)},$$

$$e_{n+1}x \mapsto e_{n+1}x = e_n(d_{n+1}x) \mapsto e_n(d_{n+1}xe_n)e_n^{-1}$$

and therefore the diagrams in (3) commute and the theorem is shown. ∎

We now establish some notations for future considerations.

Suppose (c) is a left Öre sequence for a semiprime ideal I of a left Noetherian ring A. By Theorem 3.2 there are contexts $[\Sigma_n^{(c)}, B_n^{(c)}, A_n, P_n^{(c)}]$ for $n = 1, 2, \ldots$, where $B_n^{(c)} = \{x \in A_n : dc_n x \in A_n C_n \text{ for some } d \in C(I)\}$, $\Sigma_n^{(c)} = C(I_n)c_n$, and $P_n^{(c)} = Q_n^{(c)} \otimes_{B_n^{(c)}} A_n$. We will call these contexts the contexts associated with (c). For convenience of notation, we will denote by $[\Sigma_n, B_n, A_n, P_n]$ the contexts associated with some left Öre sequence for I.

Now, if $x \in B_{n+1}$, then $dc_{n+1}x \in A_{n+1}c_{n+1}$, where $c \in C(I)$. The image of x by $A_{n+1} \to A_n$ is in $A_n c_{n+1}$. As $A_n c_{n+1} \subseteq A_n c_n$, and since $C(I)c_{n+1} \subseteq C(I)c_n$ it follows that the image of x is in B_n, thus $A_{n+1} \to A_n$ induces a unitary ring homomorphism which we denote $B_{n+1} \to B_n$.

Furthermore, it is easily verified that the maps $Q_{n+1} \to Q_n$ and $A_{n+1} \to A_n$ induce surjective maps $P_{n+1} \to P_n$. The inverse limits of the sequences $\{Q_{n+1} \to Q_n\}$, respectively $\{P_{n+1} \to P_n\}$, will be denoted by $Q = \varprojlim Q_n$, respectively $P = \varprojlim P_n$, or $Q^{(c)} = \varprojlim Q_n^{(c)}$, respectively $P^{(c)} = \varprojlim P_n^{(c)}$, if we wish to refer to the left Öre sequence for I being used.

If $\{X_{n+1} \to X_n : n = 1, 2, \ldots\}$ is a sequence of maps, then the elements of $\varprojlim X_n$ will be denoted by $(x_n) = (x_1, x_2, \ldots, x_n, \ldots)$ where x_n is the image of x_{n+1} by $X_{n+1} \to X_n$.

Thus $Q = \varprojlim Q_n$ with addition and multiplication coordinatewise is a unitary ring with the natural projections $Q \to Q_n$ being surjective ring homomorphisms. Further, P is a left Q-, right A-bimodule with addition coordinatewise, left Q-module action given by $(q_n)(p_n) = (q_n p_n)$, where $(q_n) \in Q$, $(p_n) \in P$, and right A-module action given by $(p_n)a = (p_n a)$ for $(p_n) \in P$, $a \in A$. If X is in Q_n-mod, then X is naturally in Q-mod via the homomorphism $Q \to Q_n$.

We denote the Jacobson radical of $Q_n^{(c)}$ by $J_n^{(c)}$, or just J_n and the Jacobson radical of $Q^{(c)}$ by $J^{(c)}$, or just J. The kernel of $Q \to Q_n$ will be denoted by K_n.

Theorem 3.7. *Let (c) be a left Öre sequence for a semiprime ideal I of a left Noetherian ring A. Then the following hold:*

1. *The map $Q_n \to P_n$ given by $x \mapsto x \otimes 1$ is a split monomorphism in Q-mod for $n = 1, 2, \ldots$*
2. *For $n = 1, 2, \ldots$, $J_n = Q_n I_n c_n$ and $(J_n)^n = 0$.*
3. *For $n = 1, 2, \ldots$, $K_n \supseteq J^n$, and $K_1 = J$.*
4. *The map $Q \to P$ given by $(q_n) \mapsto (q_n \otimes 1)$ is a split monomorphism in Q-mod.*

Proof. Proof of (1): Let $e_n = c_1 c_2 \cdots c_n$. Define $P_n \to Q_n$ by $x \otimes a \mapsto xae_n$, where $x \in Q_n$ and $a \in A_n$. It is easily verified that this is a well defined map and a Q-homomorphism. If $x \in Q_n$, then the composite $Q_n \to P_n \to Q_n$ maps $xe_n^{-1} \mapsto x$, so the composite is a surjective Q-homomorphism, hence an isomorphism since Q_n has finite length in Q_n-mod and hence in Q-mod; (1) follows.

Proof of (2): By 3.2, the kernel of $Q_n \to Q_1$ is J_n. Each element of Q_n is expressible as $\sigma^{-1}ac_n$ where $\sigma \in C(I)c_n$, $a \in A_n$. Thus $\sigma^{-1}ac_n$ is in the kernel of $Q_n \to Q_1$ if and only if $a \in I$, so $J_n = Q_n I_n c_n$. Since $I_n c_n Q_n \subseteq J_n$ and $(I_n)^n = 0$, (2) follows.

Proof of (3): If $(q_n) \in K_1$, then $q_1 = 0$ and $q_n \in Ker(Q_n \to Q_1) = J_1$ by 3.2. The $1 - q_n$ is a unit in Q_n with inverse u_n. Clearly $u_{n+1} \mapsto u_n$ by $Q_{n+1} \to Q_n$, so $(u_n) \in Q$. Therefore $K_1 \subseteq J$. Since $J_1 = 0$ it follows that $J \subseteq K_1$, so equality holds. Now, since $K_n \subseteq K_1 = J$ and $(J_n)^n = 0$, it follows that $J^n \subseteq K_n$, and (3) follows.

Proof of (4): Let e_n and $P_n \to Q_n$ be as in (1). Denote $H_n = Ker(P_n \to Q_n)$, $T_n = Im(Q_n \to P_n)$. then, by (1), $P_n = T_n \oplus H_n$. If $x \in Q_{n+1}$, $a \in A_{n+1}$, then denote the image of x, respectively a, under $Q_{n+1} \to Q_n$, respectively $A_{n+1} \to A_n$, by $\bar{x}$, respectively $\bar{a}$. Thus, if $x \otimes a \in H_{n+1}$, then $0 = xae_{n+1}$, so $0 = \bar{x}\bar{a}e_{n+1} = \bar{x}\bar{a}e_n c_{n_1}$. Since c_{n+1} is a unit in Q_n, $0 = \bar{x}\bar{a}e_n$, so $\bar{x} \otimes \bar{a} \in H_n$. thus $P_{n+1} \to P_n$ maps H_{n+1} to H_n, and clearly maps T_{n+1} to T_n. Let $P_n \to T_n$ be the projection with respect to the decomposition $P_n = T_n \oplus H_n$, so for each n we have a commutative diagram:

$$\begin{array}{ccccc} Q_{n+1} & \to & P_{n+1} & \to & T_{n+1} \\ \downarrow & & \downarrow & & \downarrow \\ Q_n & \to & P_n & \to & T_n \end{array}$$

Since each composite $Q_n \to P_n \to T_n$ is an isomorphism, we can extend to inverse limts and (4) follows. ∎

Suppose A is a Noetherian ring (i.e. left and right Noetherian) and N is its nilradical. Then there is a reduced rank function for finitely generated modules in A-mod and also one for finitely generated modules in mod-A. We will denote each by $\rho = \rho_N$ as there will be no confusion as to which applies.

Also, if Σ is a left and right denominator set in a ring B, then the ring of left fractions of B with respect to Σ is also a ring of right factions of B with respect to Σ, and so is called the ring of fractions of B with respect to Σ.

Lemma 3.8. *Let A be a Noetherian ring and N its nilradical. Then there is an element $c \in C(N)$ such that $cC(N)c$ is a denominator set in cAc.*

Proof. Let $d \in C(N)$ be such that $\rho(Ad)$ in A-mod is minimal and $e \in C(N)$ be such that $\rho(eA)$ in mod-A is minimal. Let $c = de \in C(N)$. Then clearly $\rho(Ac)$ and $\rho(cA)$ are minimal. Therefore $C(N)c$ is a left denominator set in Ac and so $cC(N)c$ is a left denominator set in cAc. By symmetry, $cC(N)c$ is also a right denominator set in cAc, and the result follows. ∎

Corollary 3.9. *If A is a Noetherian ring and I a semiprime ideal, then there is a sequence $\{c_n\} \subseteq C(I)$ that is simultaneously a left and right Öre sequence for I.*

Proof. Choose $d_n \in C(I)$ so that $d_n C(I_n) d_n$ is a denominator set in $d_n A_n d_n$. Define $c_n \in C(I)$ recursively by $c_1 = d_1$ and $c_{n+1} = c_n d_{n+1} c_n$. It is now evident that $\{c_n\}$ has the desired properties. ∎

Corollary 3.10. *If* (c) *is an Öre (left and right) sequence for a semiprime ideal* I *of a Noetherian ring, then* $Q_n^{(c)}$ *is a left and right Artinian ring.*

Proof. By 3.2, $Q_n^{(c)}$ is left Artinian. If S_n is a ring of right fractions of $c_n A_n$ with respect to $c_n C(I_n)$ then S_n is right Artinian. But then S_n is the ring of right fractions of $c_n A_n c_n$ with respect to $c_n C(I_n) c_n$, hence a ring of fractions of $c_n A_n c_n$ with respect to $c_n C(I_n) c_n$. Therefore S_n and $Q_n^{(c)}$ are isomorphic and the corollary follows. ∎

4. Examples

Definition 4.1. Inverse limit contexts of a left Noetherian ring A with respect to semi-prime ideal I can be classified into one of four categories:
: Class I: The ring A is left localizable at $C(I)$ (and therefore A_n is left localizable at $C(I_n)$ for all n).
: Class II: The ring A is not left localizable at $C(I)$, but A_n is left localizable at $C(I_n)$ for all $n \geq 1$
: Class III: The ring A_n is not left localizable at $C(I_n)$ for $n > 1$, but there is $c \in C(I)$ such that for each n, $\rho(A_n c)$ is minimal among $\{\rho(A_n d) : d \in C(I)\}$.
: Class IV: For any $c \in C(I)$, there is some N such that $\rho(A_n c)$ is not minimal for any $n \geq N$.

In this section we will present examples of inverse limit contexts that fit into each of categories II, III, and IV. Since Class I depicts the classical situation, we will not give an example for it. For x an element in a ring of integers, $\mathbb{Z}$, (x) will denote the ideal generated by x, and $C(x)$ will denote the elements in $\mathbb{Z}$ that are regular modulo (x). In each case we will work with the inverse limit context of the ring A with respect to the ideal I and use the notation of inverse limit contexts.

Example 4.2. Let $A = \begin{bmatrix} \mathbb{Z} & \mathbb{Z} \\ 0 & \mathbb{Z} \end{bmatrix}$ and $I = \begin{bmatrix} (2) & \mathbb{Z} \\ 0 & (3) \end{bmatrix}$, so A is left Noetherian and I is a semi-prime ideal with $A_1 \cong \begin{bmatrix} \mathbb{Z}/(2) & 0 \\ 0 & \mathbb{Z}/(3) \end{bmatrix}$. Therefore A_1 is its own classic ring of left fractions. It is easy to see that

$$C(I) = \begin{bmatrix} C(2) & \mathbb{Z} \\ 0 & C(3) \end{bmatrix}$$

is not a left denominator set in A.
For each $n \geq 1$ $I^n = \begin{bmatrix} (2^n) & \mathbb{Z} \\ 0 & (3^n) \end{bmatrix}$, so

$$A_n \cong \begin{bmatrix} \mathbb{Z}/(2^n) & 0 \\ 0 & \mathbb{Z}/(3^n) \end{bmatrix}.$$

is commutative and $C(I_n)$ is a left denominator set in A_n. Thus the inverse limit context of A with respect to I is of class II. Moreover, $A_n = Q_n$, so Q is the I-adic completion of A.

Example 4.3. Let $A = \begin{bmatrix} \mathbb{Z} & \mathbb{Z} \\ 0 & \mathbb{Z} \end{bmatrix}$ and $I = \begin{bmatrix} (2) & \mathbb{Z} \\ 0 & 0 \end{bmatrix}$. Then

$$A_1 \cong \begin{bmatrix} \mathbb{Z}/(2) & 0 \\ 0 & \mathbb{Z} \end{bmatrix}$$

is commutative and semi-prime. Its classic ring of left fractions is

$$Q_1 \cong \begin{bmatrix} \mathbb{Z}/(2) & 0 \\ 0 & \mathbb{Q} \end{bmatrix},$$

where $\mathbb{Q}$ is the rational field. In addition $C(I) = \begin{bmatrix} C(2) & \mathbb{Z} \\ 0 & C(0) \end{bmatrix}$. It is easily seen that $C(I)$ is not a left denominator set in A.

For each $n \geq 1$, $I^n = \begin{bmatrix} (2^n) & (2^{n-1}) \\ 0 & 0 \end{bmatrix}$, $A_n = \begin{bmatrix} \mathbb{Z}/(2^n) & \mathbb{Z}/(2^{n-1}) \\ 0 & \mathbb{Z} \end{bmatrix}$

and

$C(I_n) = \begin{bmatrix} C((2)/(2^n)) & \mathbb{Z}/(2^{n-1}) \\ 0 & C(0) \end{bmatrix}$. Let $c_n = \begin{bmatrix} \bar{1} & 0 \\ 0 & 2^{n-1} \end{bmatrix}$. Then $C(I_n)c$

is a left denominator set in $A_n c_n = \begin{bmatrix} \mathbb{Z}/(2^n) & 0 \\ 0 & (2^{n-1}) \end{bmatrix}$, $B_n \cong \begin{bmatrix} \mathbb{Z}/(2^n) & 0 \\ 0 & \mathbb{Z} \end{bmatrix}$

and $Q_n \cong \mathbb{Z}/(2^n) \times \mathbb{Q}$. It follows that $Q \cong \mathbb{Z}_2 \times \mathbb{Q}$, where $\mathbb{Z}_2$ is the (2)-adic completion of $\mathbb{Z}$.

Notice that $\mathbb{Z}/(2^{n-1})$, and therefore any $\mathbb{Z}/(2^n)$-submodule, X, of $\mathbb{Z}/(2^{n-1})$ is $C((2)/(2^n))$-torsion free. Let $d_n = \begin{bmatrix} \bar{u} & \bar{x} \\ 0 & m \end{bmatrix} \in C(I_n)$ with m not divisible by 2^{n-1}. Then

$$A_n d_n = \begin{bmatrix} \mathbb{Z}/(2^n) & X \\ 0 & (m) \end{bmatrix}$$

and

$$C(I_n)d_n = \begin{bmatrix} C((2)/(2^n)) & X \\ 0 & mC(0) \end{bmatrix},$$

where $X \neq 0$. It is easily seen that $C(I_n)d_n$ is not a left denominator set in $A_n d_n$. Thus for each $c \in C(I)$ there is some N such that $C(I_n)c$ is not a left denominator set in $A_n c$ for any $n \geq N$. Thus the inverse limit context of A with respect to the ideal I is of class IV.

Example 4.4. Let $A = \begin{bmatrix} \mathbb{Z} & \mathbb{Z}/(4) \\ 0 & \mathbb{Z} \end{bmatrix}$ and $I = \begin{bmatrix} (2) & \mathbb{Z}/(4) \\ 0 & 0 \end{bmatrix}$. Therefore A is left Noetherian, I is a semiprime ideal, $A_1 \cong \begin{bmatrix} \mathbb{Z}/(2) & 0 \\ 0 & \mathbb{Z} \end{bmatrix}$, and $Q_1 \cong \mathbb{Z}/(2) \times \mathbb{Q}$. Since $C(I) = \begin{bmatrix} C(2) & \mathbb{Z}/(4) \\ 0 & C(0) \end{bmatrix}$ and $\mathbb{Z}/(4)$ is $C(2)$-torsion free. It is evident that $C(I)$ is not a left denominator set in A.

Next look at $I^2 = \begin{bmatrix} (4) & (2)/(4) \\ 0 & 0 \end{bmatrix}$, with $A_2 \cong \begin{bmatrix} \mathbb{Z}/(4) & \mathbb{Z}/(2) \\ 0 & \mathbb{Z} \end{bmatrix}$ and $C(I_2) = \begin{bmatrix} C((2)/(4)) & \mathbb{Z}/(2) \\ 0 & C(0) \end{bmatrix}$. It can easily be seen that $C(I_2)$ is not a left denominator set in A_2, but if $c_2 = \begin{bmatrix} \bar{1} & 0 \\ 0 & 2 \end{bmatrix}$ then $C(I_2)c_2$ is a left denominator set in $A_2 c_2$.

However, if $k > 2$, then $I^k = \begin{bmatrix} (2^k) & 0 \\ 0 & 0 \end{bmatrix}$, with $A_k \cong \begin{bmatrix} \mathbb{Z}/(2^k) & \mathbb{Z}/(4) \\ 0 & \mathbb{Z} \end{bmatrix}$. The set $C(I_k)$ is not a left denominator set in A_k, and $C(I_k)c_2$ is not a left denominator set in $A_k c_2$, but if $c = \begin{bmatrix} \bar{1} & 0 \\ 0 & 4 \end{bmatrix}$ then $C(I_k)c$ is a left denominator set in $A_k c$. Thus $C(I_k)c$ is a left denominator set in $A_k c$ for all $k \geq 1$. It follows that the inverse limit context of A with respect to I is of class III.

5. Questions and Problems

There are many questions that arise, some of which we pose here. When A is a left Noetherian ring and (c) is a left Öre sequence for a semiprime ideal I.

1. When is $Q^{(c)}$ a left Noetherian ring?
   Remark: It is easily verified that $Q^{(c)}$ is a semiperfect ring.
2. When is $Ker(Q^{(c)}_{n+1} \to Q^{(c)}_n) = (J^{(c)}_{n+1})^n$? The inclusion $\supseteq$ is always valid.

   More specifically, is there a left Noetherian ring with nilradical N and $c \in C(N)$ such that $\rho(Ac)$ is minimal, but $(Nc \cap N^2)/(Nc)^2$ is not $C(N)$ torsion? If there is, then
   $$Ker(Q^{(c)}_3 \to Q^{(c)}_2) \neq (J^{(c)}_3)^2.$$

3. When is $Ker(Q^{(c)} \to Q^{(c)}_n) = (J^{(c)})^n$? The inclusion $\supseteq$ is always valid.
4. When is $Q^{(c)}$ $J^{(c)}$-adically complete?
5. When is $J^{(c)}$ finitely generated in $Q^{(c)}$-mod?
   This is related to
6. Is there a positive integer m such that $J^{(c)}/(J^{(c)})^2$ is a direct sum of $\leq m$ simple left $Q^{(c)}$-modules?

There are analogous questions when A is Noetherian.

References

1. Anderson, F.W., Fuller, K.R. Rings and Categories of Modules. Graduate Texts in Mathemtics. Berlin-New York: Springer Verlag, 1974.
2. Atiyah, M.F., MacDonald, I.G. Introduction to Commutative Algebra. Reading, Mass: Addison-Wesley, 1969.
3. Chatters, A.W., Goldie, A.W., Hajarnavis, C.R., and Lenagan, T.H. Reduced Rank in Noetherian Rings. J. of Alg. 61 582-89 (1979).
4. Cozzens, J.H., Sandomierski, F.L. Localization at a Semiprime Ideal of a Right Noetherian Ring. Comm. in Alg. 5(7) 707-726 (1977).

5. Eid, G.M. Classical Quotient Ring with Perfect Topologies. Ph.D. Thesis, Kent State University, 1983.
6. Jategaonkar, A.V. Localization in Noetherian Rings. London Math. Soc. Lecture Note Series, No. 98. Cambridge: Cambridge University Press, 1986.
7. Goldie, A.W. The Structure of Prime Rings under Ascending Chain Conditions. Proc. London Math Soc. (3) 8 589-609 (1958).
8. Goldie, A.W. Semi-prime Rings with Maximum Condition. Proc. London Math Soc. (3) 10 201-220 (1960).
9. Goldie, A.W. Torsion-free Modules and Rings. J. of Alg. 1 (1964) 268-287.
10. Goldman, O. Rings and Modules of Quotients. J. of Alg. 13 10-47 (1969).
11. Goodearl, K.R., Warfield, R.B. An Introduction to Noncommutative Noetherian Rings. London Math. Soc. Student Texts, No. 16. Cambridge: Cambridge University Press, 1989.
12. Hinohara, Y. Note on Non-commutative Semi-local Rings. Nagoya Math. J. 17 161-166 (1960).
13. Lambek, J., Michler, G. The Torsion Theory at a Prime Ideal of a Right Noetherian Ring. J. of Alg. 25, 364-389 (1973).
14. McConnell, J.C., Robson, J.C. Noncommutative Noetherian Rings. Wiley-Interscience Series. New York: John Wiley and Sons, 1987.
15. Morita, K. Localization at Categories of Modules I. Math Z. 114 121-144 (1970).
16. Morita, K. Localization in Categories of Modules II. J. Reine Angew. Math 242 163-169 (1970).
17. Morita, K. Localization in Categories of Modules III. Math Z. 119 313-320 (1971).
18. Morita, K. Flat Modules, Injective Modules and Quotient Rings. MathZ. 120 25-40 (1971).
19. Öre, O. Linear Equations in Non-commutative Fields. Annals of Math 32 463-477 (1931).
20. Samuel, P., Zariski, O. Commutative Algebra, Vols. I and II. Princeton, New Jersey: D. Van Nostrad Co. Inc, 1960.
21. Stenström, B. Rings and Modules of Quotients. Lecture Notes in Math., Vol. 237. Berlin-New York: Springer-Verlag, 1971.

DEPARTMENT OF MATHEMATICAL SCIENCES, KENT STATE UNIVERSITY, KENT, OHIO 44242
Current address: Mathematics Department, Clarion University of Pennsylvania, Clarion, Pennsylvania 16214
E-mail address: mmcconnell@vaxa.clarion.edu

DEPARTMENT OF MATHEMATICAL SCIENCES, KENT STATE UNIVERSITY, KENT, OHIO 44242
E-mail address: sandomie@mcs.kent.edu

PROJECTIVE DIMENSION OF IDEALS IN VON NEUMANN REGULAR RINGS

BARBARA L. OSOFSKY

Dedicated to Bruno Mueller on his retirement.

ABSTRACT. This paper is motivated by an attempt to solve an old problem of Wiegand, which asks whether the projective dimension of an ideal in a commutative von Neumann regular ring depends only on the lattice of idempotents in that ideal. We compute the projective dimension of some infinitely generated ideals in von Neumann regular rings. In previous work, this projective dimension, if computable, was either 'obvious' or the subscript of the aleph of a generating set. We give nontrivial examples which can have arbitrary preassigned projective dimension and arbitrarily large cardinality of a generating set. The paper then presents a function ℓ from the class of all nonzero submodules of projective modules over a von Neumann regular ring to the class of all ordinals. This function depends only on the lattice of cyclic submodules of M. We show that $\ell(M) = 0 \iff M$ is projective and $\ell(M) \geq \mathrm{pd}(M)$. We conjecture that $\mathrm{pd}(M) < \infty \implies \mathrm{pd}(M) = \ell(M)$ for all M. Since $\ell(M)$ is defined lattice theoretically, this would answer Wiegand's question affirmatively. Even if our conjecture is false, $\ell(M)$ seems like an interesting lattice invariant to explore.

1. INTRODUCTION

The main mathematical terms and notation in this paper will be explained later in the paper. Here we simply introduce the problem of interest.

Back in the late 60's, Roger Wiegand asked the following question:

> Let R be a commutative regular ring and J an ideal of R generated by a set $\mathcal{E}$ of idempotents. Let $\mathcal{B}$ be the Boolean algebra of all idempotents of R. Then is the projective dimension of $J = \mathcal{E}R$ as an R-module the same as the projective dimension of $\mathcal{E}\mathcal{B}$ as a $\mathcal{B}$-module?

1991 *Mathematics Subject Classification.* Primary 16E10, 16E50; Secondary 06E20, 13C05, 13D02, 13D05.

Key words and phrases. Projective dimension, von Neumann regular rings, lattices.

The author wishes to thank the University of Haifa Department of Mathematics and Computer Science for their kind hospitality during the time when most of the work on this paper was being done.

It is not difficult to see that the answer is 'yes' if J is projective. Richard Pierce showed that it is 'yes' in case either projective dimension is 1. To the best of my knowledge, that is the most that can be said in general, although the answer is known to be 'yes' in some other special cases.

Pierce used a specific projective resolution to get his result in the case of dimension 1. He was able to express projective dimension 1 in terms of idempotents in the ring essentially because there is no torsion in the first homology group of a simplicial complex. He also observed that his approach did not work for higher dimensions because you could get torsion in higher homology groups. One way of looking at this is that for higher dimensions, the problem is one of purity in abelian groups. Such problems are notoriously intractable. Another way of looking at the problem is that, for distinct primes p, the lattices of subspaces of a finite dimensional vector space over the integers modulo p are different. Clearly a new approach is needed to get further results on the Wiegand problem for larger dimensions. This paper indicates a possibility of what such an approach might be.

Section 2 of this paper presents, with sketched proofs, the basic well known results from ring and module theory which we need. Section 3 gives the major tools used to compute projective dimensions of infinitely generated modules. These tools are also in the literature, but, especially in the lower bound for projective dimension tools, in a more abstract form.

In Section 4 we extend the rather small collection of kinds of (right) ideals in a not necessarily commutative von Neumann regular ring for which the projective dimension can be computed. The results actually hold for a somewhat wider class of rings and modules. In the literature, such computed projective dimensions, if not easily seen, were essentially always given by the subscript of the aleph corresponding to a generating set. That is the case for one of our examples here, but we also get ideals in some von Neumann regular rings that can have arbitrary preassigned finite projective dimension but which require arbitrarily large generating sets.

In Section 5, we present a conjecture which implies Wiegand's question has a positive answer. This conjecture is stated in terms of a property which is of interest on its own whether or not the conjecture is true. The Pierce criterion for dimension 1, while based on a complicated property of idempotents in the commutative von Neumann regular ring $\Re$, does not appear to be neatly expressible in lattice theoretic terms. What might lattice theoretic terms look like? In *all* of the known cases where projective dimension can be determined, the upper bound is obtainable by 'adding a generator at a time'. This observation gives rise to an ordinal valued function ℓ with domain the class of submodules of projective $\Re$-modules. $\ell(M)$ is determined by the lattice of cyclic submodules of the module M. If $\ell(M) = n < \infty$, then the projective dimension of M is at most n. We conjecture that M has projective dimension n if and only if $\ell(M) = n$. Even if that turns out not to be the case, the property $\ell(M) = n$ is an intriguing extension of the concept of projectivity, which is $\ell(M) = 0$.

2. Preliminary definitions and results

In this section we fix some notation that will be in use for the rest of the paper and give a few background results. This is meant to be a reference. The section contains nothing new.

In this paper, all rings have 1 and all modules and ideals are right modules and right ideals.

Definition. A module M over a ring R is called **projective** iff any map from M to a quotient module B/C factors through B iff M is a direct summand of a free.

We first note some basic properties of von Neumann regular rings.

Definition. A ring R is **von Neumann regular** iff for all $x \in R$ there exists a $y \in R$ such that $x = xyx$.

Here are some equivalent properties.

Theorem 2.1. *Let R be a ring. Then the following are equivalent.*
  (i) *For all $x \in R$ there exists a $y \in R$ such that $x = xyx$.*
  (ii) *For all $x \in R$, there exists an $e = e^2 \in R$ with $xR = eR$.*
  (iii) *Every finitely generated submodule of a projective R-module is a direct summand.*
  (iv) *Every R-module is flat.*
  (v) *If x and y are elements of R, then $xR \cap yR$ is generated by an idempotent, and $xR + yR$ is also generated by an idempotent. Thus the cyclic right ideals of R form a lattice under $\vee =$ sum and $\wedge =$ intersection.*

Corollary 2.2. *If R is von Neumann regular, then every countably generated submodule of a projective R-module is projective.*

Proof. Let M be a countably generated submodule of a free R-module. Then M is a sum $\sum_{i=0}^{\infty} m_i R$ of cyclic submodules. Since every finitely generated submodule of a projective R-module is a direct summand by (iii), $\sum_{i=0}^{j} m_i R = \sum_{i=0}^{j-1} m_i R \oplus K_j$ where K_j must be cyclic projective since $m_j R$ is. Then M equals $\bigoplus_{i \in \omega} K_i$. □

Notation. The German letter $\mathfrak{R}$ will denote a von Neumann regular ring,

The usual letter R will be used for an arbitrary ring.

Notation. A set of idempotents in $\mathfrak{R}$ will be denoted by the script letter $\mathcal{E}$. That is,
$$\mathcal{E} = \{e_\alpha : \alpha \in \mathcal{I}\} \subseteq \mathfrak{R}$$
where $e_\alpha = e_\alpha^2$. We let $\mathcal{E}\mathfrak{R}$ be the right ideal generated by $\mathcal{E}$.

The right ideals $\{e_\alpha \mathfrak{R} : \alpha \in \mathcal{I}\}$ generate a lattice of cyclic idempotent generated right ideals in with join = sum and meet = intersection. We denote this lattice by
$$\tilde{\mathcal{E}} = \{e_\alpha \mathfrak{R} : \alpha \in \mathcal{I}\}$$
where $e_\alpha = e_\alpha{}^2$.

The idempotents in $\mathcal{E}$ commute if and only if $\tilde{\mathcal{E}} \cup \{1\}$ generates a distributive complemented lattice. Much of what we do does not require that the idempotents commute. However, on some significant occasions we will assume that they do commute, or at least some of the lattice properties behave as in the commuting idempotents case. For example, we have to take intersections at one point, and then we will use the distributivity property

$$\left(\bigcap_{i=1}^{n} e_{\alpha_i} \mathfrak{R}\right) \cap \sum_{j=1}^{m} e_{\beta_j} \mathfrak{R} = \sum_{j=1}^{m} \left(\bigcap_{i=1}^{n} e_{\alpha_i} \mathfrak{R} \cap e_{\beta_j} \mathfrak{R}\right)$$

for all $\{\alpha_i,\ \beta_j\} \subseteq \mathcal{I}$.

When looking at the projective dimension of $\mathcal{E}\mathfrak{R}$, it helps to know what projective dimension is. The most common definition is in terms of a long projective resolution, but I prefer to do things a step at a time.

Definition. A **short projective resolution** of a module M is a short exact sequence
$$0 \longrightarrow K \longrightarrow P \longrightarrow M \longrightarrow 0$$
where P is projective.

A **long projective resolution** of M is an exact sequence $\mathfrak{P}$:
$$\cdots \xrightarrow{\partial_{n+1}} P_n \xrightarrow{\partial_n} P_{n-1} \xrightarrow{\partial_{n-1}} \cdots \xrightarrow{\partial_1} P_0 \xrightarrow{\partial_0} M \longrightarrow 0$$
where each of the P_i is projective.

Definition. It is convenient to say that a module M has projective dimension -1 if and only if $M = 0$. We will adopt this convention.

A module M has projective dimension ≤ 0 if and only if M is projective.

Inductively, a module M has projective dimension $\leq n+1$ if and only if there is a short projective resolution
$$0 \longrightarrow K \longrightarrow P \longrightarrow M \longrightarrow 0$$
where K has projective dimension $\leq n$.

Then the **projective dimension** of M, denoted $\mathrm{pd}_R(M)$ or usually just $\mathrm{pd}(M)$ if the ring is clear, is defined by

$$\mathrm{pd}_R(M) = \begin{cases} \min\{k : \mathrm{pd}(M) \leq k\} & \text{if there is such a } k \\ \infty & \text{otherwise} \end{cases}$$

The first thing we observe about this definition is that, for nonzero modules, it is independent of the short projective resolution used.

Lemma 2.3 (Schanuel). *For R any ring and M an R-module, given two exact sequences*

$$0 \longrightarrow K \longrightarrow P \xrightarrow{\mu} M \longrightarrow 0$$

$$0 \longrightarrow L \longrightarrow Q \xrightarrow{\nu} M \longrightarrow 0$$

with P and Q projective, we have $P \oplus L \cong Q \oplus K$.

Proof. Let $\mathcal{M}$ be the pullback $\mathcal{M} = \{(p,q) \subseteq P \oplus Q : \mu p = \nu q\}$. The projection of $\mathcal{M}$ to the first (respectively second) component maps onto the projective module P (respectively Q). The kernel of that projection is, up to multiplication by -1, L (respectively K). Thus $\mathcal{M} \cong P \oplus L$ (respectively $\mathcal{M} \cong Q \oplus K$). □

Schanuel's Lemma implies that, if $\operatorname{pd}(M) \leq k$, then for any short projective resolution $0 \longrightarrow K \longrightarrow P \longrightarrow M \longrightarrow 0$ of M, $\operatorname{pd}(K) \leq \max\{k-1, 0\}$. By induction we extend this to any long projective resolution $\mathfrak{P}$:

$$\cdots \xrightarrow{\partial_{n+1}} P_n \xrightarrow{\partial_n} P_{n-1} \xrightarrow{\partial_{n-1}} \cdots \xrightarrow{\partial_1} P_0 \xrightarrow{\partial_0} M \longrightarrow 0$$

of M, to get $\operatorname{pd}(M) \leq n \iff \partial_n P_n$ is projective.

We now list some standard properties of projective dimension.

Lemma 2.4. *Let $0 \longrightarrow A \xrightarrow{\sigma} B \xrightarrow{\tau} C \longrightarrow 0$ be exact, and*

$$0 \longrightarrow H \longrightarrow Q \xrightarrow{\alpha} A \longrightarrow 0$$

$$0 \longrightarrow K \longrightarrow L \xrightarrow{\gamma} C \longrightarrow 0$$

exact with Q and L projective. Then there is a simultaneous short projective resolution of the initial exact sequence, that is, a commutative diagram with exact rows and columns

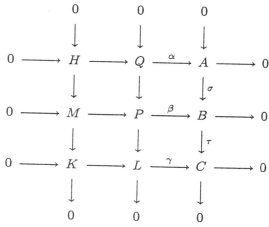

and P also projective. This extends by induction to long simultaneous projective resolutions.

Proof. Set $P = Q \oplus L$. Then since L is projective, there exists a map γ' from L to B such that $\gamma = \tau\gamma'$. Set $\beta = (\sigma\alpha, \gamma')$. The image of β contains A and maps onto C, so β must be onto. The kernel exact sequence is a relatively straightforward diagram chase. □

Lemma 2.5. Let $0 \longrightarrow A \longrightarrow B \longrightarrow C \longrightarrow 0$ be exact, and let the diagram

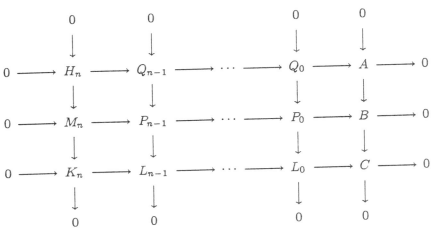

be commutative with exact row and columns, and the P_i, Q_i, and L_i all projective. Then

(a) If $\operatorname{pd}(A) \leq n$ and $\operatorname{pd}(B) \leq n$, then $\operatorname{pd}(C) \leq n+1$.
(b) If $\operatorname{pd}(B) \leq n$, then $\operatorname{pd}(C) \leq n$ if and only if H_n is a direct summand of M_n.
(c) If $\operatorname{pd}(C) \leq n+1$ and $\operatorname{pd}(B) \leq n$, then $\operatorname{pd}(A) \leq n$.

Proof. These are straightforward consequences of the definition of projective dimension and of the fact that it is independent of the projective resolution used. For example, to see (a), note that if H_n and M_n are projective, then K_n has dimension at most 1. □

Actually, considerably more can be said. If two of $\operatorname{pd}(A)$, $\operatorname{pd}(B)$, $\operatorname{pd}(C)$ are finite, then so is the third, and

(a') If $\operatorname{pd}(A) = \operatorname{pd}(B)$, then $\operatorname{pd}(C) \leq \operatorname{pd}(A) + 1$,
(b') If $\operatorname{pd}(A) < \operatorname{pd}(B)$, then $\operatorname{pd}(C) = \operatorname{pd}(B)$.
(c') If $\operatorname{pd}(A) > \operatorname{pd}(B)$, then $\operatorname{pd}(C) = \operatorname{pd}(A) + 1$.

The inequality in (a) (or (a')) is the source of the major problems arising in our computations of projective dimension.

Lemma 2.6 (Kaplansky). *Let M be a projective module over any ring R. Then M is a direct sum of countably generated (projective) modules.*

Proof. See [K]. □

In the case of von Neumann regular rings, this can be refined further.

Corollary 2.7. *A projective module M over a von Neumann regular ring $\Re$ is a direct sum of cyclic modules (isomorphic to ideals $e\Re$ of $\Re$).*

Proof. By Lemma 2.6, M is a direct sum of countably generated modules. The proof of Corollary 2.2 shows that each countably generated summand is a direct sum of cyclics. □

Our tools for computing projective dimension hold in more generality than for von Neumann regular rings. For getting lower bounds, we need coherence, so we record the appropriate properties here. These results derive from 'exact direct limits'.

Notation. *Following standard usage, we will use the Hebrew letter $\aleph$ to denote an infinite cardinal. We will identify cardinals with initial ordinals when we wish to index by the ordinal with smallest given cardinality. It is also convenient to let the notation $\aleph_{-1}$ mean 'finite' or 'finitely'.*

Definition. *A module M is $\aleph$-generated if it has some generating set of cardinality $\leq \aleph$. M is strictly $\aleph$-generated if M is $\aleph$-generated but has no generating set of cardinality $< \aleph$. M is $\aleph$-related if there is some short projective resolution*

$$0 \longrightarrow K \longrightarrow P \longrightarrow M \longrightarrow 0$$

with K $\aleph$-generated. M is $\aleph$-presented if it is both $\aleph$-generated and $\aleph$-related.

The same definitions hold if one replaces $\aleph$ with $\aleph_{-1} =$ 'finitely'.

By Schanuel's lemma, if $0 \longrightarrow K \longrightarrow P \longrightarrow M \longrightarrow 0$ is a short projective resolution with K and P $\aleph$-generated, then for any short projective resolution

$$0 \longrightarrow L \longrightarrow Q \longrightarrow M \longrightarrow 0,$$

if Q is $\aleph$-generated, then so is L since $Q \oplus K \cong P \oplus L$.

Definition. *A module M is called **coherent** if every finitely generated submodule of M is finitely related. M is $\aleph$-**coherent** if every finitely generated submodule of M is $\aleph$-related.*

Lemma 2.8. *Let R be $\aleph$-coherent as a module over itself, and let $\aleph' \geq \aleph$. Then every $\aleph'$-generated submodule of a free R-module is $\aleph'$-related.*

Proof. We first show, by induction on n, that every $\aleph$-generated submodule K of a finitely generated free $\bigoplus_{i=1}^{n} R_i$ is $\aleph$-related. The case $n = 1$ is by hypothesis.

Let $K \subseteq \bigoplus_{i=1}^{n+1} R_i$, K $\aleph$-generated. Projecting onto the last summand, we get an exact sequence

$$0 \longrightarrow K \cap \bigoplus_{i=1}^{n} R_i \longrightarrow K \longrightarrow M \longrightarrow 0$$

where M is isomorphic to an $\aleph$-generated submodule of $R_{n+1} \cong R$ and thus $\aleph$-related by hypothesis. Set $\overline{K} = K \cap \bigoplus_{i=1}^{n} R_i$. Let

$$0 \longrightarrow L \longrightarrow Q \longrightarrow K \longrightarrow 0$$

be a short projective resolution of K with Q $\aleph$-generated projective and let N denote the kernel of the composition $Q \longrightarrow K \longrightarrow M$. Then N is $\aleph$-generated and we have a commutative diagram with exact rows and columns

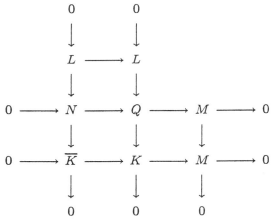

Here $L \subseteq N$ because N is the kernel of the map from Q to M. Then $\overline{K} = K \cap \bigoplus_{i=1}^{n} R_i$ is a quotient of N and hence $\aleph$-generated. By the induction hypothesis, $K \cap \bigoplus_{i=1}^{n} R_i$ is $\aleph$-related.

Map an $\aleph$-generated projective Q' onto N. The kernel L' of the composition $Q' \longrightarrow N \longrightarrow K \cap \bigoplus_{i=1}^{n} R_i$ is $\aleph$-generated by the induction hypothesis, and L' maps onto $L \subseteq N$ so L is $\aleph$-generated.

Now let K be any submodule of a free R-module and let $\{m_\alpha : \alpha \in \aleph'\}$ be a set of generators for K. Let $\mathfrak{F}$ be the set of all finite subsets of $\aleph'$. Then the cardinality of $\mathfrak{F}$ is $\aleph'$ and for any $\mathfrak{f} \in \mathfrak{F}$, the kernel of the map $\bigoplus_{\alpha \in \mathfrak{f}} R_\alpha \longrightarrow \sum_{\alpha \in \mathfrak{f}} m_\alpha R \longrightarrow 0$ is $\aleph$-generated. Now $\bigoplus_{\alpha \in \aleph'} R_\alpha \longrightarrow \sum_{\alpha \in \aleph'} m_\alpha R \longrightarrow 0$ is a map from a projective onto K, and one easily sees that its kernel is generated by the $\aleph$-generated kernels of the $\aleph'$ maps $\bigoplus_{\alpha \in \mathfrak{f}} R_\alpha \longrightarrow \sum_{\alpha \in \mathfrak{f}} m_\alpha R$. Thus this kernel is $\aleph \cdot \aleph' = \aleph'$-generated and K is $\aleph'$-related. $\square$

3. Upper and Lower Bounds on Projective Dimension

It is rare to be able to compute the projective dimension of infinitely generated modules except in very special cases. In this section we present the major known ways of computing bounds on projective dimension of infinitely generated ideals in nice rings.

A module may be given in such a way that it is obvious that it is projective, or perhaps has projective dimension 1. For example, if $\mathfrak{R} = \prod_{i \in \mathcal{I}} F_i$ where each F_i is a field, the ideal $\bigoplus_{i \in \mathcal{I}} F_i$ is projective. If $\mathcal{I} = \omega$ and if $\mathcal{X} = \{X_\alpha \subseteq \mathcal{I}\}$ is a family of almost disjoint subsets of $\mathcal{I}$, that is, distinct members of $\mathcal{X}$ have finite intersection, then the ideal generated by characteristic functions of the elements in $\mathcal{X}$ is projective if $\mathcal{X}$ is countable and has projective dimension 1 otherwise.

Moving on from these 'obvious' cases, I know of basically two tools that have been used in all successful computations of projective dimension of ideals of von Neumann regular rings so far. The first, Auslander's Lemma, has been used in computing upper bounds on projective dimensions almost from the beginning of the study of projective dimension. The second is a more complicated pair of results which appeared about a decade after Auslander's Lemma. This tool enables us to use induction in some cases to compute lower bounds (see [O67] and [P67]). It succeeds only if we can find correct induction hypotheses which are preserved when we take intersections. These tools are developed for a somewhat wider class of rings than the von Neumann regular rings. Even more generality can be found in [O78].

The tools to obtain an upper bound.

In the 50's, the Nagoya Journal of Mathematics ran a series of articles on the dimensions of rings and modules. One of those papers, [A] by Maurice Auslander, contained a beautiful lemma which immediately gave a result called the Global Dimension Theorem. There is a slick proof of the Global Dimension Theorem which is much easier to state but which uses the derived functor $\operatorname{Ext}^i_R(\cdot, \cdot)$ and machinery from derived functors. This machinery does not seem to be useful in most work with projective dimension of infinitely generated modules, but the original lemma is invaluable.

Lemma 3.1 (Auslander). *Let M be a (continuous ascending) union of submodules $\{M_\alpha : \alpha < \Omega\}$ where*

(1) Ω is an ordinal,
(2) $M = \bigcup_{\alpha < \Omega} M_\alpha$,
(3) $M_\beta \subseteq M_\alpha$ if $\beta < \alpha$, and
(4) $\operatorname{pd}_R \left(M_\alpha \big/ \bigcup_{\beta < \alpha} M_\beta \right) \leq n$ for all $\alpha < \Omega$.

Then $\operatorname{pd}_R(M) \leq n$.

Proof. If $n = 0$, then each of the quotients $M_\alpha \big/ \bigcup_{\beta < \alpha} M_\beta$ is projective and so $\bigcup_{\beta<\alpha} M_\beta$ is a direct summand of M_α. By transfinite induction one gets $M \cong \bigoplus_{\alpha<\Omega} \left(M_\alpha \big/ \bigcup_{\beta<\alpha} M_\beta \right)$.

Now assume $n > 0$. By transfinite induction we get a family of projective resolutions

$$0 \longrightarrow K_\alpha \longrightarrow P_\alpha \longrightarrow M_\alpha \longrightarrow 0$$

for $\alpha < \Omega$, with $\bigcup_{\beta<\alpha} P_\beta$ a direct summand of P_α. Once we have P_β for all $\beta < \alpha$, let Q_α be a projective mapping onto $M_\alpha \big/ \bigcup_{\beta<\alpha} M_\beta$. Use the simultaneous resolution of Lemma 2.4 to get the projective $P_\alpha = Q_\alpha \oplus \bigcup_{\beta<\alpha} P_\beta$ mapping onto M_α.

This family of short projective resolutions is directed (indeed well ordered) by the directed set Ω. Since the category of modules over a ring has exact direct limits, for each $\alpha \leq \Omega$ one gets short projective resolutions

$$0 \longrightarrow \bigcup_{\beta<\alpha} K_\beta \longrightarrow \bigcup_{\beta<\alpha} P_\beta \longrightarrow \bigcup_{\beta<\alpha} M_\beta \longrightarrow 0$$

and for $\alpha < \Omega$,

$$0 \longrightarrow K_\alpha \bigg/ \bigcup_{\beta<\alpha} K_\beta \longrightarrow P_\alpha \bigg/ \bigcup_{\beta<\alpha} P_\beta \cong Q_\alpha \longrightarrow M_\alpha \bigg/ \bigcup_{\beta<\alpha} M_\beta \longrightarrow 0.$$

Then $\bigcup_{\alpha<\Omega} K_\beta$ is the kernel of a map from the projective $\bigcup_{\alpha<\Omega} P_\beta$ onto M, and also a well ordered union where $\mathrm{pd}\left(K_\alpha \big/ \bigcup_{\beta<\alpha} K_\beta \right) \leq n-1$. By finite induction $\mathrm{pd}\left(\bigcup_{\alpha<\Omega} K_\alpha \right) \leq n-1$, so $\mathrm{pd}(M) \leq n$. □

This lemma has the immediate corollary:

Corollary 3.2 (Global Dimension Theorem). *Let M be an R-module and let $\{m_\alpha : \alpha < \Omega\}$ be a set of generators for M. Then*

$$\mathrm{pd}(M) \leq \sup_{\alpha<\Omega} \mathrm{pd}\left(m_\alpha R \bigg/ \left(m_\alpha R \cap \sum_{\beta<\alpha} m_\beta R \right) \right).$$

We now state another corollary of Auslander's Lemma 3.1.

Theorem 3.3 (A Corollary to Auslander's Lemma). *Let $\aleph_\lambda$ be an infinite cardinal or if $\lambda = -1$ an abbreviation for finite. Assume every $\aleph_\lambda$-generated submodule N of an R-module M has $\mathrm{pd}(N) \leq n$. Then for any nonnegative integer κ, any $\aleph_{\lambda+\kappa}$-generated submodule N' of M has $\mathrm{pd}(N') \leq \kappa + n$.*

Proof. We use induction on κ. If $\kappa = 0$, this is true by hypothesis.

Now assume the upper bound is correct for the nonnegative integer κ. Then if $N \subseteq M$ is generated by $\{m_\alpha : \alpha < \aleph_{\lambda+\kappa+1}\}$ we express N as a well ordered union of submodules

$$\left\{ N_\alpha = \sum_{\beta < \alpha} m_\beta R : \alpha < \aleph_{\lambda+\kappa+1} \right\}$$

with each N_α having at most $\aleph_{\lambda+\kappa}$ generators. $N = \bigcup_{\alpha < \aleph_{\lambda+\kappa+1}} N_\alpha$ is a well ordered union of submodules of projective dimension $\leq \kappa + n$. By Lemma 2.5(a), each quotient $\sum_{\beta \leq \alpha} m_\beta R \big/ \sum_{\beta < \alpha} m_\beta R$ has dimension $\leq \kappa + n + 1$, so by Auslander's Lemma $\operatorname{pd}(N) \leq n + \kappa + 1$. □

Corollary 3.4. *Let M be a module and κ a nonnegative integer. Let $\Omega \geq \aleph_\kappa$ be an ordinal and $\{m_\alpha : \alpha < \Omega\} \subseteq M$ a set of generators for M. Assume every finitely generated (respectively countably generated) submodule of M is projective, and for all $\alpha \geq \aleph_\kappa$, $m_\alpha R \cap \sum_{\beta < \alpha} m_\beta R$ is $\aleph_{\kappa-1}$-generated.*

Then $\operatorname{pd}(M) \leq \kappa+1$ ($\leq \kappa$ if countably generated ideals are projective).

Proof. Use Theorem 3.3 to get $\operatorname{pd}\left(\sum_{\beta < \aleph_\kappa} m_\beta R\right) \leq \kappa + 1$ (or $\leq \kappa$) and any $\aleph_{\kappa-1}$-generated submodule of M has $\operatorname{pd}(M) \leq \kappa$ (or $\leq \max(0, \kappa - 1)$). For all $\alpha \geq \aleph_\kappa$ we have an isomorphism

$$\sum_{\beta \leq \alpha} m_\beta R \bigg/ \sum_{\beta < \alpha} m_\beta R \cong m_\alpha R \bigg/ m_\alpha R \cap \sum_{\beta < \alpha} m_\beta R \,.$$

For N any $\aleph_{\kappa-1}$-generated submodule of M and $m \in M$, we have an exact sequence

$$0 \longrightarrow mR \cap N \longrightarrow mR \oplus N \longrightarrow mR + N \longrightarrow 0$$

with both $\operatorname{pd}(mR \oplus N)$ and $\operatorname{pd}(mR + N)$ at most κ (or $\leq \max(0, \kappa - 1)$). By Lemma 2.5(b), $\operatorname{pd}(mR \cap N) \leq \kappa$ (or $\leq \max(0, \kappa - 1)$). Auslander's Lemma completes the proof. □

The only cases where I can compute the projective dimension of an ideal in a von Neumann regular ring $\mathfrak{R}$ are where the ideal is built up from a sum of ideals with the property that the upper bound in one of the above theorems is also a lower bound.

The tools to obtain a lower bound.

A version of the basic results used to compute lower bounds for infinitely generated ideals is given in general form in [O78], although in that reference a specific consequence, rather than the general tools, is highlighted. Here we develop the results comprising the general tool for $\aleph$-coherent rings.

Theorem 3.5. *Let R be an $\aleph$-coherent ring. Let M be a submodule of a free R-module, assume $\operatorname{pd}(M) \leq n < \infty$, and let $\{m_\beta : \beta \in \mathcal{I}\}$ be a set of generators for M. Then there exist a directed set $\mathcal{D}$ and a function $\alpha \mapsto M_\alpha$ from $\mathcal{D}$ to the set of submodules of M such that:*

(a) *If $\alpha < \beta \in \mathcal{D}$ then $M_\alpha \subseteq M_\beta$.*
(b) *Each M_α is generated by some subset $\{m_\beta : \beta \in \mathcal{I}_\alpha\}$ with $\mathcal{I}_\alpha \subseteq \mathcal{I}$ and $\operatorname{card}(\mathcal{I}_\alpha) \leq \aleph$.*
(c) *Every $\aleph$-generated submodule of M is contained in some M_α.*
(d) *Every chain $\mathcal{C}$ in $\mathcal{D}$ of cardinality $\leq \aleph$ has an upper bound γ in $\mathcal{D}$ with $\bigcup_{\alpha \in \mathcal{C}} M_\alpha = M_\gamma$.*
(e) *For all $\alpha \in \mathcal{D}$, $\operatorname{pd}(M/M_\alpha) \leq n$.*

Proof. We first start with a construction which will then be applied to prove the theorem. Fix a long projective resolution

$$0 \to P_n \to \cdots \to P_0 \to M \to 0$$

of M. Using this resolution, we show:

Let $L \subseteq M$ be $\aleph$-generated. Then there exists a module L' with $L \subseteq L' \subseteq M$, L' generated by a subset of at most $\aleph$ elements in $\{m_\beta : \beta \in \mathcal{I}\}$, and $\operatorname{pd}(M/L') \leq n$.

Since each P_k is projective, by Kaplansky's Theorem (Lemma 2.6), P_k is a direct sum of countably generated modules, say $P_k = \bigoplus_{i \in \mathcal{I}_t} C_i^{(k)}$. We obtain a complex

$$0 \to T_{n,0} \to T_{n-1,0} \to \cdots \to T_{0,0} \to L_0 \to 0$$

where $L \subseteq L_0 \subseteq M$ and each $T_{i,0}$ is of the form $\bigoplus_{i \in \mathcal{I}_{i,k}} C_i^{(k)}$ where $\operatorname{card}(\mathcal{I}_{i,k}) \leq \aleph$. L is contained in the submodule generated by some subset $X \subseteq \{m_\beta : \beta \in \mathcal{I}\}$ of cardinality $\leq \aleph$. X is in the image of the sum S_0 of a set of at most $\aleph$ of the $C_i^{(0)}$. Since this sum is $\aleph \cdot \aleph_0 = \aleph$-generated, its image in M is $\aleph$-generated and hence $\aleph$-related. The kernel is then contained in some sum S_1 of at most $\aleph$ of the $C_i^{(1)}$. Continue down the resolution in this manner until you reach a sum S_n of at most $\aleph$ of the $C_i^{(n)}$. Now go forward along the resolution. Let $T_{n,0} = S_n$. The image of $T_{n,0} + S_{n-1}$ is contained in some sum $T_{n-1,0}$ of at most $\aleph$ of the $C_i^{(n-1)}$. Once we have $T_{j,0}$ as a sum of at most $\aleph$ of the $C_i^{(j)}$ we let $T_{j-1,0}$ be a sum of at most $\aleph$ of the $C_i^{(j-1)}$ which contains both the image of $T_{j,0}$ in P_{j-1} and S_{j-1}. Let L_0 be generated by some subset of $\{m_\beta\}$ of cardinality at most $\aleph$, with L_0 containing the image of $T_{0,0}$. Once we have

$$0 \to T_{n,h} \to T_{n-1,h} \to \cdots \to T_{0,h} \to L_h \to 0$$

for a nonnegative integer h, we repeat the above procedure with the additional property that each S_i contains the kernel of the map $T_{i,h} \to T_{i-1,h}$

(or L_h). This kernel must be $\aleph$-generated since the $T_{i,h}$ are $\aleph$-generated and all the images are contained in frees.

By induction we get such a complex for every nonnegative integer h, and the kernel of each outgoing map at stage h is contained in the image of the incoming map at stage $h+1$, so the complex

$$0 \to \bigcup_{h=0}^{\infty} T_{n,h} \to \bigcup_{h=0}^{\infty} T_{n-1,h} \to \cdots \to \bigcup_{h=0}^{\infty} T_{0,h} \to \bigcup_{h=0}^{\infty} L_h \to 0$$

is exact. By construction each $\bigcup_{h=0}^{\infty} T_{j,h}$ is projective and $\aleph$-generated, and $\bigcup_{h=0}^{\infty} T_{n,h}$ is a direct summand of P_n. By Lemma 2.5(b),

$$\mathrm{pd}\left(M \Big/ \bigcup_{h=0}^{\infty} L_h\right) \leq n.$$

Note that $\bigcup_{h=0}^{\infty} L_h$ is generated by a subset of $\{m_\beta : \beta \in \mathcal{I}\}$.

We are now ready to prove the theorem. Set

$$\mathcal{D} = \{\alpha = \langle 0 \to T_n \to T_{n-1} \to \cdots \to T_0 \to M_\alpha \to 0 \rangle\}$$

which can arise in this manner. That is, a sequence

$$\langle 0 \to T_n \to T_{n-1} \to \cdots \to T_0 \to M_\alpha \to 0 \rangle$$

is in $\mathcal{D}$ provided it is exact, M_α is generated by a set $\{m_\beta : \beta \in \mathcal{I}_\alpha\} \subseteq \{m_\beta : \beta \in \mathcal{I}\}$ with $\mathrm{card}\,(\mathcal{I}_\alpha) \leq \aleph$, and each T_i is a sum of some set of at most $\aleph$ of the $C_j^{(i)}$. Order $\mathcal{D}$ by containment at every slot in the sequence. To show that $\mathcal{D}$ is directed, we will actually show more—namely that every subset of $\mathcal{D}$ of cardinality at most $\aleph$ has an upper bound. Let

$$\left\{\langle 0 \to T_n^{(\alpha)} \to T_{n-1}^{(\alpha)} \to \cdots \to T_0^{(\alpha)} \to M_\alpha \to 0 \rangle : \alpha \in \mathfrak{J}\right\} \subseteq \mathcal{D}$$

have cardinality at most $\aleph$. In the construction of the first part of this proof, take the initial S_i to be $\sum_\alpha T_i^{(\alpha)}$ and proceed with the construction. This will clearly give an upper bound. The other required conditions are equally clear. $\square$

Note that this proof implies that the set

$$\{M_\alpha : \alpha \in \mathcal{D}\}$$

as a set of submodules of M ordered by inclusion is a directed set, but it does *not* imply that it is closed under unions of chains. It is likely that you will have some modules $M_\alpha \subseteq M_\beta$ without α and β being comparable in $\mathcal{D}$.

The proof of the above theorem has an immediate consequence which is a form of converse to Auslander's Lemma for modules which are not countably generated.

Theorem 3.6. Let R be an $\aleph_1$-coherent ring and $\aleph > \aleph_0$. Let M be a strictly $\aleph$-generated submodule of a free R-module, and let $\{m_\alpha : \alpha < \aleph\}$ be a set of generators for M. Assume $pd(M) \leq n < \infty$. Then there is a well ordered (by inclusion) family $\{M_\beta : \beta < \Lambda\}$ of submodules of M such that

(a) $M = \bigcup_{\beta < \Lambda} M_\beta$.
(b) Each M_β is generated by some subset $S_\beta \subset \{m_\alpha : \alpha < \aleph\}$ where $card(S_\beta) < \aleph$ and S_β contains the set $\{m_\gamma : \gamma < \lambda(\beta)\}$ for some $\lambda(\beta) < \aleph$.
(c) $pd\left(M_\beta / \bigcup_{\gamma < \beta} M_\gamma\right) \leq n$ for all $\beta < \Lambda$.

Proof. Modify the proof of Theorem 3.5 by taking $\mathcal{D}$ as all sequences

$$\langle 0 \to T_n \to T_{n-1} \to \cdots \to T_0 \to M_\alpha \to 0 \rangle$$

where each module in the sequence is generated by fewer than $\aleph$ elements. Let Λ be the cofinality of $\aleph$, that is, the smallest ordinal for which there is a $1-1$ order preserving unbounded function $\lambda : \Lambda \to \aleph$. Let $\beta \in \Lambda$, and assume we have a chain of elements

$$\{\langle 0 \to T_n \to T_{n-1} \to \cdots \to T_0 \to M_\gamma \to 0 \rangle : \gamma < \beta\} \subset \mathcal{D}$$

with each entry in every sequence generated by at most $card(\lambda(\beta))$ elements. The union of this chain is in $\mathcal{D}$, and every module in this union, which is an exact sequence, is $card(\lambda(\beta))$-generated. Then let

$$\langle 0 \to T_n \to T_{n-1} \to \cdots \to T_0 \to M_\beta \to 0 \rangle$$

be an upper bound of $\{\langle 0 \to T_n \to T_{n-1} \to \cdots \to T_0 \to M_\gamma \to 0 \rangle : \gamma < \beta\}$ such that $M_\beta \supseteq \{m_\alpha : \alpha < \lambda(\beta)\}$ and M_β is $card(\lambda(\beta))$-generated. As in Theorem 3.5, the quotients have the correct projective dimension. □

How does the above give a lower bound on projective dimension? One takes intersections as in the following theorem, and hopes that enough properties of the original situation hold that one can do induction.

Theorem 3.7. Let R be any ring, and let $K, L \subseteq M$ be R-modules. Assume $pd(M)$, $pd(M/K)$, and $pd(K+L)$ are all $\leq n$. Then if $pd(L) < n$ we have $pd(L \cap K) < n$.

Proof. In the exact sequence

$$0 \longrightarrow K + L \longrightarrow M \longrightarrow M/(K+L) \longrightarrow 0$$

we have the kernel and middle module of projective dimension at most n. By Lemma 2.5(a), $pd(M/(K+L)) \leq n+1$. Then in the exact sequence

$$0 \longrightarrow (K+L)/K \longrightarrow M/K \longrightarrow M/(K+L) \longrightarrow 0$$

we have middle module of projective dimension $\leq n$ and the cokernel of projective dimension $\leq n+1$. By Lemma 2.5(c), $\operatorname{pd}((K+L)/K) \leq n$. Then in the exact sequence

$$0 \longrightarrow L/(K \cap L) \longrightarrow L \longrightarrow (K+L)/K \longrightarrow 0$$

we have the cokernel with projective dimension $\leq n$ and the middle with projective dimension $< n$, so again by Lemma 2.5(c), $\operatorname{pd}(L/(K \cap L)) \leq \operatorname{pd}(L) < n$. □

4. Computing projective dimensions

We now give some situations where we can actually compute projective dimension of modules over a von Neumann ring.

Our first lower bound result uses the idea of independence of idempotents.

Definition. A set $\mathcal{E} = \{e_\alpha : \alpha \in \mathfrak{J}\}$ of commuting idempotents is called **independent** if for any disjoint pair $\mathfrak{f}$ and $\mathfrak{g}$ of finite subsets of $\mathfrak{J}$,

$$\prod_{\alpha \in \mathfrak{f}} e_\alpha \prod_{\beta \in \mathfrak{g}} (1 - e_\beta) \neq 0.$$

In a ring R in which every finite sum or finite intersection of idempotent generated cyclic right ideals is generated by an idempotent, a family $\tilde{\mathcal{E}} = \{e_\alpha R : \alpha \in \mathfrak{J}\}$ of cyclic ideals is called **independent** if for any disjoint pair $\mathfrak{f}$ and $\mathfrak{g}$ of finite subsets of $\mathfrak{J}$, $\bigcap_{\alpha \in \mathfrak{f}} e_\alpha R$ is not contained in $\sum_{\beta \in \mathfrak{g}} e_\beta R$. This definition agrees with the previous one if the idempotents $\{e_\alpha : \alpha \in \mathfrak{J}\} \subseteq R$ happen to commute.

A family $\tilde{\mathcal{E}} = \{e_\alpha R : \alpha \in \mathfrak{J}\}$ of idempotent generated right ideals in a ring R is **almost independent** provided

(1) either any finite sum or finite intersection of idempotent generated cyclic right ideals of R is generated by an idempotent or the idempotents e_α commute,
(2) for any disjoint pair $\mathfrak{f}$ and $\mathfrak{g}$ of finite subsets of $\mathfrak{J}$, $\bigcap_{\alpha \in \mathfrak{f}} e_\alpha R \neq 0 \Rightarrow \bigcap_{\alpha \in \mathfrak{f}} e_\alpha R$ is not contained in $\sum_{\beta \in \mathfrak{g}} e_\beta R$, and
(3) $\left(\bigcap_{i=1}^n e_{\alpha_i} \mathfrak{R}\right) \cap \sum_{j=1}^m e_{\beta_j} \mathfrak{R} = \sum_{j=1}^m \left(\bigcap_{i=1}^n e_{\alpha_i} \mathfrak{R} \cap e_{\beta_j} \mathfrak{R}\right)$ for all $\{\alpha_i, \beta_j\} \subseteq \mathfrak{J}$.

Theorem 4.1. Let $\tilde{\mathcal{E}} = \{e_\alpha R : \alpha \in \mathfrak{J}\} \cup \{f_\beta R : \beta \in \mathfrak{J}'\}$ be an almost independent family of idempotent generated ideals in a ring R. Also assume that $\{e_\alpha R : \alpha \in \mathfrak{J}\}$ is actually independent and of cardinality $\aleph_\kappa$ for some nonnegative integer κ. Let $M = \sum_{\varepsilon R \in \tilde{\mathcal{E}}} \varepsilon R$. Then $\operatorname{pd}(M) \geq \kappa$.

Proof. If $\kappa = 0$, there is nothing to prove. Now assume that $\kappa > 0$ and the theorem is true whenever $\operatorname{card}(\{e_\alpha R : \alpha \in \mathfrak{J}\}) < \aleph_\kappa$.

Let $\text{pd}(M) = \lambda < \kappa$. Then let $\tilde{\mathcal{E}}'$ be any subset of $\{e_\alpha R : \alpha \in \mathcal{I}\}$ of cardinality $\aleph_\lambda$. By Theorem 3.5, there is some L generated by a subset $\mathcal{F}$ of $\tilde{\mathcal{E}}$ with $\mathcal{F} \supseteq \tilde{\mathcal{E}}'$, $\text{card}(\mathcal{F}) = \aleph_\lambda$, and $\text{pd}(M/L) \leq \lambda$. Then there is some $eR \in \{e_\alpha R : \alpha \in \mathcal{I}\} \setminus \mathcal{F}$, and the set

$$\left\{ \varepsilon R = eR \cap e'R : e'R \in \tilde{\mathcal{E}}' \right\}$$

is independent of cardinality $\aleph_\lambda$. Moreover, the set

$$\tilde{\mathcal{F}} = \{\varepsilon' R : \varepsilon' R = eR \cap \varepsilon_1 R, \ \varepsilon_1 R \in \mathcal{F}\}$$

is almost independent. By Theorem 3.3, $\text{pd}(L + eR) \leq \lambda$ so by Theorem 3.7, $\text{pd}(eR \cap L) < \lambda$. But by definition of almost independent, $eR \cap L$ is generated by $\tilde{\mathcal{F}}$, and this contradicts the induction hypothesis. $\square$

We can now put together appropriate upper and lower bounds to get ideals generated by commuting idempotents (or ideals in a von Neumann regular ring) with small projective dimension and large strict generation that are not trivially a direct sum of modules with small projective dimension. I believe this is the first time that such a construction has appeared for arbitrary finite projective dimension. To avoid the kind of technicalities found in the definition of almost independent idempotent generated ideals, we will give an example with commuting idempotents.

Example 1. *Let Ω be an ordinal $> \aleph_\kappa$ for κ a nonnegative integer. Let $\mathfrak{B}$ be a free Boolean ring on a set $\{\varepsilon_\alpha : \alpha < \Omega\}$, that is,*

$$\mathfrak{B} = \mathbb{Z}/2\mathbb{Z}\left[\{\varepsilon_\alpha : \alpha < \Omega\}\right]$$

where the $\{\varepsilon_\alpha : \alpha < \Omega\}$ is a set of commuting, independent idempotents. For each α with $\aleph_\kappa \leq \alpha < \Omega$, let I_α be a set of ordinals $< \alpha$ such that $\text{card}(\alpha \setminus I_\alpha) \leq \aleph_{\kappa-1}$. Now let R be a ring containing a set $\{e_\alpha : \alpha < \Omega\}$ of commuting idempotents. These idempotents generate a Boolean ring under the operations of $e \boxplus f = e(1-f) + f(1-e)$ and ring multiplication. There is a homomorphism from $\mathfrak{B}$ to this Boolean ring taking ε_α to e_α. Assume the kernel of this homomorphism is generated by $\{\varepsilon_\alpha \varepsilon_\beta : \beta \in I_\alpha, \aleph_\kappa \leq \alpha < \Omega\}$. Then $\text{pd}\left(\sum_{\alpha<\Omega} e_\alpha R\right) = \kappa$.

Proof. By hypothesis, there are no relations on $\{e_\alpha : \alpha < \aleph_\kappa\}$ so that is an independent set of idempotents. Theorem 4.1 gives κ as a lower bound for the projective dimension of $\sum_{\alpha<\Omega} e_\alpha R$, and Corollary 3.4 gives κ as an upper bound. $\square$

The next theorem concerns a situation somewhat opposite to the above one.

Theorem 4.2. *Let κ be a nonnegative integer, R any ring, and*
$$\tilde{\mathcal{E}} = \{e_\alpha R : \alpha < \Omega\}$$
where Ω is an ordinal, $e_\alpha = e_\alpha^2 \in R$, and $e_\beta R \subsetneq e_\alpha R$ if $\beta < \alpha$. Then $\operatorname{pd}\left(\bigcup_{\alpha<\Omega} e_\alpha R\right) = \kappa$ if and only if the cofinality of Ω is $\aleph_\kappa$.

Proof. Note that, for any idempotents e and f, if $ef = f$ then $g = e - fe$ is an idempotent with $fg = 0$ and $eR = fR \oplus gR$ so
$$\sum_{i=0}^\infty e_{\alpha_i} R = \bigoplus_{i=0}^\infty \left(e_{\alpha_i} - e_{\alpha_{i-1}} e_{\alpha_i}\right) R$$
is projective where $\alpha_i < \alpha_{i+1}$ for all $i \in \omega$ (here $e_{\alpha_{-1}} = 0$). Thus the ideal generated by a countable subset of $\tilde{\mathcal{E}}$ is projective because the subset is well ordered by inclusion and has countable cofinality (ω or 1).

If the cofinality of Ω is at most $\aleph_\kappa$, that is, if $M = \bigcup_{\alpha<\Omega} e_\alpha R$ is $\aleph_\kappa$-generated, then by Theorem 3.3 $\operatorname{pd}(M) \leq \kappa$.

Now assume $\operatorname{pd}(M) \leq \kappa$. If $\kappa = 0$, M must be projective. Then it is a direct sum of countably generated modules. Let $C = \sum_{i=0}^\infty c_i R$ be a strictly $\aleph_0$-generated direct summand. By the well ordering of $\tilde{\mathcal{E}}$, for each $i \in \omega$ there is a smallest $e_{\alpha_i} R$ with $e_{\alpha_i} R \supseteq \sum_{j=0}^i c_j R$. If $\{e_{\alpha_i} R\}$ did not generate M, $\{e_{\alpha_i} R\}$ would be bounded above by some $e_\alpha R$. Then C would be a direct summand of $e_\alpha R$. Since a direct summand of a cyclic is cyclic, C would be cyclic, contradicting the assumption that it is strictly $\aleph_0$-generated. Thus C cannot be bounded above by any $e_\alpha R$, so $\{e_{\alpha_i} R\}$ generates M.

Now assume $\kappa > 0$ and any well ordered union N of idempotent generated cyclic modules with $\operatorname{pd}(N) < \kappa$ is $\aleph_\lambda$-generated for some $\lambda < \kappa$. If M is not $\aleph_\kappa$-generated, by Theorem 3.5 there is some ascending chain $\left\{\sum_{\beta<\alpha_\nu} e_\beta R\right\}$ of order type $\aleph_\kappa$ with $\operatorname{pd}\left(M \Big/ \sum_\nu \sum_{\beta<\alpha_\nu} e_\beta R\right) \leq \kappa$. Since the ideals $e_\alpha R$ are linearly ordered, the ideal $\sum_\nu \sum_{\beta<\alpha_\nu} e_\beta R$ must be strictly $\aleph_\kappa$-generated. Then there is some α for which $e_\alpha \notin \sum_\nu \sum_{\beta<\alpha_\nu} e_\beta R$. By linear ordering,
$$e_\alpha R \cap \sum_\nu \sum_{\beta<\alpha_\nu} e_\beta R = \sum_\nu \sum_{\beta<\alpha_\nu} e_\beta R.$$
Since $\operatorname{pd}(e_\alpha R) = 0 < \kappa$, we may apply Theorem 3.7, which gives
$$\operatorname{pd}\left(e_\alpha R \cap \sum_\nu \sum_{\beta<\alpha_\nu} e_\beta R\right) < \kappa,$$
a contradiction. Hence M must be $\aleph_\kappa$-generated. $\square$

Since our previous example was for commutative rings, we will give one in the not necessarily commutative case. We will also give a commutative example.

Example 2. *Let $\mathfrak{R}$ be the ring of all linear transformations on a right $\aleph_\kappa$-dimensional vector space V over a division ring D. Then $\mathfrak{R}$ is von Neumann regular, and any cyclic right ideal $\lambda\mathfrak{R}$ is determined completely by the image of λ. Since V is a well-ordered union of $\aleph_\kappa$ subspaces, $\mathfrak{R}$ has an ideal $\bigcup_{\alpha<\aleph_\kappa} e_\alpha \mathfrak{R}$ where $\alpha < \beta \implies \text{image}(e_\alpha) \subset \text{image}(e_\beta)$. The e_α commute only if you carefully select their kernels. Then for any such ideal, $pd\left(\bigcup_{\alpha<\aleph_\kappa} e_\alpha \mathfrak{R}\right) = \kappa$.*

The full linear rings used in Example 2 actually have ideals with larger projective dimension. You can at a minimum find an independent set of commuting idempotents in such a ring with cardinality $2^{\aleph_\kappa}$.

Example 3. *Let $\mathfrak{B}$ be the Boolean algebra generated by the set of all initial segments $s(\alpha) = \{\beta : \beta < \alpha\}$ and their complements, where $\alpha \in \aleph_\kappa$ is not a limit ordinal. This is the Boolean algebra of all clopen sets of $\aleph_\kappa$ under the order topology. Then the global dimension of $\mathfrak{B} = \kappa + 1$.*

Proof. Since the cardinality of $\mathfrak{B}$ is $\aleph_\kappa$, by Theorem 3.3, every ideal has dimension $\leq \kappa$, so by the Global Dimension Theorem, the global dimension of $\mathfrak{B}$ is at most $\kappa + 1$. Since the ideal generated by $\{s(\alpha) : \alpha \in \aleph_\kappa\}$ is an ascending union of $\aleph_\kappa$ cyclics, it must have projective dimension κ so the global dimension is at least $\kappa + 1$. □

We now have theorems which enable us to compute the projective dimension of some messy ideals. Now let us look at some ways of combining ideals of known projective dimension to get other ideals of known projective dimension.

Lemma 4.3. *Let $K, N \subseteq M$ be modules over a ring R, with*

$$pd(K) = k < \infty$$
$$pd(N) = n \leq k$$
$$pd(K \cap N) = m < \infty$$

Then

$$pd(K+N) \begin{cases} \leq k+1 & \text{if } m = k \\ = k & \text{if } m < k \\ = m+1 & \text{if } m > k \end{cases}.$$

Proof. Note that the projective dimension of a direct sum is the supremum of the projective dimensions of the summands. Then this lemma is an immediate consequence of Lemma 2.5 and the fact that there is an exact sequence

$$0 \longrightarrow K \cap N \longrightarrow K \oplus N \longrightarrow K + N \longrightarrow 0. \quad \square$$

Theorem 4.4. *Let $\kappa < \omega$, and let $\{M_\alpha : 0 \leq \alpha \leq \kappa < \omega\}$ be a finite family of submodules of a module M over an arbitrary ring R. Assume $pd(M_0) = n < \infty$ and for all α with $1 \leq \alpha \leq \kappa$, $pd(M_\alpha) \leq n$ and $pd\left(M_\alpha \cap \sum_{\beta=0}^{\alpha-1} M_\beta\right) \leq n - 1$. Then $pd(\sum_{\alpha=0}^{\kappa} M_\alpha) = n$.*

Proof. Use finite induction and apply Lemma 4.3. □

This gives us a way of making somewhat messier ideals of a given projective dimension from perhaps less messy ones. We give a couple of examples.

Corollary 4.5. *Let $\kappa < \omega$, and let M_0 be an ideal with projective dimension κ in a von Neumann regular ring $\mathfrak{R}$. For example, M_0 could be an ascending union of cyclic submodules $M = \bigcup_{\beta < \aleph_\kappa} e_\beta \mathfrak{R}$ where $e_\beta \mathfrak{R} \subsetneq e_\alpha \mathfrak{R}$ if $\beta < \alpha$ as in Theorem 4.2. Let M_1 be any ideal of $\mathfrak{R}$ with $pd(M_1) \leq \kappa$ such that $M_1 \cap M_0$ is an $\aleph_{\kappa-1}$-generated ideal. Then $pd(M_0 + M_1) = \kappa$.*

Proof. Apply Theorem 3.3 and Theorem 4.4. □

Corollary 4.6. *Let M be an ideal in a von Neumann regular ring $\mathfrak{R}$ with $pd(M) = \kappa > 0$, and let $K \subseteq M$ have $pd(M/K) = \kappa$. Then for any countably generated ideal $C \subseteq M$ we have $pd(K + C) = \kappa \iff pd(K \cap C) < \kappa$.*

Proof. Apply Theorem 4.4 for $\Leftarrow$ and Theorem 3.7 for $\Rightarrow$. □

Theorem 4.4 leads to an interesting question. What if, in a von Neumann regular ring $\mathfrak{R}$, you have a sum of ideals indexed by some infinite ordinal Ω with $pd(M_\beta) = \kappa < \infty$ (or $pd(M_0) = \kappa$ and $pd(M_\alpha) \leq \kappa$) for all $\beta < \Omega$ and $pd\left(M_\beta \cap \sum_{\alpha < \beta} M_\alpha\right) \leq \kappa - 1$? By Lemma 2.5 and Auslander's lemma, $pd\left(\sum_{\alpha < \Omega} M_\alpha\right) \leq \kappa$. Can this dimension actually be less than κ? If $\kappa = 0$, the answer is clearly no. What if $\kappa \geq 1$? One can immediately reduce this question to looking at the smallest ordinal Ω for which such a sum exists so that every sum over an initial segment has dimension κ. This minimum Ω must be a regular cardinal. But there is no restriction on the cardinalities of generating sets for the M_α or on the properties of M_0 that might help us show that its projective dimension is $\geq \kappa$. Note that the condition on projective dimension of intersections with smaller sums is an essential part of this question. Otherwise it would be easy to reduce the dimension—just look at the embedding of M in a free module (which is projective).

Conjecture 1. *Let M be a submodule of a free module over a von Neumann regular ring $\mathfrak{R}$. Assume M is a well ordered sum $M = \sum_{\alpha < \Omega} M_\alpha$ such that for all $\alpha < \Omega$, $pd(M_\alpha) = \kappa < \infty$ and $pd\left(M_\alpha \cap \sum_{\beta < \alpha} M_\beta\right) \leq \kappa - 1$. Then $pd(M) = k$.*

There are close connections between this conjecture and the major conjecture of this paper, which appears in the next section.

5. A LATTICE THEORETIC DIMENSION

The work on projective dimension in this paper gives rise to an ordinal valued invariant that one can define on bounded complemented distributive lattices. Such lattices are precisely the lattices of idempotent generated ideals in some commutative von Neumann regular ring. We will define this invariant on more than just the ideals of the ring.

Definition. *Let $\mathfrak{R}$ be a von Neumann regular ring. We define a function ℓ from the class of all submodules of free $\mathfrak{R}$-modules to the class of all ordinals $\cup \{-1\}$ by:*

(1) $\ell(M) = -1$ if and only if $M = 0$.
(2) $\ell(M) \leq \alpha$ if and only if there exists an ordinal Ω and some set $\{m_\beta : \beta < \Omega\}$ of generators for M such that, for all $\beta < \Omega$,

$$\ell\left(m_\beta \mathfrak{R} \cap \sum_{\gamma < \beta} m_\gamma \mathfrak{R}\right) < \alpha.$$

(3) $\ell(M) = \inf\{\alpha : \ell(M) \leq \alpha\}$.

Let us look at some properties of this definition. The modules M looked at in the rest of this section will be submodules of projective $\mathfrak{R}$-modules.

Proposition 5.1. $\ell(M) \leq 0$ *if and only if M is projective.*

Proof. This merely says that $\ell(M) \leq 0$ if and only if M is a direct sum of cyclic submodules. Since M is a submodule of a free $\mathfrak{R}$-module, this is equivalent to projectivity. □

Proposition 5.2. *Let κ be an ordinal. Then $\ell(M) \leq \kappa$ if M is $\aleph_\kappa$-generated. Hence $\ell(M)$ is defined for all M.*

Proof. If M is countably generated, then it is projective and $\ell(M) = \text{pd}(M) = 0$. Now assume the result for all $\kappa' < \kappa$ and let M be $\aleph_\kappa$-generated. Index a set of generators $\{m_\alpha : \alpha < \aleph_\kappa\}$ for M by the smallest ordinal of cardinality $\aleph_\kappa$. Then each $\sum_{\beta < \alpha} m_\beta \mathfrak{R}$ is generated by fewer than $\aleph_\kappa$ elements for $\alpha < \aleph_\kappa$. By the coherence of $\mathfrak{R}$, $m_\alpha \mathfrak{R} \cap \sum_{\beta < \alpha} m_\beta \mathfrak{R}$ is also generated by fewer than $\aleph_\kappa$ elements. By the induction hypothesis, $\ell\left(m_\alpha \mathfrak{R} \cap \sum_{\beta < \alpha} m_\beta \mathfrak{R}\right) < \kappa$, so by definition $\ell(M) \leq \kappa$. □

Proposition 5.3. $\ell(M) \geq \text{pd}(M)$ *if $\ell(M) < \omega$.*

Proof. If $\ell(M) = 0$, we have seen that M is projective. Now let $\ell(M) = n \geq 1$, and assume $\ell(M') \geq \text{pd}(M')$ if $\ell(M') < n$. Let $\{m_\alpha : \alpha < \Omega\}$ be a set of generators for M with $\ell\left(m_\alpha \mathfrak{R} \cap \sum_{\beta < \alpha} m_\beta \mathfrak{R}\right) \leq n - 1$. By the induction hypothesis, $\text{pd}\left(m_\alpha \cap \sum_{\beta<\alpha} m_\beta \mathfrak{R}\right) \leq \ell\left(m_\alpha \cap \sum_{\beta<\alpha} m_\beta \mathfrak{R}\right) \leq n-1$. By a standard isomorphism theorem,

$$\sum_{\beta\leq\alpha} m_\beta \mathfrak{R} \Big/ \sum_{\beta<\alpha} m_\beta \mathfrak{R} \cong m_\alpha \mathfrak{R} \Big/ m_\alpha \mathfrak{R} \cap \sum_{\beta<\alpha} m_\beta \mathfrak{R}.$$

Since we are working with submodules of a free $\mathfrak{R}$-module, $m_\alpha \mathfrak{R}$ is projective, so

$$\text{pd}\left(m_\alpha \mathfrak{R} \Big/ m_\alpha \mathfrak{R} \cap \sum_{\beta<\alpha} m_\beta \mathfrak{R}\right) \leq n.$$

Since

$$M = \bigcup_{\alpha<\Omega} \left(\sum_{\beta<\alpha} m_\beta \mathfrak{R}\right),$$

Auslander's Lemma completes the proof. □

Except that it is contrary to usual terminology, one could let $\text{pd}(M) = -1$ iff M is finitely generated projective. Then one would let $\ell(M) = -1$ if and only if M is finitely generated. Then Propositions 5.1, 5.2, and 5.3 would be true for all ordinals κ and for $\kappa = -1$.

Conjecture 2. $\ell(M) = \text{pd}(M)$ *if either of the two is finite.*

We conclude with some observations concerning this conjecture.

Remark 5.1. *If Conjecture 2 is true, then Wiegand's conjecture is true. This is because $\ell(M)$ for M an ideal of $\mathfrak{R}$ is defined strictly in terms of the lattice structure of direct summands, and not in terms of the additive structure.*

Remark 5.2. *If $\text{pd}(M) = n$ and M is $\aleph_n$-generated, $\ell(M) = \text{pd}(M)$.*

Remark 5.3. *Assume Conjecture 2 is false. Let n be the smallest positive integer for which there is a submodule M of a free $\mathfrak{R}$-module with $\text{pd}(M) = n < \ell(M)$. Find such an M with smallest possible generating set. Then by Theorem 3.6, M is a directed union of submodules M_α each of which has $\ell(M_\alpha) = \text{pd}(M_\alpha)$ since they are generated by fewer elements than M. The quotients $M_\alpha / \bigcup_{\beta<\alpha} M_\beta$ all have projective dimension $\leq n$. If we could show that $\ell(M_\alpha) \leq n$ and $\ell\left(\bigcup_{\beta<\alpha} M_\beta\right) \leq n$ implies $\ell\left(M_\alpha / \bigcup_{\beta<\alpha} M_\beta\right) \leq n$ whenever that quotient is embeddable in a free, we might be able to prove the conjecture by starting at the first kernel in a projective resolution of an ideal rather than the ideal itself.*

Remark 5.4. It is tempting to try and modify the proof of Theorem 3.5 to prove Conjecture 2. This runs into the problem that the ring $\mathfrak{R}$ may contain two projective ideals J and K such that $\text{pd}\,(\mathfrak{R}/J)$ and $\text{pd}\,(\mathfrak{R}/K)$ are both $= 1$, but $\text{pd}\,(\mathfrak{R}/(J+K))$ is 2. Thus even in the case of projective dimension 1, the fact that nice kernels associated with projective ideals are direct summands of $\partial_1 P_1$ does not mean that the sum of the ideals will have a corresponding kernel which is a direct summand of $\partial_1 P_1$.

Remark 5.5. It is not clear how to do the following, but looking at it might give insights into the conjecture. In Example 1 one has a set of relations $\{e_\alpha e_\beta = 0\}$ where, for some specific ordering Ω of the indexing set, we can compute simultaneously both $\ell(M)$ and $\text{pd}\,(M)$. Just giving which pairs (e_α, e_β) have product $e_\alpha e_\beta = 0$ and knowing that in the specific ordering Ω, for any $\alpha \geq \aleph_\kappa$ this set of pairs will have $\{\beta : e_\alpha \cdot e_\beta = 0 \text{ and } \beta < \alpha\}$ with complement in α of cardinality $< \aleph_\kappa$, can one reconstruct an order similar to the specific order Ω. That is, if I choose a different ordering of the set of idempotents, how can you find an ordering in which one can compute an upper bound for $\ell(M)$ and $\text{pd}\,(M)$ simultaneously?

Remark 5.6. Since the only way currently known to get upper bounds on projective dimension of specific ideals actually compute an upper bound for $\ell(M)$, disproving the conjecture would require new techniques for getting upper bounds. Conjecture 1 is an indication that new methods for getting lower bounds may also be needed.

We do have some evidence that the Conjecture 2 might be true, at least in the commutative case. The following proposition 'just' reduces a proof of the conjecture to handling unions. This is actually the first proof in this section where commutativity of the idempotents matters.

Proposition 5.4. *If $\mathfrak{R}$ is commutative and $I \subseteq J$ are ideals of $\mathfrak{R}$ with $\text{pd}\,(I) \leq n$ and $\text{pd}\,(J/I) \leq n$, then for all $e = e^2 \in J$, $\text{pd}\,(J/(I + e\mathfrak{R})) \leq n$.*

Proof. Multiplication by e gives a map from J/I to $(I + e\mathfrak{R})/I$ which splits the exact sequence

$$0 \longrightarrow (I + e\mathfrak{R})/I \longrightarrow J/I \longrightarrow J/(I + e\mathfrak{R}) \longrightarrow 0. \quad \square$$

Another possible attack on the Wiegand problem goes as follows. Let $\langle \mathfrak{B}, +, \cdot \rangle$ be any Boolean rng (not necessarily with identity), that is, $e^2 = e$ for all $e \in \mathfrak{B}$. Form the semigroup ring $\mathbb{Z}\,[\langle \mathfrak{B} \cup \{1\}, \cdot \rangle]$ and mod out the ideal generated by

$$\{e_1 + e_2 - (e_1 + e_2) : e_1, e_2 \in \mathfrak{B},\ e_1 \cdot e_2 = 0\}.$$

The result is a commutative ring whose idempotents have the same lattice structure as $\mathfrak{B}$ but the ring has no additive torsion. Call this ring $\mathbb{Z}\langle \mathfrak{B} \rangle$, that is,

$$\mathbb{Z}\langle \mathfrak{B} \rangle = \mathbb{Z}\,[\langle \mathfrak{B} \cup \{1\}, \cdot \rangle] / \langle \{e_1 + e_2 - (e_1 + e_2) : e_1, e_2 \in \mathfrak{B},\ e_1 \cdot e_2 = 0\} \rangle.$$

Proposition 5.5. *Let R be any ring, and let $\mathcal{E} = \{e_\alpha : \alpha \in \mathfrak{I}\}$ be a set of commuting idempotents in R which generate the Boolean ring $\langle \mathfrak{B}, \boxplus, \boxtimes \rangle$ where $e \boxplus f = e+f-2ef$ and $e \boxtimes f = ef$. Then $\mathrm{pd}_R(\mathcal{E}R) \leq \mathrm{pd}_{\mathbb{Z}\langle\mathfrak{B}\rangle}(\mathcal{E}\mathbb{Z}\langle\mathfrak{B}\rangle)$.*

Proof. The inclusion $\mathcal{E}\mathbb{Z}\langle\mathfrak{B}\rangle \hookrightarrow \mathbb{Z}\langle\mathfrak{B}\rangle$ is pure, so there is a pure exact projective resolution

$$\cdots \longrightarrow P_n \longrightarrow P_{n-1} \longrightarrow \cdots \longrightarrow P_0 \longrightarrow \mathcal{E}\mathbb{Z}\langle\mathfrak{B}\rangle \longrightarrow 0$$

where the P_i are $\mathbb{Z}\langle\mathfrak{B}\rangle$-projective. Then R is a left $\mathbb{Z}\langle\mathfrak{B}\rangle$-module in the obvious way, and by purity we get an exact R-projective resolution

$$\cdots \to P_n \otimes_{\mathbb{Z}\langle\mathfrak{B}\rangle} R \to P_{n-1} \otimes_{\mathbb{Z}\langle\mathfrak{B}\rangle} R \to \cdots \to P_0 \otimes_{\mathbb{Z}\langle\mathfrak{B}\rangle} R \to \mathcal{E}R \to 0.$$

If the image of any $P_n \longrightarrow P_{n-1}$ is projective, then so is the image of $P_n \otimes_{\mathbb{Z}\langle\mathfrak{B}\rangle} R \to P_{n-1} \otimes_{\mathbb{Z}\langle\mathfrak{B}\rangle} R$. □

All of our computations of projective dimensions of special kinds of ideals in von Neumann regular rings work for this ring $\mathbb{Z}\langle\mathfrak{B}\rangle$ also. The definition of $\ell(M)$ also makes sense for this ring, especially if we restrict to pure submodules of free $\mathbb{Z}\langle\mathfrak{B}\rangle$-modules. I suspect any proof or disproof of Wiegand's question or of Conjecture 2 will also give the same result for $\mathbb{Z}\langle\mathfrak{B}\rangle$.

References

[A] M. Auslander, *On dimensions of rings and modules, III: Global dimension*, Nagoya Math. J. **9** (1955), 67–77.

[K] I. Kaplansky, *Projective modules/*, Ann. of Math. **68** (1958), 372–377.

[O67] B. L. Osofsky, *Global dimension of valuation rings*, Trans. Amer. Math. Soc. **127** (1967), 136–149.

[O70] B. L. Osofsky, *Homological dimension and cardinality*, Trans. Amer. Math. Soc. **151** (1970), 641–649.

[O78] B. L. Osofsky, *Projective dimension of "nice" directed unions*, J. Pure and Applied Algebra **13** (1978), 179–219.

[P67] R. S. Pierce, *The global dimension of Boolean rings*, Journ. of Alg. **7** (1967), 91–99.

[P76] R. S. Pierce, *The global dimension of commutative regular rings*, Houston J. Math **2** (1976), 97–110.

[W] R. Wiegand, *Some topological invariants of Stone spaces*, Mich. Math. Jour. **16** (1969), 289–296.

HOMOLOGICAL PROPERTIES OF COLOR LIE SUPERALGEBRAS

KENNETH L. PRICE

ABSTRACT. Let $\mathcal{L} = \mathcal{L}_+ \oplus \mathcal{L}_-$ be a finite dimensional color Lie superalgebra over a field of characteristic 0 with universal enveloping algebra $U(\mathcal{L})$. We show that $\text{gldim}(U(\mathcal{L}_+)) = \text{lFPD}(U(\mathcal{L})) = \text{rFPD}(U(\mathcal{L})) = \text{injdim}_{U(\mathcal{L})}(U(\mathcal{L})) = \dim(\mathcal{L}_+)$. We also prove that $U(\mathcal{L})$ is Auslander-Gorenstein and Cohen-Macaulay and thus that it has a QF classical quotient ring.

1. INTRODUCTION

Color Lie superalgebras are graded over an Abelian group G and generalize Lie superalgebras. Background information on Lie superalgebras and color Lie superalgebras can be found in [7] and [13] and [1] and [14], respectively. We will show, in a forthcoming paper [12], that the ideal structure of the enveloping algebra of a color Lie superalgebra can be very different from that of the enveloping algebra of a Lie superalgebra. By contrast, the results in this note illustrate that certain homological properties are the same. Our main strategy is to pass to the case where the grading group G is finitely generated, so that the color Lie superalgebra is determined by a Lie superalgebra and a 2-cocycle defined on G.

In §2 we define color Lie superalgebras and state a theorem due to Scheunert (see [14]). In §3 we calculate the finitistic and injective dimensions of the enveloping algebra of a finite dimensional color Lie superalgebra over a field of characteristic 0. The enveloping algebra of a finite dimensional color Lie superalgebra may have infinite global dimension (this is known for ordinary Lie superalgebras, see [2], Proposition 5). However, for a finite dimensional color Lie superalgebra $\mathcal{L}$ which is *positively graded*, i.e., $\mathcal{L} = \mathcal{L}_+$, we prove that, in analogy with the nongraded case, $\text{gldim}(U(\mathcal{L})) = \dim \mathcal{L}$. In particular, Theorem 3.1 generalizes [8], Proposition 2.3.

In Theorem 3.2, we show that the enveloping algebra $U(\mathcal{L})$ of a color Lie superalgebra is Auslander-Gorenstein and Cohen-Macaulay. It thus follows from [9], Theorem 1.4 that $U(\mathcal{L})$ has a (right and left) quasi-Frobenius (QF) classical quotient ring.

2. COLOR LIE SUPERALGEBRAS

Throughout k denotes a field of charecteristic $\neq 2$, G denotes an Abelian group, and all algebras are associative k-algebras with 1. We call a map $\gamma : G \times G \to k^\times$

1991 *Mathematics Subject Classification.* Primary 17B35; Secondary 16S30.

($k^\times = k\backslash\{0\}$) a *skew-symmetric bicharacter* on G if it satisfies

(1) $\gamma(f, gh) = \gamma(f, g)\gamma(f, h)$ and $\gamma(gh, f) = \gamma(g, f)\gamma(h, f)$ for any $f, g, h \in G$.
(2) $\gamma(g, h)\gamma(h, g) = 1$ for any $g, h \in G$.

Note that (2) implies that $\gamma(g, g) = \pm 1$ for any $g \in G$. Set $G_\pm = \{g \in G : \gamma(g, g) = \pm 1\}$, then G_+ is a subgroup of G and $[G : G_+] \leq 2$. In [14], Lemma 2, it is shown that if G is finitely generated then the skew-symmetric bicharacters on G are completely determined by the subgroup G_+ and the 2-cocycles on G with coefficients in $k^\times$.

Generally, we shall be discussing objects which are graded over G so let G-**vec** denote the category of G-graded vector spaces and graded linear maps. For any $V \in G$-**vec**, set $\partial x = g$ if $0 \neq x \in V_g$. Let γ be a skew-symmetric bicharacter on G. We will shorten our notation by writing $\gamma(x, y)$ instead of $\gamma(\partial x, \partial y)$ for homogeneous $0 \neq x \in V$, $0 \neq y \in W$ and $V, W \in G$-**vec**. Note that G-**vec** is naturally contained in $\mathbb{Z}_2$-**vec** via the decomposition $G = G_+ \dot\cup G_-$ and the group homomorphism $\pi : G \to \mathbb{Z}_2$ given by $\pi(G_+) = \bar{0}$ and $\pi(G_-) = \bar{1}$.

Definition. A (G, γ)-*color Lie superalgebra* is a pair $(\mathcal{L}, \langle,\rangle)$ such that $\mathcal{L} \in G$-**vec** and $\langle,\rangle : \mathcal{L} \otimes_k \mathcal{L} \to \mathcal{L}$ is a graded bilinear map which satisfies the following for any homogeneous $x, y, z \in \mathcal{L}$.

γ-*skew-symmetry* $\langle x, y \rangle = -\gamma(x, y)\langle y, x \rangle$
γ-*Jacobi identity* $\gamma(z, x)\langle x, \langle y, z \rangle\rangle + \gamma(y, z)\langle z, \langle x, y \rangle\rangle + \gamma(x, y)\langle y, \langle z, x \rangle\rangle = 0$

Example. (1) A *Lie superalgebra* is a $(\mathbb{Z}_2, \chi)$-color Lie superalgebra where $\chi(\bar{i}, \bar{j}) = (-1)^{ij}$ for any $i, j \in \mathbb{Z}$.
(2) If A is a G-graded algebra then there is a color Lie superalgebra structure A^- defined on A subject to the condition that $\langle a, b \rangle = ab - \gamma(a, b)ba$ for any nonzero homogeneous $a, b \in A$.

We shall now describe how to construct color Lie superalgebras from Lie superalgebras as in [14]. In as much as Lie superalgebras are $\mathbb{Z}_2$-graded, we must consider G-gradings on Lie superalgebras which respect the $\mathbb{Z}_2$-grading.

Definition. A Lie superalgebra $(L, [,])$, $L = L_{\bar{0}} \bigoplus L_{\bar{1}}$, is G-*graded* if $L = \bigoplus_{g \in G} L_g$, $[L_f, L_g] \subseteq L_{fg}$ for any $f, g \in G$, and $L_h \subseteq L_{\bar{0}}$ or $L_h \subseteq L_{\bar{1}}$ for each $h \in G$.

Starting with a Lie superalgebra L there are many choices of groups G and G-gradings on L. Let $G(L)$ be the smallest subgroup of G which grades L, that is, $G(L)$ is generated by $\{g \in G : L_g \neq 0\}$. We cannot always pass to the case that $G = G(L)$ since, for example, there are G-graded representations of L which are not $G(L)$-graded.

Let $G(L)_+$ be the subgroup of $G(L)$ which is generated by $\{g \in G : U(L)_g \subseteq U(L)_{\bar{0}}\}$. Then $[G(L) : G(L)_+] \leq 2$. Set $G_- = G \backslash G_+$; then L is a (G, γ_0)-color Lie superalgebra where γ_0 is the skew-symmetric bicharacter on G defined below.

$$\gamma_0(g, h) = \begin{cases} -1 & \text{if } g, h \in G_- \\ 1 & \text{otherwise} \end{cases}$$

To continue our construction of a color Lie superalgebra from a Lie superalgebra, we need the notion of a 2-cocycle.

Definition. A *2-cocycle* on G is a map $\sigma : G \times G \to k^\times$ which satisfies
$$\sigma(f,gh)\sigma(g,h) = \sigma(f,g)\sigma(fg,h)$$
for any $f, g, h \in G$.

If σ is a 2-cocycle on G then there is a skew-symmetric bicharacter γ defined on G by $\gamma(g,h) = \gamma_0(g,h)\sigma(g,h)\sigma(h,g)^{-1}$ for any $g,h \in G$. Moreover, there is a (G,γ)-color Lie superalgebra $(L^\sigma, [,]^\sigma)$ which has the same vector space structure as L but bracket $[,]^\sigma : L^\sigma \times L^\sigma \to L^\sigma$ defined subject to the condition that $[x,y]^\sigma = \sigma(\partial x, \partial y)[x,y]$ for any homogeneous $x,y \in L$.

By setting $f = h = e$ in the definition of 2-cocycle, e the identity element of G, we obtain $\sigma(e,e) = \sigma(g,e) = \sigma(e,g)$ for any $g \in G$. Note that there is no loss in generality by assuming that $\sigma(e,e) = 1$ since we can replace σ with $\sigma' = \sigma(e,e)^{-1}\sigma$ and have $L^\sigma \cong L^{\sigma'}$ as (G,γ)-color Lie superalgebras.

Theorem 2.1 summarizes the results from [14] we are interested in. For the reader's convenience we will go over some basic definitions.

Definition. Let $\mathcal{L}$ be a (G,γ)-color Lie superalgebra.

(1) A linear map $\phi : \mathcal{L}_1 \to \mathcal{L}_2$ between (G,γ)-color Lie superalgebras $(\mathcal{L}_1, \langle,\rangle_1)$ and $(\mathcal{L}_2, \langle,\rangle_2)$ is called a *homomorphism* if $\phi(\langle x,y\rangle_1) = \langle \phi(x), \phi(y)\rangle_2$ for any $x,y \in \mathcal{L}_1$.

(2) A *G-graded representation of $\mathcal{L}$* is a pair (V, ρ) where $V \in G$-**vec** and $\rho : \mathcal{L} \to End_k(V)^-$ is a homomorphism of color Lie superalgebras ($End_k(V)$ is a G-graded algebra since $V \in G$-**vec**).

(3) For any G-graded algebra A, a graded map $\phi : \mathcal{L} \to A^-$ is called *compatible* if it is a homomorphism of (G,γ)-color Lie superalgebras.

(4) The *universal enveloping algebra* of $\mathcal{L}$ is a G-graded algebra $U(\mathcal{L})$ and a compatible map $\iota : \mathcal{L} \to U(\mathcal{L})^-$ which satisfies the property that for any compatible map $\phi : \mathcal{L} \to A$ there is a unique (graded) algebra homomorphism $\Phi : U(\mathcal{L}) \to A^-$ such that $\Phi \circ \iota = \phi$.

Theorem 2.1 (Scheunert). *Let $\mathcal{L}$ be a (G,γ)-color Lie superalgebra and L and L^σ as above.*

(1) *If G is finitely generated, then any color Lie superalgebra can be obtained as an L^σ for appropriately chosen L and 2-cocycle σ.*

(2) *The enveloping algebra $U(L^\sigma)$ is obtained from $U(L)$ by defining a new multiplication $*$ on $U(L)$ subject to the condition that for any homogeneous $x,y \in U(L)$ we have $x * y = \sigma(\partial x, \partial y)xy$.*

(3) *For a graded representation $\rho : L \to End_k(V)$ of L, there is a graded representation $\rho^\sigma : L^\sigma \to End_k(V)$ of L^σ which is obtained from ρ subject to the condition that for any homogeneous $x \in L^\sigma$ and $v \in V$, $\rho^\sigma(x)(v) = \sigma(\partial x, \partial v)\rho(x)(v)$. This defines a category equivalence between the categories of graded representations of L and L^σ.*

We are particularly interested in parts (2) and (3) of Theorem 2.1. More generally, consider an arbitrary G-graded algebra $\mathcal{R}$ and left $\mathcal{R}$-modules V and W. An $\mathcal{R}$-module homomorphism $\phi : V \to W$ is called *graded* if $\phi(V_h) \subseteq W_h$ for each $h \in G$. Let $_{\mathcal{R}}\mathcal{M}^G$ denote the category whose objects are G-graded left $\mathcal{R}$-modules and morphisms are all graded $\mathcal{R}$-module homomorphisms from V to W, denoted $\mathrm{Hom}_{\mathcal{R}}^0(V,W)$, where $V, W \in {}_R\mathcal{M}^G$. In particular, ${}_k\mathcal{M}^G = G\text{-}\mathbf{vec}$.

Lemma 2.2. *Let σ be a 2-cocycle on G and R a G-graded k-algebra.*

(1) *There is a G-graded k-algebra R^σ with the same vector space structure as R and whose multiplication $*$ is obtained from the multiplication of R subject to the condition $r*s = \sigma(\partial r, \partial s)rs$ for any homogeneous $r, s \in R$.*

(2) *For any $M \in {}_R\mathcal{M}^G$ there corresponds $M^\sigma \in {}_{R^\sigma}\mathcal{M}^G$ such that M^σ has the same vector space structure as M but the R^σ-module structure. is obtained from the R-module structure of M subject to the condition $r.m = \sigma(\partial r, \partial m)rm$ for any homogeneous $r \in R$ and $m \in M$.*

(3) *The functor $^\sigma : {}_R\mathcal{M}^G \to {}_{R^\sigma}\mathcal{M}^G$ defined as in (2) is a category equivalence. In particular, $\mathrm{Hom}_R^0(V, W) = \mathrm{Hom}_{R^\sigma}^0(V^\sigma, W^\sigma)$. Moreover, if $V \in {}_R\mathcal{M}^G$ is graded free, then so is $V^\sigma \in {}_{R^\sigma}\mathcal{M}^G$.*

(4) *For any $V, W \in {}_R\mathcal{M}^G$ and $\phi \in \mathrm{Hom}_R^0(V, W)$ we have $\ker(\phi^\sigma) = (\ker \phi)^\sigma = \ker \phi$ and $\mathrm{im}(\phi^\sigma) = (\mathrm{im}\,\phi)^\sigma = \mathrm{im}\,\phi$.*

Proof. It is easy to prove (1)-(3) directly. For (4), first note that $\ker \phi$, $\mathrm{im}\,\phi$, $\ker(\phi^\sigma)$ and $\mathrm{im}(\phi^\sigma)$ are G-graded submodules. For any homogeneous $x \in V$ we have $x \in \ker(\phi^\sigma) \Leftrightarrow x \in \ker(\phi) \Leftrightarrow x \in (\ker \phi)^\sigma$. This proves that $\ker(\phi^\sigma) = (\ker \phi)^\sigma = \ker \phi$. The proof that $\mathrm{im}(\phi^\sigma) = (\mathrm{im}\phi)^\sigma = \mathrm{im}\,\phi$ is similar. □

3. Homological Properties of $U(\mathcal{L})$

Throughout this section, γ is a skew-symmetric bicharacter on G and $G_\pm = \{g \in G : \gamma(g,g) = \pm 1\}$. For any $V \in G\text{-}\mathbf{vec}$, set $V_\pm = \oplus_{g \in G_\pm} V_g$. If $V = V_+$ (respectively $V = V_-$) then V is called *positively* (respectively *negatively*) *graded*. We shall be primarily interested in the case where $\mathcal{L}$ is an Abelian color Lie superalgebra. If $\mathcal{L}$ is a positively graded, finite dimensional and Abelian (G, γ)-color Lie superalgebra, then $U(\mathcal{L}) \cong k^\gamma[\theta_1, \theta_2, \ldots, \theta_n]$, the *color polynomial ring* in $n = \dim \mathcal{L}$ variables (see [1]). If $\mathcal{L}$ is a negatively graded, finite dimensional and Abelian (G, γ)-color Lie superalgebra, then $U(\mathcal{L}) \cong \Lambda^\gamma(\mathcal{L})$, the *color exterior algebra (or color Grassmann algebra) of $\mathcal{L}$* (see [1]).

Example. (1) Choose $q \in k^\times$. Let γ be the skew-symmetric bicharacter on $G = \mathbb{Z}^2$ defined by $\gamma(g, h) = q^{g_1 h_2 - g_2 h_1}$ for any $g = (g_1, g_2)$, $h = (h_1, h_2) \in G$. Then $G = G_+$. Consider the Abelian (G, γ)-color Lie superalgebra with homogeneous basis $\{x, y\}$ such that $\partial x = (1, 0)$ and $\partial y = (0, 1)$. Then $\mathcal{L}$ is positively graded and $U(\mathcal{L}) \cong k[x, y : xy = qyx]$, the so-called *quantum plane*.

(2) Choose $q \in k^\times$. Let γ be the skew-symmetric bicharacter on $G = \mathbb{Z}^2$ defined by $\gamma(g, h) = (-1)^{g_1 h_1 + g_2 h_2} q^{g_1 h_2 - g_2 h_1}$ for any $g = (g_1, g_2)$, $h = (h_1, h_2) \in G$. Then $G_+ = \{(i, j) \in G : i + j \in 2\mathbb{Z}\}$ and $G_- = \{(i, j) \in G : i + j \in 1 + 2\mathbb{Z}\}$. Consider the Abelian (G, γ)-color Lie superalgebra with homogeneous basis $\{x, y\}$ such that $\partial x = (1, 0)$ and $\partial y = (0, 1)$. Then $\mathcal{L}$ is negatively graded and $U(\mathcal{L}) \cong k[x, y : xy = qyx]/I$, where I is the ideal generated by $\{x^2, y^2\}$.

If $V, W \in G\text{-}\mathbf{vec}$ then the tensor product $V \otimes W \in G\text{-}\mathbf{vec}$ has grading defined by $(V \otimes W)_g = \oplus_{h \in G}(V_h \otimes W_{gh^{-1}})$. For G-graded algebras A and B there is a γ-graded tensor product $A \widehat{\otimes}_k B$ with multiplication defined subject to the condition that $(a\widehat{\otimes}b)(a'\widehat{\otimes}b') = \gamma(b,a')aa'\widehat{\otimes}bb'$ for any nonzero homogeneous $a, a' \in A$ and $b, b' \in B$. The γ-graded tensor product is essential to the study of enveloping algebras of color Lie superalgebras. For example, the enveloping algebra of a color Lie superalgebra is not a Hopf algebra in the usual sense; it is only a Hopf algebra with respect to the γ-graded tensor product (see [1], §3.2.9 for details).

For any Noetherian ring R, the left (respectively right) finitistic dimension of R, denoted lFPD(R) (respectively rFPD(R)), is the supremum of the projective dimensions of left (respectively right) R-modules of finite projective dimension. The following theorem generalizes Proposition 2.3 of [8]. As in [8] and [9], we assume that char $k = 0$ in all that follows.

Theorem 3.1. *Let $\mathcal{L} = \mathcal{L}_+ \oplus \mathcal{L}_-$ be a finite dimensional color Lie superalgebra. Then $\mathrm{gldim}(U(\mathcal{L}_+)) = lFPD(U(\mathcal{L})) = rFPD(U(\mathcal{L})) = \mathrm{injdim}_{U(\mathcal{L})}(U(\mathcal{L})) = \dim(\mathcal{L}_+)$.*

Proof. Let $U = U(\mathcal{L})$ denote the universal enveloping algebra of $\mathcal{L}$ and set $n = \dim \mathcal{L}_+$. We first show that lFPD(U) = rFPD(U) = injdim$_U(U) = n \geq \mathrm{gldim}(U(\mathcal{L}_+))$. Then we show that $\mathrm{gldim}(U(\mathcal{L}_+)) \geq n$, that is, there exists a $U(\mathcal{L}_+)$-module with projective dimension at least n. As in the case of ordinary Lie algebras, the existence of such a module will be demonstrated using the *Chevalley-Eilenberg complex* (see Chapter 7 of [16] for background).

1) With respect to the standard filtration, $gr(U)$ is Noetherian. The filtration is defined by $U^{-1} = \{0\}$, $U^0 = k$ and, for $m > 0$, U^m is spanned by all monomials of length $\leq m$. By [1], Proposition 3.2.8, U is Noetherian and the associated graded algebra $gr(U)$ is isomorphic to $k^\gamma[\theta_1, \theta_2, \ldots, \theta_n]\widehat{\otimes}\Lambda^\gamma(\mathcal{L}_-)$, where $\Lambda^\gamma\mathcal{L}_-$ is the color exterior algebra determined by the vector space $\mathcal{L}_-$ with an Abelian color Lie superalgebra structure. Thus $gr(U)$ is an iterated Ore extension of $\Lambda = \Lambda^\gamma(\mathcal{L}_-)$ without derivations, that is, $gr(U) \cong \Lambda[\theta_1; \alpha_{\partial\theta_1}][\theta_2; \alpha_{\partial\theta_2}]\cdots[\theta_n; \alpha_{\partial\theta_n}]$. Therefore, $gr(U)$ is Noetherian by Theorem I.2.10 of [11].

2) injdim$_U U \leq$ injdim$_{gr(U)} gr(U) = n$ and gldim$U(\mathcal{L}_+) \leq$ gldim$gr(U(\mathcal{L}_+)) = n$ Since $U^m = 0$ when $m < 0$, the filtration $\{U^m\}$ satisfies the *closure condition* of [3], §1.2.11. Therefore, we may apply Theorem 1.3.12 of [3] to obtain injdim$_U(U) \leq$ injdim$_{gr(U)} gr(U)$. Moreover, gldim$U(\mathcal{L}_+) \leq$ gldim$(gr(U(\mathcal{L}_+))$ by 7.6.18 of [11]. For any ring R and automorphism σ of R, one can show that, $R[t;\sigma]$, we have gldim$(R[t;\sigma]) = $ gldim$(R) + 1$ and injdim$_{R[t;\sigma]}(R[t;\sigma]) = $ injdim$_R(R) + 1$ by similar methods to the proofs in the case that σ is the identity on R. Since $gr(U(\mathcal{L}_+)) \cong k^\gamma[\theta_1, \theta_2, \ldots, \theta_n]$ we have gldim$(gr(U(\mathcal{L}_+))) = n = \dim \mathcal{L}_+$. By [6], Corollary 6.3, Λ is a Frobenius algebra; hence injdim$_{gr(U))}(gr(U)) = n = \dim \mathcal{L}_+$ as $gr(U) \cong \Lambda^\gamma[\theta_1, \theta_2, \ldots, \theta_n]$.

3) lFPD(U) = rFPD(U) = injdim$_U(U) = n \geq$ gldim$(U(\mathcal{L}_+))$. Since gldim$(U(\mathcal{L}_+))$ is finite and $U(\mathcal{L})$ is free over $U(\mathcal{L}_+)$ on both sides, we may apply Theorem 1.4 of [4], to obtain gldim$(U(\mathcal{L}_+)) \leq$ lFPD$(U(\mathcal{L}))$. By Proposition 2.1 of [8], we have injdim$_U(U) = $ lFPD(U) = rFPD(U).

The rest of the proof is dedicated to showing that if $\mathcal{L}$ is a positively graded finite dimensional (G, γ)-color Lie superalgebra then $n = \dim \mathcal{L} \leq$ gldim$(U(\mathcal{L}))$.

4) Pass to the case that $\mathcal{L} = L^\sigma$ for some Lie algebra L. The modules we will consider are graded over the subgroup $G(\mathcal{L})$ of G generated by $\{g \in G : \mathcal{L}_g \neq 0\}$. Therefore, we can pass to the case that G is finitely generated

by assuming that $G = G(\mathcal{L})$. By Theorem 2.1(1), there is a G-graded Lie algebra L with bracket $[,]$ and a 2-cocycle σ on G such that $\mathcal{L} = L^\sigma$.

5) The Chevalley-Eilenberg complex $V_*(L) \xrightarrow{\varepsilon} k$ is a complex in $_{U(L)}\mathcal{M}^G$.

We must show that $V_*(L) \xrightarrow{\varepsilon} k$ is a *graded complex*, i.e., the modules are graded and all differentiations are graded maps. Let ΛL denote the exterior algebra of L and $\Lambda^i L$ denote the ith exterior algebra of L. Then $\Lambda L = \oplus_{i=1}^n \Lambda^i L$ can be G-graded so that the homogeneous component of degree $g \in G$ is $(\Lambda L)_g = \sum_{i=1}^n (\Lambda^i L)_g$ and $(\Lambda^i L)_g = \sum_{g_1 g_2 \cdots g_i = g} L_{g_1} \wedge L_{g_2} \wedge \cdots \wedge L_{g_i}$ where the sum is indexed by all appropriate $g_1, g_2, \ldots, g_i \in G$.

We have $V_i(L) = U(L) \otimes_k \Lambda^i L$ which is a graded free left $U(L)$-module such that $V_i(L)_g = \sum_{h \in G} (U(L)_h) \otimes_k (\Lambda^i L)_{hg^{-1}}$ so we only need to check that the *augmentation map* $\varepsilon : V_0(L) \to k$ is graded and, for $1 \leq i \leq n$, $d_i : V_i(L) \to V_{i-1}(L)$ is graded. The augmentation map $\varepsilon : V_0(L) = U(L) \to k$ is graded since it is induced from the Lie algebra map $L \to k$ which sends everything to zero, a graded map. It is also easy to see that $d_1 : V_1(L) \to V_0(L)$ is graded since it is just the product map $u \otimes x \longmapsto ux$. For $1 < i \leq n$, the map $d_i : V_i(L) \to V_{i-1}(L)$ is defined by the formula $d_i(u \otimes x_1 \wedge x_2 \wedge \cdots \wedge x_i) = \theta_1 + \theta_2$ where θ_1 and θ_2 are defined as follows.

$$\theta_1 = \sum_{j=1}^i (-1)^{j+1} u x_i \otimes x_1 \wedge x_2 \wedge \cdots \wedge \hat{x}_j \wedge \cdots \wedge x_i$$

$$\theta_2 = \sum_{l<m} (-1)^{l+m} u \otimes [x_l, x_m] \wedge x_1 \wedge x_2 \wedge \cdots \wedge \hat{x}_l \wedge \cdots \wedge \hat{x}_m \wedge \cdots \wedge x_i$$

Therefore d_i is graded since if $u, x_1, x_2, \ldots, x_i$ are homogeneous elements then $u \otimes x_1 \wedge x_2 \wedge \cdots \wedge x_i, \theta_1$ and θ_2 are all homogeneous of the same degree.

6) The Chevalley-Eilenberg complex can be defined in $_{U(\mathcal{L})}\mathcal{M}^G$.

By parts (2), (3) and (4) of Lemma 2.2, there is a projective resolution $V_*(\mathcal{L})$ of $k = k^\sigma$ over $U(\mathcal{L})$ such that $V_i(\mathcal{L}) = V_i(L)^\sigma$. Note that the modules $V_i(L)$ are graded free $U(L)$-modules hence the $V^i(\mathcal{L})$ are graded free $U(\mathcal{L})$-modules by Theorem 2.2(3).

7) $\operatorname{gldim}(U(\mathcal{L})) \geq n$

As in [16], Exercise 7.7.2, there is a graded representation $\rho : L \to End_k(\Lambda^n L)$ defined as below for nonzero $y \in L$ and $x_1, x_2, \ldots, x_n$ a (homogeneous) basis of L.

$$\rho(y)(x_1 \wedge x_2 \wedge \cdots \wedge x_n) = \sum_{i=1}^n x_1 \wedge x_2 \wedge \cdots \wedge [y, x_i] \wedge \cdots \wedge x_n$$

Let $M = \Lambda^n L$ be the (graded) $U(L)$-module defined by the above action. Then according to [16], $\operatorname{Ext}_{U(L)}^n(k, M) \cong \ker d_n^* / \operatorname{im} d_{n-1}^* \cong k$. By parts (2) and (4) of Lemma 2.2, and 6) above, we have $\ker d_n^* = (\ker d_n^*)^\sigma = (\ker(d_n^*)^\sigma)$ and $(\operatorname{im} d_{n-1}^*) = (\operatorname{im} d_{n-1}^*)^\sigma = (\operatorname{im}(d_{n-1}^*)^\sigma)$. This implies that

$$\operatorname{Ext}_{U(\mathcal{L})}^n(k, M^\sigma) \cong (\operatorname{Ext}_{U(L)}^n(k, M))^\sigma \cong k$$

therefore $pd_{U(\mathcal{L})}(M^\sigma) \geq n$ and thus $\operatorname{gldim}(U(\mathcal{L})) \geq n$ as desired.

The result follows from 3) and 7) above. $\square$

Theorem 3.2. *Let $\mathcal{L}$ be a finite dimensional color Lie superalgebra. Then $U(\mathcal{L})$ is Auslander-Gorenstein and Cohen-Macaulay and thus has a quasi-Frobenius classical quotient ring.*

Proof. We have Λ is Auslander-Gorenstein and Cohen-Macaulay since Λ is a Frobenius algebra (see [6], Corollary 6.3). As $\mathrm{gr}(U(\mathcal{L}))$ is an iterated Ore extension of Λ where each iteration is of the form $R[x; \sigma]$, it follows follows from ([10], Lemma (ii), p. 184) and Theorem 4.2 of [5] that $\mathrm{gr}(U(\mathcal{L}))$ is Auslander-Gorenstein and Cohen-Macaulay. We thus conclude that $U(\mathcal{L})$ is Auslander-Gorenstein and Cohen-Macaulay by [3], Theorem 1.4.1, [15], Lemma 4.4 (note that Theorem 3.1 is needed so that $\mathrm{injdim}_{U(\mathcal{L})} U(\mathcal{L}) < \infty$). The last statement now follows from [9], Theorem 1.4. □

Acknowledgements

The author is indebted to Professor James Kuzmanovich for many helpful conversations concerning this material.

REFERENCES

1. Yu. A. Bahturin, A. A. Mikhalev, V. M. Petrogradsky and M. V. Zaicev, *Infinite Dimensional Lie Superalgebras*, DeGruyter, Berlin, 1992.
2. E. J. Behr, 'Enveloping Algebras of Lie Superalgebras', *Pacific J. Math.* **130** (1987), 9-25.
3. J-E Björk, 'Filtered Noetherian Rings', in *Noetherian Rings and Their Applications"*, Math. Surveys and Monographs, vol. **24**, 58-97, Amer. Math. Soc., Providence, RI, 1987.
4. B. Cortzen, 'Finitistic Dimension of Ring Extensions', *Comm. Algebra* **10** (1982), 993-1001.
5. E. K. Ekström, 'The Auslander Condition on Graded and Filtered Noetherian Rings', in *Sem. Dubreil-Malliavin*, Lecture Notes in Math. **1404**, Springer-Verlag (1989), 220-245.
6. D. Fischman, S. Montgomery and H.-J. Schneider, 'Frobenius Extensions of Subalgebras of Hopf Algebras', to appear in *Trans. Amer. Math. Soc.*
7. V. G. Kac, 'Lie Superalgebras', *Adv. in Math.* **26** (1977), 8-96.
8. E. Kirkman, J. Kuzmanovich and L. Small, 'Finitistic Dimensions of Noetherian Rings', *Journal of Algebra* **147(2)** (1992), 350-364.
9. E. Kirkman and J. Kuzmanovich, 'Minimal Prime Ideals in Enveloping Algebras of Lie Superalgebras', *Proc. Amer. Math. Soc.* **124(6)**, 1693-1702.
10. T. Levasseur and J. T. Stafford, 'The Quantum Coordinate Ring of the Special Linear Group', *J. Pure Appl. Alg.* **86** (1993), 181-186.
11. J. C. McConnell and J. C. Robson, *Noncommutative Noetherian Rings*, Wiley, Chichester, England, 1987.
12. K. L. Price, 'Primeness Criterion for Enveloping Algebras of Color Lie Superalgebras', manuscript in preparation.
13. M. Scheunert, *The Theory of Lie Superalgebras: An Introduction*, Springer Lecture Notes in Mathematics (716), 1979.
14. M. Scheunert, 'Generalized Lie Algebras', *J. Math. Phys.* **20(4)** (1979), 712-720.
15. J. T. Stafford and J. J. Zhang, 'Homological Properties of (Graded) Noetherian PI Rings', *J. Algebra* **168** (1994), 988-1026.
16. C. A. Weibel, *An Introduction to Homological Algebra*, Cambridge Studies in Advanced Mathematics (38), Cambridge University Press, 1994.

DEPARTMENT OF MATHEMATICS, UNIVERSITY OF WISCONSIN, MILWAUKEE, WISCONSIN 53201
E-mail address: price@csd.uwm.edu

INDECOMPOSABLE MODULES OVER ARTINIAN RIGHT SERIAL RINGS

SURJEET SINGH

ABSTRACT. Let R be an artinian ring with Jacobson radical J such that $J^2 = 0$ and R/J is a direct product of matrix rings over finite dimensional division rings. The structure of R is determined, in case every indecomposable right R-module is uniform. Furthermore, all indecomposable right or left modules over such a ring are determined.

Let R be an artinian ring. It is well known that every indecomposable right R-module is uniserial iff R is generalized uniserial (see Faith [3], Theorem 25.2.6). Every uniserial module is uniform. A right serial exceptional ring is not generalized uniserial, but every indecomposable right module over it is uniform [1]. This motivates one to determine the structure of a ring R, with the property that every indecomposable right R-module is uniform. Let R be an artinian ring with radical J, such that R/J is a direct product of matrix rings over finite dimensional division rings. Theorem (3.5) gives the structure of R, in case $J^2 = 0$, and every finitely generated right R-module is uniform. If $J^2 = 0$ and every finitely generated, indecomposable right R-module is uniform, Theorem (3.6) shows that every right R-module is a direct sum of finitely generated, uniform R-modules; it is evident from these results that this ring of bounded representation type. By Dlab and Ringel [2], if R is a local artinian ring of bounded representation type, with $J^2 \neq 0$ and R/J a field, then R is a serial ring. Example 3 in section 3 gives an indecomposable, artinian ring R, with $J^2 \neq 0$ and every indecomposable right R-module uniform, but R is not generalized uniserial. Theorem (4.6) determines the structure of left R-modules, in case R satisfies the hypothesis of Theorem (3.5). In section 2, some constructions of finite length, indecomposable modules over arbitrary rings are discussed.

1. PRELIMINARIES

All the rings considered here are with unity $1 \neq 0$, and the modules are unital. For any ring R, $J(R)$ (or simply J) denotes the Jacobson radical of R. Let M be any module. Then $E(M), End(M), J(M)$, and $soc(M)$ denote the injective hull, the ring of endomorphisms, the Jacobson radical and the socle of M, respectively. If M is of finite (composition) length, then $d(M)$ denotes its composition length. A module in which the lattice of its submodules is linearly ordered under inclusion is called a *uniserial* module. A module that is a direct sum of uniserial modules is called a *serial* module. A ring R is called a right (left) serial ring, if $R_R (_RR)$ is a serial module. If a ring is right as well as left artinian, it is called an artinian ring. An

---

1991 Mathematics Subject Classification. 16P20.

artinian ring which is left and right serial is traditionally called a *generalized uniserial* ring. Let k be any non-negative integer. Then $soc^k(M)$ is defined as follows: $soc^0(M) = 0$, for $k \geq 1, soc(M/soc^{k-1}(M)) = soc^k(M)$ $soc^{k-1}(M)$. A module M is said to be of *Loewy length* k, if k is the smallest non-negative integer, such that $M = soc^k(M)$. Further $M^{(n)}$ denotes a direct sum of n copies of M. For any subdivision ring D' of a division ring $D, [D:D']_l([D:D']_r)$ denotes the left (right) dimension of D over D'. For the basic results on rings and modules, we refer to Faith [3].

2. INDECOMPOSABLE MODULES

Throughout, R is any ring unless otherwise stated.

LEMMA 2.1. Let A_1 and A_2 be two uniform right R-modules with $S = soc(A_1) \cong soc(A_2) \neq 0$. Let $\omega : S \to A_2$ be an R-monomorphism, $N = \{(s, -\omega s) : s \in S\}$. Then $M = (A_1 \times A_2)/N$ is uniform if and only if given any $(a_1, a_2) \notin N, f : a_1 R \to a_2 R$, given by $f(a_1 r) = -a_2 r$, does not define an R homomorphism extending ω.

Proof. Now $B = \{\overline{(s,0)} : s \in S\}$ is a simple submodule of M. Suppose that M is uniform. Consider $(a_1, a_2) \notin N$, such that $f : a_1 R \to a_2 R, f(a_1 r) = -a_2 r$, is a function extending ω. Let for some r, $\overline{(a_1, a_2)}r \in B$. Then $a_1 r - s = s'$, $a_2 r = -\omega s'$ for some $s, s' \in S$. Then $f(a_1 r - s) = f(s')$ yields $-a_2 r - \omega s = \omega s' = -a_2 r, -\omega s = 0, s = 0$. Thus $B \cap \overline{(a_1, a_2)}R = 0$. So M is not uniform. This is a contradiction. Conversely, let the given hypothesis hold. Consider $(a_1, a_2) \notin N$. Let for some $r \in R, a_1 r = 0$, but $a_2 r \neq 0$. Then for some $r_0, a_1 r_0 = 0$, and $0 \neq a_2 r_0 \in soc(A_2)$. Then for some $s \in S, a_2 r_0 = \omega(s)$, and $0 \neq \overline{(a_1, a_2)}r_0 = \overline{(s, 0)} \in B$. Let $a_2 r = 0$, whenever $a_1 r = 0$. We get the R-homomorphism $\eta : a_1 R \to a_2 R$, given by $\eta(a_1 r) = -a_2 r$. For some $a_1 r_0 \in S, \eta(a_1 r_0) \neq \omega(a_1 r_0)$. Then $\overline{(a_1, a_2)}r_0 \neq 0$ and it is in B. Hence M is uniform.

Let K_R be any uniform, finite length module of Loewy length at least $2, S = soc(K), D = End(S), T = End(K)$ and $D' = T/J(T)$. We have an embedding $\sigma : D' \to D$, such that for any $\overline{\eta} \in D', [\sigma(\overline{\eta})](a) = \eta(a), a \in S$. With this embedding, we shall regard D', a subdivision ring of D, and call (D, D') a *division ring pair associate* (in short drpa) of K. Throughout A_R is a uniserial module with $d(A) = 2, drpa(D, D')$ and $S = soc(A)$.

LEMMA 2.2. Let A_R be quasi-projective and $\omega_1 = I, \omega_2, \cdots, \omega_n$ be any n non-zero members of D. Then $M = A^{(n)}/N$, where $N = \{(\omega_1 x_1, \omega_2 x_2, \cdots, \omega_n x_n) : x_i \in S, \Sigma x_i = 0\}$, is uniform if and only if there do not exist $\eta_1, \eta_2, \cdots, \eta_n \in D'$, not all zeros, such that $\Sigma_i \omega_i^{-1} \eta_i = 0$ on S.

Proof. For $n = 1$, M is uniform. So the result holds for $n = 1$. Observe that in general $d(N) = n - 1$ and $d(M) = n + 1$. To apply induction, let $n > 2$ and result hold for $n - 1$. Let $M' = A^{(n-1)}/L$, where

$$L = \{(\omega_1 x_1, \omega_2 x_2, \cdots, \omega_{n-1} x_{n-1}) : x_i \in S, \Sigma x_i = 0\}.$$

Then $\bar{S} = \{\overline{(\omega_1 s, 0, \cdots, 0)} : s \in S\} \subseteq M'$. We get a monomorphism $\lambda : S \to M', \lambda(\omega_n s)) = \overline{(\omega_1 s, 0, \cdots, 0)}$. Let $T = \{(s, -\lambda(s)) : s \in S\}$. Then $d(T) = 1$, gives $d((A \times M')/T) = n+1 = d(M)$. Define $\mu : A \times M' \to M$ as follows. For any $x_n \in A$ and $y = \overline{(x_1, x_2, \cdots, x_{n-1})} = (x_1, x_2, \cdots, x_{n-1}) + L \in M'$,

$$\mu(x_n, y) = (x_1, x_2, \cdots, x_n) + N$$

This is well defined. For any $s \in S, \mu(s, -\lambda(s)) = (-\omega_1 \omega_n^{-1} s, 0, \cdots, 0, s) + N = (\omega_1(-\omega_n^{-1} s), 0, \cdots, 0, \omega_n \omega_n^{-1} s) + N = 0$. So μ induces an epimorphism $\bar{\mu}$ from $(A \times M')/T$ to M, which is an isomorphism. Thus $M \cong (A \times M')/T$. Let $\omega_1^{-1}, \omega_2^{-1}, \cdots, \omega_n^{-1}$ be right linearly independent over D'. By the induction hypothesis, M' is uniform. Suppose λ extends to some $\eta : A \to M'$. As A is quasi-projective, we get endomorphisms $\eta_i, 1 \leq i \leq n-1$, of A such that $\eta(a) = \overline{(\eta_1 a, \eta_2 a, \cdots, \eta_{n-1} a)}, a \in A$. Then in $\lambda(\omega_n(s)) = \eta(\omega_n(s))$, by replacing s by $\omega_n^{-1} s$, we get $\overline{(\omega_1 \omega_n^{-1} s, 0, \cdots, 0)} = \overline{(\eta_1 s, \eta_2 s, \cdots, \eta_{n-1} s)}$. So there exist $r_i \in S$ such that $\Sigma_1^{n-1} r_i = 0, \omega_1 \omega_n^{-1} s - \eta_1 s = r_1$ and $-\eta_i s = \omega_i r_i$ for $2 \leq i \leq n-1$. This gives $\omega_n^{-1} = \Sigma_1^{n-1} \omega_i^{-1} \eta_i$. This is a contradiction. Hence, by (2.1), M is uniform.

Conversely, let M be uniform, but $\omega_1^{-1}, \omega_2^{-1}, \cdots, \omega_n^{-1}$ be not right linearly independent over D'. As M' embeds in M, it is uniform. By the induction hypothesis $\omega_1^{-1}, \omega_2^{-1}, \cdots, \omega_{n-1}^{-1}$ are right linearly independent over D'. Then $\omega_n^{-1} = \Sigma_1^{n-1} \omega_i^{-1} \eta_i$, for some $\eta \in D'$. Then $\nu : A \to M', \eta(a) = \overline{(\eta_1 a, \eta_2 a, \cdots, \eta_{n-1} a)}$ extends λ. This contradicts (2.1) and the results follow.

LEMMA 2.3. Let A_R be quasi-projective and $E = E(A)$. Consider any submodule K of E of the form, $K = A_1 + A_2 + \cdots, + A_n, A_1 = A \cong A_i$, such that $A_j \not\subset \Sigma_{i \neq j} A_i$. Let $\lambda_i \in End(E)$ such that $\lambda_i(A) = A_i$, and $\lambda_1 = I$. If $\omega_i = \lambda_i \mid S$, then $\omega_1, \omega_2, \cdots, \omega_n$ are right linearly independent over D', and $K \cong A^{(n)}/L$, where $L = \{(\omega_1^{-1} s_1, \omega_2^{-1} s_2, \cdots, \omega_n^{-1} s_n) : s_i \in S, \Sigma s_i = 0\}$.

Proof. Suppose for some $\eta_i \in D'$, not all zeros on $S, \Sigma \omega_i \eta_i = 0$ on S. Then $\Sigma \lambda_i \eta_i = 0$ on S. For some $j, \eta_j(A) = A$. This gives $A_j = \eta_j(A) \subseteq \Sigma_{i \neq j} \lambda_i \eta_i(A) + S \subseteq \Sigma_{i \neq j} A_i$. This is a contradiction. Hence $\omega_1, \omega_2, \cdots, \omega_n$ are right linearly independent over D'. Consequently $M = A^{(n)}/L$ is uniform by (2.2). Then $\lambda : M \to K, \lambda \overline{(a_1, a_2, \cdots, a_n)} = a_1 + \omega_2 a_2 + \cdots + \omega_n a_n$, is an isomorphism.

LEMMA 2.4. Let A_R be quasi-projective, $\omega_1 = I, \omega_2, \cdots, \omega_n$ be any n members of D right linearly independent over D'. For each i, let $\lambda_i \in End(E)$ be an extension of ω_i with $\lambda_1 = I$, and $A_i = \lambda_i(A)$. Then $A_j \not\subset \Sigma_{i \neq j} A_i$.

Proof. In the notation of the last lemma $M = A^{(n)}/L$ is a uniform module with $d(M) = n+1$ and $M \cong K$. This proves the lemma.

The following is immediate from (2.3) and (2.4).

COROLLARY 2.5. Let $e^2 = e \in R$, such that eR is uniserial and $d(eR) = 2$. For $E = E(eR), D = End(S), D' = eRe/eJe$ and a positive integer n, $[D : D']_r \geq n$ if and only if the homogeneous component of $soc(E/eJ)$ determined by the simple module eR/eJ is of length at least n.

LEMMA 2.6. Let M_R be a finite length, quasi-projective module and S be a simple submodule of M, such that S is not contained in a (proper) summand of M. Then M/S is indecomposable.

Proof. Let $T = End(M), L = End(M/S)$ and $K = \{\sigma \in T : \sigma(S) \subseteq S\}$. As M is quasi-projective, L is a homomorphic image of K. Consider any $\sigma \in K$, such that σ is not invertible. Then for some positive integer n, $M = \ker \sigma^n \oplus \text{range } \sigma^n$. If $\sigma^n(S) = 0$, then $S \subseteq \ker \sigma^n$, so by the hypothesis, range $\sigma^n = 0, \sigma^n = 0$. If $\sigma^n(S) \neq 0$, then $S \subseteq \text{range } \sigma^n$ yields σ is an automorphism of M; which is a contradiction. Thus every non-unit member of K is nilpotent, so is for L. As L is semiprimary, it must be a local ring. Hence M is indecomposable.

LEMMA 2.7. Let K_R by any finite length uniform module of composition length at least 2, $S = soc(K)$ and (D, D') be the drpa of K. Let $\omega_1 = I, \omega_2, \cdots, \omega_n$ be any n non-zero member of D. Then $L = \{(\omega_1 x, \omega_2 x, \cdots, \omega_n x) : x \in S\}$ is not contained in a summand of $K^{(n)}$ if and only if $\omega_1, \omega_2, \cdots, \omega_n$ are left linearly independent over D'.

Proof. Clearly $L \cong S$. Let L be not contained in a summand of $K^{(n)}$. On the contrary, let the given members of D be left linearly dependent over D'. Without lost of generality, we suppose that $\omega_n = \Sigma_1^{n-1} \eta_i \omega_i, \eta_i \in End(K)$. For $1 \leq i \leq n-1$, let B_i be the submodule of $K^{(n)}$ consisting of those $(x_1, x_2, \cdots, x_n)$ such that $x_j = 0$, for $j \neq i, n$ but $x_n = \eta_i x_i$. Then $\oplus \Sigma_1^{n-1} B_i$ is a summand of $K^{(n)}$ containing L. This is a contradiction. Hence the given members of D are left linearly independent over D'. Conversely let the given hypothesis hold, but L be contained in a summary of $K^{(n)}$. There exists a non-trivial idempotent endomorphism $\eta = [\eta_{ij}]_{n \times n}, \eta_{ij} \in End(K)$, of $K^{(n)}$, such that $(I - \eta)z = 0$ for $z \in L$. So for any $x \in S, \omega_i x - \Sigma_j \eta_{ij} \omega_j x = 0$. The hypothesis gives that $1 - \eta_{ii} = 0$ and $-\eta_{ij} = 0$ on S. Consequently, $I - \eta \in J(End(K^{(n)}))$. Hence $I = \eta$. This is a contradition. This completes the proof.

PROPOSITION 2.8. Let R be an artinian right serial ring. If M is an indecomposable, non-uniform, right R-module of smallest composition length, then for some finite length, uniform right R-modules $A_1, A_2, \cdots, A_k, A_{k+1}$, with $k \geq 1, A_{k+1}$ uniserial, $z_i \in A_i, z_i R$ simple, $z_i R \cong z_j R$, we have $M \cong (A_1 \oplus A_1 \oplus \cdots \oplus A_{k+1})/L$, where $L = \{(z_1, z_2, \cdots, z_{k+1})r : r \in R\}$ a simple module. Further for $i \neq j$, the mapping $\lambda_{ji} : z_i R \to z_j R, \lambda_{ji}(z_i r) = z_j r$, cannot be extended to an R-homomorphism $\sigma_{ji} : A_i \to A_j$.

Proof. Clearly M is not local. So for some submodules N and A of M, with $d(N) = d(M)-1$ and A local, $M = N+A$. As R is right serial, A is uniserial. The hypothesis on M gives $N = \oplus \Sigma_1^k B_j$, for some uniform modules B_j's. Further, $N \cap A = zR$ is maximal in A, $z = \Sigma_j z_j, z_j \in B_j$. If some z_j say $z_1 = 0$, then $M = B_1 \oplus K$ with $K = \Sigma_2^k B_j + A$. This is a contradiction. Hence $z_j \neq 0$ for every j. Put $B_{k+1} = A$ and $z_{k+1} = -z$. Then $M \cong (\oplus \Sigma_1^{k+1} B_j)/T$, where $T = \{(z_1 r, z_2 r, \cdots, z_{k+1} r) : r \in R\}$. Let uR be maximal in zR with $u = zs$. Let $u_j = z_j s, A_j = B_j u_j R$ and $w_j = z_j + u_j R$. Then $M \cong (\oplus \Sigma_1^{k+1} A_j)/L$, where $L = \{(w_1 r, w_2 r, \cdots, w_{k+1} r) : r \in R\}$. If the last part does not hold, then L will be contained in a summand of $\oplus \Sigma_1^{k+1} A_j$ and hence M will not be indecomposable. This proves the result.

3. ARTINIAN RIGHT SERIAL RINGS

Throughout R is an artinian right serial ring with $J^2 = 0$, unless otherwise stated. Let S be a simple right R-module. Then either S is injective or there exists an indecomposable idempotent $e \in R$ such that $d(eR) = 2$ and $S \cong eJ$. This observation, together with (2.2) and (2.4) gives the following.

PROPOSITION 3.1. Let S be a simple right R-module, $E = E(S)$. If for some indecomposable idempotent $e \in R, S \cong eJ, D = End(S), D' = eRe/eJe$ and $[D : D']_r = n < \infty$, then the homogeneous component of the socle of E/S, determined by the simple module eR/eJ, is of length n.

LEMMA 3.2. If every finitely generated right R-module is uniform, then for any indecomposable idempotent $e \in R$ with $eJ \neq 0$, the drpa (D, D') of eR satisfies $[D : D']_l \leq 2$.

Proof. For some indecomposable idempotent $e \in R$, let $[D : D']_l \geq 3$. Let $1, \omega_2, \omega_3$ in D be left linearly independent over D'. Then $S = \{(x, \omega_2 x, \omega_3 x) : x \in eJ\}$ is a simple submodule of $T = eR \oplus eR \oplus eR$. By (2.6) and (2.7), $M = T/S$ is indecomposable. As $eR \oplus eR$ embeds in M, M is not uniform. Hence $[D : D']_l \leq 2$.

LEMMA 3.3. Let every indecomposable right R-module be uniform, e and e' be two non-isomorphic indecomposable idempotents in R such that $0 \neq eJ \cong e'J$. Then eR and $e'R$ are quasi-injective.

Proof. Let eR be not quasi-injective. Then by Lemma (3.2) the drpa (D, D') of eR satisfies $[D : D']_l = 2$. Consider $E = E(eR)$. We get two submodules A and B of E, such that $A = eR, B \cong e'R, S = soc(A) = soc(B)$. Let $T = A_1 \oplus A_2 \oplus A_3, A = A_1 = A_2, A_3 = B$. There exists $\omega \in D$, such that $1, \omega$ are left linearly independent over D'. Then $S' = \{(x, \omega x, x) : x \in S\}$ is a simple submodule of T. Suppose for some proper idempotent endomorphism $\eta = [\eta_{ij}]$ of $T, (I - \eta)(S) = 0$. Then for any $x \in S, x = \eta_{11} x + \eta_{12} \omega x + \eta_{13} x, \omega x = \eta_{21} x + \eta_{22} \omega x + \eta_{23} x, x = \eta_{31} x + \eta_{32} \omega x + \eta_{33} x$. As $A \cong B, \eta_{13}, \eta_{23}, \eta_{31}, \eta_{32}$ are all zeros on S. So that $(1 - \eta_{11})x - \eta_{12} \omega x = 0, (1 - \eta_{22}) \omega x - \eta_{21} x = 0, (1 - \eta_{33})x = 0$. The left linear independence of $1, \omega$ over D' gives that $1 - \eta_{11}, 1 - \eta_{22}, \eta_{12}, \eta_{21}$ and $-\eta_{33}$ are zeros on S. So $I - \eta \in J(End(T))$, hence $I - \eta = 0$. Thus S' is not contained in any summand of T. Consequently $M = T/S$ is indecomposable, by (2.6). As $A \oplus A$ embeds in M, M is not uniform. This is a contradiction. So $[D : D']_l = 1$. Hence $D = D'$, as a consequence eR is quasi-injective. Similarly $e'R$ is quasi-injective.

LEMMA 3.4. If every indecomposable right R-module is uniform, then there do not exist three pairwise non-isomorphic indecomposable idempotents e_1, e_2 and e_3 in R such that $0 \neq e_1 J \cong e_2 J \cong e_3 J$.

Proof. Suppose the contrary. Then for $e = e_1$, $E = E(eR)$ has three submodules A_1, A_2 and A_3, such that $A_i \cong e_i R$. Consider $T = A_1 \oplus A_2 \oplus A_3$. Then $S = \{(x, x, x); x \in eJ\}$ is a simple submodule of T. On similar lines as in (3.3), we get that S is not contained in any summand of T, and that T/S is a non-uniform, indecomposable module. This is a contradiction. This proves the result.

We now prove the main theorem. First of all observe that given two indecomposable idempotents e and f in R with $eJf \neq 0$, as R is right serial, for the drpa (D, D') of eR, $[D : D']_l$ is precisely equal to the dimension of eRf as a left vector space over $D' = eRe/eJe$.

THEOREM 3.5. Let R be an artinian ring such that $J^2 = 0$ and R/J is a direct product of matrix rings over finite dimensional division rings. Then every finitely generated, indecomposable, right R-module is uniform if and only if R is a right serial ring satisfying the following conditions.

(i) For any indecomposable idempotent $e \in R$, if $eJ \neq 0$, then $[D : D']_l \leq 2$, for drpa (D, D') of eR.

(ii) If $e_1, e_2, \cdots, e_k$ are $k \geq 2$, pairwise non-isomorphic, indecomposable idempotents in R, such that $0 \neq e_1 J \cong e_j J$ for every i, then $k = 2$, and $e_1 R, e_2 R$ both are quasi-injective.

Proof. Let every indecomposable, right R-module be uniform. Consider any indecomposable idempotent $e \in R$. As eR is uniform, by the hypothesis, eR is uniserial. Hence R is right serial. That it satisfies conditions (i) and (ii) follows from (3.2), (3.3) and (3.4).

Conversely let R be a right serial ring, satisfying conditions (i) and (ii). Consider any simple right R-module S, such that $E = E(S) \neq S$. There exists an indecomposable idempotent $e \in R$, such that $S = eJ$. Let (D, D') be the drpa of eR. Then $[D : D']_\ell \leq 2$. As D is finite dimensional, by Jacobson ([4], Proposition 3, p. 158), $[D : D']_\ell = [D : D']_r$. So by (3.1), the homogeneous component of E/S determined by the simple module eR/eJ is of length atmost two. Let this composition length be 2. Then (ii) gives $E/S \cong eR/eJ \oplus eR/eJ$, and there exists no indecomposable idempotent f

not isomorphic to e, such that $S \cong fJ$. Let $[D : D']_l = 1$. Then eR is quasi-injective. Again by using condition (ii), we get that E/S of length atmost two. Thus any uniform right R-module is of length atmost three, and it must be one of the following types: (a) a simple module, (b) a projective module, and (c) an injective module A with $d(A) \leq 3$. Let the converse be not true. We can find an indecomposable, non-uniform right R-module M of smallest length. Then M is not local. By (2.8), there exist uniform modules $A_i, 1 \leq i \leq k+1$, with A_{k+1} uniserial, such that $M \cong M' = (\oplus \Sigma_1^{k+1} A_i)/L$, and $L = \{(z_1 r, z_2 r, \cdots, z_{k+1} r) : r \in R\}$ is a simple submodule; further $z_i R = soc(A_i)$. The last part of (2.8) gives that no A_i is simple, so $d(A_i) = 2$ for every i. The last part of this proposition also gives that if two of the A_i's are isomorphic, then they cannot be quasi-injective. Further as $k \geq 2$, (ii) gives that A_i's are isomorphic. Let $A = A_1, E = E(A)$. We can regard every $z_i = \lambda_i z_1$, for some $\lambda_i \in End(E)$, with $\lambda_1 = I, M' = A^{(k+1)}/L$ and $L = \{(\omega_1 z, \omega_2 z, \cdots, \omega_{k+1} z) : z \in S\}$ where $\omega_i = \lambda_i \mid S$. Let (D, D') be the drpa of A. As $k+1 \geq 3, \omega_i$ are left linearly dependent over D'. By (2.7), L is contained in a summand of $A^{(k+1)}$. Thus M is not indecomposable. This is a contradiction. Hence the result follows.

THEOREM 3.6. Let R be an artinian ring with $J^2 = 0$ and R/J a direct product of matrix rings over finite dimensional division rings. If every finitely generated, indecomposable, right R-module is uniform, then every right R-module is a direct sum of finitely generated, uniform right R-modules.

Proof. Let M be any reduced right R-module. Let $N = \oplus \Sigma_{i \in I} A_i$ be a maximal direct sum of finitely generated, uniform, non-simple submodules A_i's in M. As seen in (3.5), every A_i is projective and $d(A_i) = 2$. Consider any finitely generated, uniform, non-simple submodule of M, not contained in N. Then $d(A) = 2, S = A \cap N$ is a simple submodule, and for some finite subset $X = \{1, 2, \cdots, n\}$ of $I, S = A \cap \Sigma_1^n A_i$. Consider $T = \oplus \Sigma_1^n A_i$. Then, as $K = T + A$ is finitely generated, $K = \oplus \Sigma_1^m L_i$, for some uniform modules L_i's. As K is a homomorphic image of $T \oplus A, m \leq n+1$. As $d(K) = 2n+1, m = n+1$. Thus, only one L_i, say L_1 is simple. As $L_1 \not\subset J(K)$, we get $L_1 \not\subset T$ and $K = T \oplus L_1$. Hence $M = N + soc(M)$. This completes the proof.

It should be observed that the only place in the proof of (3.5) where the finite dimensionality of a division ring D has been used is to conclude that $[D : D']_l = [D : D']_r$ for a division subring D' of D, satisfying $[D : D']_\ell \leq 2$. It is very easy to construct examples of rings that are not serial, not local, but they satisfy the conditions in (3.5).

EXAMPLE 1. Let K be any finite dimensional division ring admitting a division subring F with $[K : F] = 2$. Then the matrix ring $R = \begin{bmatrix} F & K \\ 0 & K \end{bmatrix}$

is a right serial ring with $J^2 = 0$, such that every indecomposable, right R-module is uniform.

EXAMPLE 2. Let D be an finite dimensional division ring. Then $R = \begin{bmatrix} D & 0 & D \\ 0 & D & D \\ 0 & 0 & D \end{bmatrix}$ is a right serial ring with $J^2 = 0$, such that every indecomposable right R-module is uniform.

It is clear from Theorem (3.5) that under the hypothesis of that theorem, the ring R is of bounded representation type. If R is a local artinian ring, such that R/J is a field, $J^2 \neq 0$ and R is of bounded representation type, then by Dlab and Ringel [1], R is serial. The following example shows that, if R is not local, $J^2 \neq 0$ and every indecomposable right R-module is uniform, then R need not be generalized uniserial.

EXAMPLE 3. Let S be any commutative, local, artinian, principal ideal ring with $J(S)^3 = 0$, but $J(S)^2 \neq 0$. Further let $F = S/J(S)$. Consider the ring $R = \begin{bmatrix} F & F \\ 0 & S \end{bmatrix} sx$. Then R is a right serial, but not generalized uniserial. For $J = J(R), J^2 \neq 0$, but $J^3 = 0$. We claim that every finitely generated, indecomposable right R-module is uniform. Now $soc(e_1 R) \cong soc(e_2 R), e_2 R e_1 = 0$, both $e_1 R$ and $e_2 R$ are quasi-injective. So if for $K = soc(e_1 R), E = E(K)$, then $E = A + B$, where $A \cong e_1 R, B \cong e_2 R, A \cap B = K$ and $d(E) = 5$. This R admit only one more indecomposable injective module, namely $e_1 R/e_1 J$. Every submodule of E is quasi-injective and is of the form $C + D, C \subseteq A, D \subseteq B$. Suppose there exists an indecomposable, non-uniform right R module M of smallest length. Then by (2.8), $M \cong (\Sigma_1^{k+1} A_j)/L$, where A_j's are uniform, A_{k+1} is uniserial with $d(A_{k+1}) = 2, z_j R = soc(A_j)$ and $L = \{(z_1 r, z_2 r, \cdots, z_{k+1} r) : r \in R\}$ is a simple module. No A_j is simple, so every A_j embeds in E, and hence is quasi-injective. As A_{k+1} is uniserial, either it is isomorphic to $e_1 R$ or to a submodule of $e_2 R$. Let $k \geq 2$. If more than one A_j's are non-uniserial, then they are comparable, leading to a contradiction to the last part of (2.8). Thus, except for one, say may be A_1, every other A_j embeds in A or B.

CASE 1. $A_{k+1} \cong A$. The last part of (2.8) gives that every A_j for $j \neq k+1, 1$, is non-isomorphic to A, so it embeds in B. If A_1 were not uniserial, then A would embed in A_1, so it would contain a copy of A_{k+1}, this once again contradicts the last part of (2.8). Thus A_1 also embeds in B. As $k \geq 2$, once again, we get a contradiction. Hence $k = 1$, and $M \cong (A_1 \times A_2)/L$, which obviously embeds in E. Thus M is uniform.

CASE 2. A_{k+1} embeds in B. Then no A_j for $j \neq k+1$, embeds in B. So every $A_j, 1 \leq j \leq k$, contains a copy of A. Thus once again, we get $k = 1$. If A_1 is not uniserial, it contains a copy of the submodule of B of length 2, and so of A_{k+1}. This again gives a contradiction. Hence A_1 embeds in A, once again M embeds in E, and it is uniform. Observe that this ring is of bounded representation type.

4. LEFT MODULES OVER RIGHT SERIAL RINGS

Throughout R is an artinian right serial ring with $J^2 = 0$, R/J is a direct product of matrix rings over finite dimensional division rings, and every indecomposable right R-module uniform. Further R does not have any non-trivial central idempotent.

LEMMA 4.1. Let e be an indecomposable idempotent in R, such that eR is quasi-injective but not injective. If for some indecomposable idempotent $f \in R, eJf \neq 0$, then there exists an indecomposable idempotent e' not isomorphic to e, such that $eJ \cong e'J, Jf = Rex \oplus Re'y$ for some $0 \neq x \in eJf, 0 \neq y \in e'Jf$.

Proof. Consider $E = E(eR)$. As seen in (3.5), $d(E) = 3$. So $E = eR + B, d(B) = 2$. Then $B \cong e'R$ for some indecomposable idempotent e'. By (3.5), $e'R$ is also quasi-injective, $e'Jf \neq 0$ and for any indecomposable idempotent e_3 not isomorphic to e and e', $e_3Jf = 0$. Consequently $Jf = ReJf \oplus Re'Jf$. However, $eJf = eRex$ and $e'Jf = e'Re'y$ for some x, y. Hence $Jf = Rex \oplus Re'y$. Clearly e and e' are not isomorphic. This proves the result.

LEMMA 4.2. Let e be an indecomposable idempotent in R, such that eR is injective. If for some indecomposable idempotent $f, eJf \neq 0$, then $Jf = Rex$, for some $x \in eJf$.

Proof. In this case $eJf = eRex$, for some $x \in eJf$, and $e'Jf = 0$, for any indecomposable idempotent e' not isomorphic to e. This gives $Jf = Rex$.

LEMMA 4.3. Let e be an indecomposable idempotent such that eR is not quasi-injective. If for some indecomposable idempotent $f, eJf \neq 0$, then $Jf = Rex \oplus Rexz$, for some x in eRf and some z in fRf, with $xz \neq 0$.

Proof. Let $D = End(eJ)$. Then for $D' = eRe/eJe, [D : D'] = 2$. Also $eJ = xfR$, for some $x \in eJf$. This gives $eJf = eRex \oplus eRexz$, for some $z \in fRf$ with $xz \neq 0$. Further, by (3.5), $e'Jf = 0$ for any indecomposable idempotent e'. Hence $Jf = Rex \oplus Rexz$.

PROPOSITION 4.4. For any indecomposable idempotent f in R, the following hold.

(i) $d(Rf) \leq 3$, and

(ii) If Rf is neither simple nor injective, then $d(Rf) = 3$.

Proof. Suppose that Rf is not simple. Then for some indecomposable idempotent $e, eJf \neq 0$. By Lemma (4.1), (4.2) and (4.3), $d(Rf) \leq 3$. By Singh [5], every uniform, injective left R-module E is uniserial, so $d(E) \leq 2$. This gives (ii).

PROPOSITION 4.5. Any indecomposable left R-module M is local and has $d(M) \leq 3$.

Proof. If M is projective, the result follows from Proposition (4.4). We use Auslander-Bridger duality (see [1]) to discuss the case, when M is not projective. Let A be an indecomposable, right R-module, which is not projective. Then either A is simple or injective. Let A be simple. Then for some indecomposable idempotent e with $eJ \neq 0$, $A = eR/eJ$. For some $x \in eJ$ and some indecomposable idempotent f, $eJ = xfR$. Then dual of A, $A^* = Rf/Rexf$, satisfies $d(A^*) \leq 2$, by (4.4). Let A be injective, but not simple. Then $d(A) = 3$. So $A = B + C$, with $d(B) = d(C) = 2$. Then for some indecomposable idempotents e, e' and f, we have $B \cong eR, C \cong e'R, eJf \neq 0$ and $eJ \cong e'J$. Then for some $x \in e'Jf, y \in e'Jf$, we have $eJ = xfR$ and $e'J = yfR$. Suppose B and C are not isomorphic, then they are quasi-injectives and $A \cong (eR \oplus e'R)/L$, where $L = \{(xr, -yr) : r \in R\}$. Then $A^* \cong Rf/(Rex \oplus Rey)$, a simple module. Let B and C be isomorphic. Then for the drpa (D, D') of eR, $[D : D'] = 2$. Then $eJf = eRex \oplus eRexz$, for some $x \in eRf$ and $z \in fRf$, with $xz \neq 0$. Then $A \cong (eR \oplus eR)L$, where $L = \{(xr, xzr) : r \in R\}$. In this case $A^* \cong Rf/(Rx \oplus Rxz)$, a simple module. This proves the result.

Notice that any indecomposable, left R-module M, which is neither simple, nor projective, has $d(M) = 3$. By using this fact, and arguments similar to those used in (3.6), we get the following:

THEOREM 4.6. Let R be an artinian ring with $J^2 = 0$ and R/J a direct product of matrix rings over finite dimensional division rings. If every finitely generated, indecomposable, right R-module is uniform, then every indecomposable, left R-module is local and of composition length atmost 3, and every left R-module is a direct sum of indecomposable modules.

ACKNOWLEDGEMENTS. This paper was partially supported by the Kuwait University Research Grant No. XM126.

6. REFERENCES

[1] V. Dlab and C.M. Ringel, *Decomposition of Modules over right uniserial rings*, Math. Z. **129** (1972), 207-230.

[2] V. Dlab and C.M. Ringel, *The structure of balance rings*, Proc. London Math. Soc. (3), **26** (1973), 446-462.

[3] C. Faith, *Algebra II, Ring Theory, Grundleheren der Mathematischen Wissenschaften*, **19**, Springer-Verlag, 1976.

[4] N. Jacobson, *Structure of Rings*, Second Edition, Colloquium Publications, **37**, American Mathematical Society, 1956.

[5] S. Singh, *Artinian right serial rings*, Proc. Amer. Math. Soc. (to appear).

CURRENT ADDRESS: Kuwait University, Faculty of Science, Department of Mathematics, P.O. Box 5969, Safat 13060, Kuwait.

NONSINGULAR EXTENDING MODULES

PATRICK F. SMITH

ABSTRACT. In this paper, it is shown that if R is a semiprime right Goldie ring, then any nonsingular extending right R-module is the direct sum of an injective module and a finite number of uniform modules.

Throughout this note, all rings are associative with identity and all modules are unital right modules. Let R be a ring and let M be an R-module. A submodule K of M is called *closed (in M)* if K has no proper essential extension in M. For example, direct summands of M are closed. The module M is called an *extending* module if every closed submodule is a direct summand. Among examples of extending modules, we would mention semisimple modules, injective modules and uniform modules. For a discussion of extending modules, see [2] or [9].

Note in particular that module M is extending if and only if every submodule is essential in a direct summand of M. It follows that if A is an ideal of R and M is an injective (R/A)-module, then M is an extending R-module.

Müller, in a series of papers with different collaborators (see, for example, [6]-[11]), has investigated extending modules in general and also particular types of extending modules, as well as various dual notions (see [9] for a good account of this work). In particular, Kamal and Müller [6] studied torsion-free extending modules over commutative domains and proved that such modules are finite direct sums of injective modules and uniform modules. The purpose of this note is to generalize this fact.

Recall that if M is a module over an arbitrary ring, then M is said to have *finite uniform dimension* if the injective hull $E(M)$ is a finite direct sum of indecomposable injective modules. It is well known that a module M has finite uniform dimension if and only if M does not contain a direct sum of an infinite number of non-zero submodules (see [2] or [3]).

For any element m in an M-module M, we set $r(m) = \{r \in R : mr = 0\}$. Note that $r(m)$ is a right ideal of R. We shall say that an R-module M has an *indecomposable decomposition* if M is the direct sum of indecomposable submodules.

Lemma 1. *The following statements are equivalent for a nonsingular extending R-module M.*

(i) M *has an indecomposable decomposition.*

(ii) *Every finitely generated submodule of M has finite uniform dimension.*

1991 Mathematics Subject Classification. 16D50, 16D99.

(iii) Every cyclic submodule of M has finite uniform dimension.

(iv) R satisfies the ascending chain condition on right ideals of the form $r(m)$ where $m \in M$.

Proof. $(i) \Rightarrow (ii)$. There exist an index set I and indecomposable submodules $M_i (i \in I)$ of M such that $M = \oplus_I M_i$. By [2, 7.1] (or see [9, Proposition 2.7]), M_i is a uniform module for each $i \in I$. If L is a finitely generated submodule of M, then $L \subseteq \oplus_J M_i$, for some finite subset J of I, and hence L has finite uniform dimension.

$(ii) \Rightarrow (iii)$. Clear.

$(iii) \Rightarrow (iv)$. Let $m \in M$. Suppose that $r(m)$ is essential in a right ideal A of R. Let $a \in A$. There exists an essential right ideal E of R such that $aE \subseteq r(m)$. It follows that $maE = 0$ and hence $ma = 0$, i.e., $a \in r(m)$. Thus $r(m) = A$. We have proved that $r(m)$ is a closed submodule of the R-module R, for each $m \in M$. Moreover, $R/r(m) \cong mR$ gives that the R-module $R/r(m)$ has finite uniform dimension. Now (iv) follows by [2, 5.10].

$(iv) \Rightarrow (i)$. By [12, Lemma 3] (or see [2, 8.2] or [9, Theorem 2.19]).

Corollary 2. Let R be a ring with finite right uniform dimension and let M be a nonsingular extending R-module. Then M has an indecomposable decomposition.

Proof. As we saw in the proof of Lemma 1, for each $m \in M$ the right ideal $r(m)$ is a closed submodule of the R-module R. Hence R satisfies the ascending chain condition on right ideals of the form $r(m)$ where $m \in M$ (see [2, 5.10]). Apply Lemma 1.

Lemma 3. Let U and V be nonsingular uniform R-modules such that the module $U \oplus V$ is extending. Then U embeds in V or U in V-injective.

Proof. Let $M = U \oplus V$, $M_1 = U \oplus 0$ and $M_2 = 0 \oplus V$, so that $M = M_1 \oplus M_2$. Suppose that U does not embed in V, i.e., M_1 does not embed in M_2. Let N be any non-zero submodule of M such that $N \cap M_1 = 0$. There exists a direct summand K of M such that N is essential in K. Let K' be a submodule of M such that $M = K \oplus K'$. Note that K and K' are both uniform submodules of M.

Since $K \cap M_1 = 0$, it follows that M_1 embeds in K'. If $K' \cap M_1 = 0$, then K' embeds in M_2, so that M_1 embeds in M_2, a contradiction. Thus $K' \cap M_1 \neq 0$. Hence $K' \cap M_1$ is an essential submodule of both K' and M_1. Now $(K' + M_1)/M_1 \cong K'/(K' \cap M_1)$ is a singular submodule of the nonsingular module M/M_1, so that $K' \subseteq M_1$. Similarly, $M_1 \subseteq K'$. Thus, $K' = M_1$ and hence $M = K \oplus M_1$. By [4, Lemma 5] (or see [2, 7.5]), M_1 is M_2-injective, i.e., U is V-injective.

The next lemma is [1, Theorem 1]. We give its proof for completeness.

Lemma 4. Let $M_i (i \in N)$ by nonsingular uniform R-modules such that for each $i \in N$ there exists a monomorphism $\phi_i : M_i \to M_{i+1}$, which is not an isomorphism. Then the R-module $M = \oplus_{i \in N} M_i$ is not extending.

Proof. The monomorphism $\phi_1^{-1} : \phi_1(M_1) \to E(M_1)$ lifts to a monomorphism $\theta_1 : M_2 \to E(M_1)$ and clearly $M_1 = \theta_1(\phi_1(M_1)) \underset{\neq}{\subset} \theta_1(M_2) \subseteq E(M_1)$. Since $M_2 \cong \theta_1(M_2)$ we can suppose without loss of generality that $M_1 \underset{\neq}{\subset} M_2 \subseteq E(M_1)$. Note that $E(M_1) = E(M_2)$. Repeating this argument for the submodules M_2, M_3, we can suppose without loss of generality that $M_2 \underset{\neq}{\subset} M_3 \subseteq E(M_1)$. Proceeding in this way, we can suppose that

$$M_1 \underset{\neq}{\subset} M_2 \underset{\neq}{\subset} M_3 \underset{\neq}{\subset} \cdots \subseteq E(M_1).$$

Define $\chi : M \to E(M_1)$ by $\chi(m_1, m_2, m_3, \cdots) = m_1 + m_2 + m_3 + \cdots$ for all $(m_1, m_2, m_3, \cdots) \in M$. Let $K = \ker \chi$. Then M/K is isomorphic to a submodule of the nonsingular module $E(M_1)$ so that K is a closed submodule of M. Suppose that K is a direct summand of M, i.e., $M = K \oplus K'$ for some submodule K' of M. Then $K' \cong M/K$, so that K' is uniform. Let $0 \neq x \in K'$. Then $x \in N = M_1 \oplus \cdots \oplus M_n$ for some positive integer n. Since K'/xR is singular and M/N is nonsingular, it follows that $K' \subseteq N$. Let $y \in M_{n+1}$. Then $y = y_1 + y_2$ for some $y_1 \in K$ and $y_2 \in K' \subseteq N$. Thus $y = \chi(y) = \chi(y_1) + \chi(y_2) = \chi(y_2) \in M_1 + \cdots + M_n = M_n$.

It follows that $M_{n+1} = M_n$, a contradiction. Thus K is not a direct summand of M and M is not an extending module.

The next result is well known, but we again include a proof for completeness.

Lemma 5. Let R be a prime right Goldie ring. Then an R-module M is injective if and only if there exists a non-zero nonsingular R-module P such that M is P-injective.

Proof. The necessity is clear. Conversely, suppose that there exists a non-zero nonsingular module P such that M is P-injective. Let $0 \neq p \in P$. Then $r(p)$ is not an essential right ideal of R. There exists a uniform right ideal U of R such that $U \cap r(p) = 0$. Then U embeds in the R-module $R/r(p) \cong pR$. Thus M is U-injective [9, Proposition 1.3].

There exists a positive integer n and independent uniform right ideals $U_1 (1 \leq i \leq n)$ of R such that $U_1 \oplus \cdots \oplus U_n$ is an essential right ideal of R (see [3]). Let $1 \leq i \leq n$. Since R is prime, it follows that $UU_i \neq 0$. Let $u \in U$ such that $uU_i \neq 0$ and define $\phi_i : U_i \to U$ by $\phi_i(x) = ux (x \in U_i)$. It is not difficult to check that because R is right nonsingular, the mapping ϕ_1 is an R-monomorphism. Thus M is U_i-injective for all $1 \leq i \leq n$ and hence M is $(U_1 \oplus \cdots \oplus U_n)$-injective. By [3, Proposition 5.9] there exists an element $c \in U_1 \oplus \cdots \oplus U_n$ with $r(c) = 0$. Then $R \cong cR$ and M is R-injective. By Baer's Lemma, M is an injective R-module.

For an arbitrary ring R, we shall call an R-module M reduced if M contains no non-zero injective submodule.

Lemma 6. Let R be a prime right Goldie ring and let M be a nonsingular reduced extending R module. Then M has finite uniform dimension.

Proof. Suppose that $M \neq 0$. By Corollary 2 there exists an index set I and uniform submodules $M_i (i \in I)$ such that $M = \bigoplus_{i \in I} M_i$. Suppose that I is infinite. For all distince i, j in I, $M_i \oplus M_j$ is extending (see, for example, [2, 7.1] or [9, Proposition 2.7]) and hence M_i embeds in M_j by Lemmas 3 and 5. By Lemma 4, there exists an infinite subset J of I such that $M_i \cong M_j$ for all $i, j \in J$.

Let $N = M_i$ for any $i \in J$. Suppose that $\phi : N \to N$ is a non-zero homomorphism. Then ϕ is monomorphism because M is nonsingular. Suppose that ϕ is not an isomorphism. Then for each $i, j \in J$, ϕ induces a monomorphism $\theta_{ij} : M_i \to M_j$ which is not an isomorphism, contradicting Lemma 4. Thus ϕ is an isomorphism. It follows that the ring $End M$ of endomorphisms of M is a division ring, and by [2, 7.3], N is quasi-injective. By Lemma 5, N is injective, a contradiction. Therefore, M has finite uniform dimension.

An element c is a ring R is called *regular* if $cr \neq 0$ and $rc \neq 0$ for all $0 \neq r \in R$. Given an ideal A of R we set $C(A) = \{c \in R : c + A \text{ is a regular element of the ring } R/A\}$. We shall need the next result later.

Lemma 7. Let R be a semiprime right Goldie ring. Then R has only a finite number of minimal prime ideals $P_i (1 \leq i \leq k)$ and $C(0) = C(P_1) \cap \cdots \cap C(P_k)$.

Proof. By [3, Proposition 6.1 and Lemma 6.4].

For any ring R, an R-module M is called *torsion-free* if $mc \neq 0$ for all $0 \neq m \in M$ and regular $c \in R$. The module M is called *divisible* if $M = Mc$ for all regular $c \in R$. It is well known that injective modules are divisible.

Lemma 8. Let R be a semiprime right Goldie ring.

(i) An R-module M is nonsingular if and only if M is torsion-free.

(ii) A nonsingular R-module M is injective if and only if M is divisible.

(iii) for any ideal A of R with $C(0) \subseteq C(A)$, any divisible (R/A)-module is a divisible R-module.

Proof.

(i) By [3, Proposition 5.9].

(ii) By [3, Proposition 6.12].

(iii) Clear.

Lemma 9. Let R be a prime right Goldie ring and let M be a nonsingular extending R-module. Then $M = M_1 \oplus \cdots \oplus M_n$ for some positive integer n and R-modules $M_i (1 \leq i \leq n)$ such that for each $1 \leq i \leq n$, M_i is injective or uniform.

Proof. By Zorn's Lemma, M contains a maximal divisible submodule N and, by Lemma 8, N is injective. Thus $M = N \oplus N'$ for some reduced submodule N'. By [2, 7.1] N' is an extending module. Apply Lemma 6.

Theorem 10. Let R be a semiprime right Goldie ring and let M be a nonsingular extending R-module. Then $M = M_1 \oplus \cdots \oplus M_n$ for some positive integer n and R-modules $M_i (1 \leq i \leq n)$ such that for each $1 \leq i \leq n$, M_i is injective or uniform.

Proof. Let $P_i (1 \leq i \leq k)$ be the minimal prime ideals of R for some positive integer k (Lemma 7). Let $K = \{m \in M : mP_1 = 0\}$. Clearly K is a submodule of M. Moreover, the (R/P_1)-module K is nonsingular for if $m \in K$ and $m(E/P_1) = 0$ for some essential right ideal E/P_1 of R/P_1, where E is a right ideal containing P_1, then E is an essential right ideal of R and $mE = 0$ so that $m = 0$.

Next we show that K is a closed submodule of the R-module M. Let L be an essential extension of K in M. Let $x \in L$. There exists an essential right ideal F of R such that $xF \subseteq K$. By [3, Proposition 5.9], there exists a regular element c of R such that $c \in F$ and hence $xcP_1 = 0$. For any p in P_1 there exists $r \in R, c' \in C(0)$ such that $cr - pc'$ [3, Theorem 5.10]. Because $c \in C(P_1)$ (Lemma 7), it follows that $r \in P_1$ and hence xpc' $xcr = 0$. But $c'R$ is essential in R, so that $xp = 0$. Hence $xP_1 = 0$ and $x \in K$. Thus $K' = L$.

By hypothesis, there exists a submodule K' of M such that $M = K \oplus K'$. Since $(P_2 \cap \cdots \cap P_n)P_1 = 0$, it follows that $M(P_2 \cap \cdots \cap P_n) \subseteq K$ and hence $K'(P_2 \cap \cdots \cap P_n) = 0$. Let $S = P_2 \cap \cdots \cap P_n$. By the above argument K' is a nonsingular (R/S)-module. Clearly K and K' are extending modules over the rings R/P_1 and R/S, respectively. By Lemma 9 and induction on k, K is a finite direct sum of nonsingular injective (R/P_1)-modules and uniform (R/P_1)-modules and K' is a finite direct sum of nonsingular injective (R/S)-modules and uniform (R/S)-modules. The result now follows by Lemma 8.

Theorem 10 fails for semiprime rings R which are not right Goldie even if they are commutative (see [5, Corollary 5.2]). For commutative rings, we have the following extension of Theorem 10.

Corollary 11. Let R be a commutative ring with prime radical N such that R/N is a Goldie ring. Let M be a nonsingular extending R-module. Then M is a finite direct sum of injective (R/N)-modules and uniform R-modules.

Proof. Let $a \in N$. For any non-zero r in R, there exists a positive integer k such that $ra^{k-1} \neq 0$ and $ra^k = 0$. Thus $0 \neq ra^{k-1} \in rR \cap r(a)$. It follows that $r(a)$ is an essential ideal of R. For any $m \in M_1$ $(ma)r(a) = 0$ implies that $ma = 0$. Thus $MN = 0$ and M is a nonsingular extending (R/N)-module. The result follows by Theorem 10.

This brings us to our final generalization of the theorem of Kamal and Müller [6, Theorem 3].

Corollary 12. Let R be a commutative ring which satisfied the ascending chain condition on annihilators and let M be a nonsingular extending R-module. Then M is a finite direct sum of injective (R/N)-modules and uniform R-modules, where N is the prime radical of R.

Proof. By [13, Lemma 1.12 and 1.16], the ring R/N is Goldie. Apply Corollary 11.

Theorem 10 and its corollaries have (partial) converses because if M is a nonsingular module over an arbitrary ring R such that $M = M_1 \oplus M_2$ for some injective module M_1 and extending module M_2, then M is extending by [4, Theorem 4]. For example, Corollary 11 can be restated thus: Let R be a commutative ring with prime radical N such that R/N is a Goldie ring. Then a nonsingular R-module M is extending if and only if $M = M_1 \oplus M_2$ for some injective (R/N)-module M_1 and extending module M_2 with M_2 having finite uniform dimension.

References

[1] J. Clark and Nguyen Viet Dung, *On the decomposition of nonsingular CS-modules*, Canad. Math. Bull., to appear.

[2] Nguyen Viet Dung, Dinh van Huynh, P.F. Smith and R. Wisbauer, *Extending modules*, Pitman Research Notes in Mathematics **313** (Longman, Harlow, 1994).

[3] K.R. Goodearl and R.B. Warfield Jr. *An introduction to noncommutative Noetherian rings*, London Math. Soc. Student Texts **16** (Cambridge Univ. Press, Cambridge, 1989).

[4] A. Harmanci and P.F. Smith, *Finite direct sums of CS-modules*, Houston J. Math **19** (1993), 523-532.

[5] A. Harmanci, P.F. Smith, A. Tercan and Y. Tiras, *Direct sums of CS-modules*, Houston J. Math. **22** (1996), 61-71.

[6] M.A. Kamal and B.J. Müller, *Extending modules over commutative domains*, Osaka J. Math. **25** (1988), 531-538.

[7] M.A. Kamal and B.J. Müller, *The structure of extending modules over Noetherian rings*, Osaka J. Math. **25** (1988), 539-551.

[8] M.A. Kamal and B.J. Müller, *Torsion free extending modules*, Osaka J. Math. **25** (1988), 825-832.

[9] S.H. Mohamed and B.J. Müller, *Continuous and discrete modules*, London Math. Soc. Lecture Note **147** (Cambridge Univ. Press, Cambridge, 1990).

[10] B.J. Müller and S.T. Rizvi, *On the decomposition of continuous modules*, Canad. Math. Bull. **25** (1982), 296-301.

[11] B.J. Müller and S.T. Rizvi, *Direct sums of indecomposable modules*, Osaka J. Math. **21** (1984), 365-374.

[12] M. Okado, *On the decomposition of extending modules*, Math Japonica **29** (1984), 939-941.

[13] L.W. Small, *Orders in Artinian rings*, J. Algebra **4** (1966), 13-41.

CURRENT ADDRESS: Department of Mathematics, University of Glasgow, Glasgow, C12 8QW, Scotland, UK.

RIGHT HEREDITARY, RIGHT PERFECT RINGS ARE SEMIPRIMARY

Mark L. Teply

ABSTRACT. If R is a right hereditary, right perfect ring, then every maximal ideal of R is idempotent, and every ideal of R has a stationary power. Consequently, every right hereditary, right perfect ring must be semiprimary.

Let R be a ring with identity element. An ideal I of R is called *right T-nilpotent* if, for each sequence $x_1, x_2, \cdots$ of elements of I, there exist an integer m depending on the sequence such that $x_m x_{m-1} \cdots x_2 x_1 = 0$. Then R is called *right perfect* if the Jacobson radical, which we denote by Rad R, is right T-nilpotent and $R/\text{Rad } R$ is a semisimple artinian ring. For such rings, every nonzero right R-module has a maximal submodule. Further results on right perfect rings can be found in [1], [2] or [4]. A right perfect ring R is called *semiprimary* is Rad R is nilpotent. The ring R is called *right hereditary* if every right ideal of R is projective. The properties of hereditary rings, as well as the homological formulas that are used in this paper, can be found in [3] or [8].

In this paper we relate these three widely-studied classes of rings by proving the theorem: every right hereditary, right perfect ring is semiprimary. While this theorem has never been published before, J. Kuzmanovich and the author long ago realized that the theorem could be obtained as a consequence of the work on piecewise domains in [5]. In this note, we offer a direct proof of the theorem that is independent of theory of piecewise domains. A small part of our proof of this result involves elementary torsion-theoretical techniques. The necessary properties of torsion classes generated by simple modules can be found in [4] and [7]. The rest of the proof only uses the standard properties of right perfect rings and homological algebra. We use $E(X)$ to denote the injective hull of an R-module X.

Lemma 1. Let R be a right hereditary, right perfect ring, and let $\mathcal{S} = \{S_1, S_2, \cdots, S_n\}$ be any set of nonisomorphic simple right R-modules. Then there exists a simple module $S_j \in \mathcal{S}$ such that $Ext_R^1(S_i, S_j) = 0$ for all $S_i \in \mathcal{S}$.

Proof. Let $\mathcal{T}$ denote the smallest torsion class of right R-modules that contains $\mathcal{S}$. Since R is right perfect, the $\mathcal{T}$-torsion submodule $\mathcal{T}(E(S_1))$ of $E(S_1)$ has a maximal submodule H. Hence, there exists $S_j \in \mathcal{S}$ such that $S_j \cong \mathcal{T}(E(S_1))/H$. Since R is right hereditary, then $E(S_1)/H$ is injective. Thus, since $\mathcal{T}$ is closed under extensions of one member of $\mathcal{T}$ by another

---

1991 Mathematics Subject Classification. 16E60, 16L30.

member of $\mathcal{T}$, we have $\mathcal{T}(E(S_j)) \cong S_j$. Hence, $Ext_R^1(S_i, S_j) = 0$ for all $S_i \in \mathcal{S}$.

Lemma 2. Let R be a right hereditary, right perfect ring, and let $\mathcal{S} = \{S_1, S_2, \cdots, S_n\}$ be any set of nonisomorphic simple right R-modules. Then there exists an ordering of the members of $\mathcal{S}$ such that $Ext_R^1(S_i, S_j) = 0$ whenever $i \geq j$.

Proof. By properly arranging our choice of subscripts, we can use Lemma 1 to obtain $Ext_R^1(S_i, S_1) = 0$ for all $S_i \in \mathcal{S}$. Suppose that we have inductively chosen $S_1, S_2, \cdots, S_k$ in $\mathcal{S}$ such that $Ext_R^1(S_i, S_j) = 0$ whenever $j \leq k$ and $i \geq j$. By applying Lemma 1 to the subset $\{S_{k+1}, S_{k+2}, \cdots, S_n\}$ of $\mathcal{S}$ and arranging our subscripts properly, we obtain $Ext_R^1(S_i, S_{k+1}) = 0$ for all $i \geq k+1$. Thus the result follows by induction.

Proposition 3. If R is a right hereditary, right perfect ring, then every maximal ideal of R is idempotent.

Proof. Let S be the simple module annihilated by the maximal ideal M. Let $x \in R$, but $x \notin M^2$. Choose a right ideal $K \supseteq M^2$ maximal with respect to $x \notin K$. Then $S \cong (xR + K)/K$ is an essential submodule of R/K and $(R/K)M^2 = 0$. Since $Ext_R^1(S, S) = 0$ by Lemma 2, then $K \supseteq M$. So $x \notin M^2$ implies $x \notin M$; i.e., $M = M^2$.

Lemma 4. Let R be a right hereditary, right perfect ring, and let $\{S_1\ S_2\ \cdots\ S_n\}$ be a set of nonisomorphic simple right R-modules such that $Ext_R^1(S_i, S_j) = 0$ whenever $i \geq j$. Let M_i be the maximal ideal that annihilates $S_i (1 \leq i \leq n)$. Then $M_1, M_2, \cdots, M_n$ is idempotent; so $\bigcap_{i=1}^n M_i$ has a stationary power.

Proof. By Proposition 3, each $M_i (1 \leq i \leq n)$ is idempotent. Let $1 \leq t \leq n$. Inductively, assume that, whenever $0 \leq k < t$ and $j + k \leq n$, any product $M_j M_{j+1} \cdots M_{j+k}$ of any $k+1$ consecutive maximal ideals is idempotent. We wish to show that the product $M_j M_{j+1} \cdots M_{j+t}$ of any $t+1$ consecutive maximal ideals is idempotent. Let $P = M_j M_{j+1} \cdots M_{j+t}$. If $PM_j = P$, then $P^2 = PM_j M_{j+1} \cdots M_{j+t} = PM_{j+1} \cdots M_{j+t} = M_j(M_{j+1} \cdots M_{j+t})^2 = P$ by our induction hypothesis.

Now assume that $PM_j \neq P$, and seek a contradiction. Note that $P/PM_j \cong \bigoplus S_j$. Let $P' = M_j M_{j+1} \cdots M_{j+t-1}$. Consider the exact sequence

$$(*) \quad 0 \longrightarrow P/PM_j \longrightarrow P'/PM_j \longrightarrow P'/P \longrightarrow 0.$$

Since M_{j+t} is maximal and R is right perfect, then $P'/P \cong \bigoplus S_{j+t}$.

We wish to show that $(*)$ splits. Note that $Ext_R^1(P'/P, P/PM_j) \cong Ext_R^1(\bigoplus S_{j+t}, \bigoplus S_j) \cong \pi\ Ext_R^1(S_{j+t}, \bigoplus S_j)$. So, to show that $(*)$ splits, it is sufficient to show that $Ext_R^1(S_{j+t}, \bigoplus S_j) = 0$. Note that

$$(**) \quad Hom_R(S_{j+t}, (\pi S_j)/(\bigoplus S_j)) \longrightarrow Ext^1_R(S_{j+t}, \bigoplus S_j)$$

$$\longrightarrow Ext^1_R(S_{j+t}, \pi S_j)$$

is exact. Now $Ext^1_R(S_{j+t}, \pi S_j) \cong \pi\ Ext^1_R(S_{j+t}, S_j) = 0$ by hypothesis. Since $S_{j+t}M_{j+t} = 0, ((\pi S_j)/(\bigoplus S_j))M_j = 0$, and $M_{j+t} + M_j = R$, then

$$Hom_R(S_{j+t}, (\pi S_j)/(\bigoplus S_j)) = 0\ .$$

Hence $Ext_R(S_{j+t}, \bigoplus S_j) = 0$ by the exactness of $(**)$.

We now can write $P'/PM_j = (P/PM_j) \bigoplus P"$, where $P" \cong P'/P$. Since P' is idempotent by our induction hypothesis and since

$$(P/PM_j)P' \subseteq (P/PM_j)M_j = 0\ ,$$

then $P'/PM_j = (P'/PM_j)P' = (P/PM_j)P' \oplus (P"P') = P"P'$. It follows from the directness of the sum that $P/PM_j = 0$, which contradicts our assumption that $PM_j \neq P$.

By induction, $M_1 M_2 \cdots M_n$ must be idempotent.

Since $Rad(R/M_1 M_2 \cdots M_n) = (\bigcap_{i=1}^n M_i)/M_1 M_2 \cdots M_n$ is nilpotent of index $p \leq n$ for some p, then $(\bigcap_{i=1}^n M_i)^p \subseteq M_1 M_2 \cdots M_n$. But since $M_1 M_2 \cdots M_n$ is idempotent, then $M_1 M_2 \cdots M_n \subseteq (\bigcap_{i=1}^n M_i)^p$. Thus, $(\bigcap_{i=1}^n M_i)^p$ is a stationary power of $\bigcap_{i=1}^n M_i$.

Proposition 5. If R is a right hereditary, right perfect ring, then every ideal of R has a stationary power.

Proof. Let I is an ideal, let $M_1, M_2, \cdots, M_n$ be the set of maximal ideals that contain I, and let $K = \bigcap_{i=1}^n M_i$. Let S be the set of n simple right R-modules annihilated by $M_1, M_2, \cdots, M_n$. By Lemmas 2 and 4, K has a stationary power; say K^P is stationary. If $I^P \neq K^P$, then K^P/I^P is a nonzero (right) ideal of the right perfect ring R/I^P, and hence K^P/I^P must have a maximal submodule N/I^P. But then $Rad(R/I^P) = K/I^P$, so that $K^P/I^P = K^{P+1}/I^P = (K^P/I^P)(K/I^P) \subseteq N/I^P$, which is a contradiction. Therefore, $K^P = I^P$, and hence I^P is a stationary power of I.

Using the same method of proof as Proposition 5, we now obtain our main result.

Theorem. A right hereditary, right perfect ring is semiprimary.

Proof. By Proposition 5, $Rad\ R$ has a stationary power, say $(Rad\ R)^P$. If $(Rad(R))^P \neq 0$, then $(Rad\ R)^P$ has a maximal submodule N. But then $(Rad\ R)^{P+1} \subseteq N$, which contradicts the fact that $(Rad\ R)^P$ is stationary. Therefore, $(Rad\ R)^P = 0$ and R is semiprimary.

We also remark that the results of this paper require the "right hereditary" hypothesis. Let F be a field, and let R be the ring of all upper triangular countably infinite matrices with entries in F that have only finitely many nonzero off-diagonal entries and constant diagonal. Then R is right perfect, but $Rad\ R$ has no stationary power.

REFERENCES

[1] F. Anderson and K. Fuller, *Rings and Categories of Modules*, Graduate Texts in Math. **13**, Springer-Verlag, New York - Heidelberg - Berlin, 1973.

[2] H. Bass, *Finitistic dimension and a homological generalization of semiprimary rings*, Trans. Amer. Math. Soc. **95** (1960), 466-488.

[3] H. Cartan and S. Eilenberg, *Homological Algebra*, Princeton University Press, Princeton, 1956.

[4] J. Golan, *Localization of noncommutative rings*, Marcel Dekker Pure and Applied Mathematics Monograph # 30, New York, 1975.

[5] R. Gordon and L.W. Small, *Piecewise Domains*, J. Algebra **23** (1972), 553-564.

[6] K. Goodearl, *Ring Theory: Nonsingular rings and modules*, Marcel Dekker Pure and Applied Mathematics Monograph # 33, New York, 1976.

[7] J.P. Jans, *Some Aspects of Torsion*, Pacific J. Math. **15** (1965), 1249-1259.

[8] J. Rotman, *An Introduction to Homological Algebra*, Academic Press, San Diego, 1979.

CURRENT ADDRESS: Department of Mathematical Sciences, University of Wisconsin-Milwaukee, Milwaukee, WI 53201-0413 USA.

E-MAIL ADDRESS: mlteply@csd.uwm.edu

ON THE ENDOMORPHISM RING OF A DISCRETE MODULE: A THEOREM OF F. KASCH

Julius M. Zelmanowitz

Dedicated to Bruno Mueller in appreciation of his many contributions to ring theory.

ABSTRACT. Using results from the monograph of Mohamed and Mueller on discrete modules, it is shown that the endomorphism ring of a discrete module, modulo its radical, is a direct product of full linear rings.

In [1], Mohamed Hasan Ali and this author showed that the endomorphism ring of a discrete module, modulo its radical, is a continuous ring. As that paper was going to press, S. Mohamed and B. Mueller observed that much more is true; namely, modulo its radical, the endomorphism ring of a discrete module is a direct product of full linear rings. This fact follows from two observations, only one of which is available in the archival literature. First of all, a discrete module has a decomposition as a direct sum of modules with local endomorphism rings which complements direct sums [3; Theorem 4.15 and Corollary 5.5], hence is a so-called "Harada-module." Secondly, F. Kasch had shown, in unpublished lecture notes dating from the early 1980s [2], that the endomorphism ring of a Harada-module, modulo its radical, is a direct product of full linear rings. The demonstration of this last theorem relies on the use of the "Total" of a module, a concept pioneered by Kasch. In fact, Kasch shows that for any module which decomposes as a direct sum of modules with local endomorphism rings, its endomorphism ring, modulo its Total, is a direct product of full linear rings [2; Corollary 3.4.3]. Furthermore, such a decomposition complements direct summands if and only if the Total equals the radical of the endomorphism ring [2; Theorem 4.1.3].

In this brief note, we will show how Kasch's result, for the special case of a discrete module, can be readily derived from material in the definitive monograph of Mohamed and Mueller [3]. That is, we give a demonstration of the following special case of Kasch's theorem.

Theorem. *If M is a discrete R-module with $S = End_R M$ and $J(S)$ is the Jacobson radical of S, then $S/J(S)$ is isomorphic to a direct product of full linear rings.*

For this result in its full generality, the only reference known to the author remains the important work [2].

Typeset by $\mathcal{A}_{\mathcal{M}}\mathcal{S}$-TEX

Recall that a module M is discrete if it satisfies the following conditions:

(D$_1$) For every submodule N of M, there is a decomposition $M = M_1 \oplus M_2$ with $M_1 \leq N$ and $N \cap M_2 << M$ (here $N << M$ means that N is a small submodule of M);

(D$_2$) If N is a submodule of M with M/N isomorphic to a direct summand of M, then N is a direct summand of M.

We will be using the following properties of discrete modules, sometimes without explicit mention.

(i) A direct summand of a discrete module is discrete [3; Lemma 4.7].

(ii) If M is discrete and $S = \text{End}_R M$ then $J(S) = \{\alpha \in S | M\alpha << M\}$ [3; Theorem 5.4].

(iii) If $A \oplus B$ is a discrete module and A and B have no nonzero isomorphic direct summands, then $Af << B$ for every $f \in \text{Hom}_R(A, B)$ [3; page 84].

(iv) A discrete module is a direct sum of hollow modules, each of which has a local endomorphism ring [3; Theorem 4.15 and Corollary 5.5].

(v) A decomposition of a discrete module M as a direct sum of hollow modules, $M = \oplus_{i \in I} M_i$, complements direct summands and is locally semi-T-nilpotent (that is, for any family of non-isomorphisms $f_{i_k} \in \text{Hom}_R(M_{i_k}, M_{i_{k+1}})$ with distinct $i_k \in I$ and any $x \in M_{i_1}$ there exists $n = n(x)$ such that $xf_1 \ldots f_n = 0$) [3; Theorem 2.26 and Theorem 4.15].

Unless otherwise indicated, M denotes a discrete left R-module with endomorphism ring S acting as right operators, and $J(S)$ is the Jacobson radical of S. From (iv), we may write $M = \underset{i \in I}{\oplus} M_i$ where the M_i are the homogeneous components of M with respect to a decomposition of $M = \underset{a \in A}{\oplus} M_a$ as a direct sum of hollow modules; that is, for each $i \in I$, there exists $a_i \in A$ such that $M_i = \oplus \{M_a | a \in A \text{ and } M_a \cong M_{a_i}\}$. Set $S_i = \text{End}_R M_i$, and for any subset $F \subseteq I$, set $M_F = \underset{i \in F}{\oplus} M_i$ and $S_F = \text{End}_R M_F$. The proof of the theorem will be completed by showing that

(a) $S/J(S) \cong \prod_{i \in I} S_i/J(S_i)$, and that
(b) $S_i/J(S_i)$ is isomorphic to a full linear ring for each $i \in I$.

We adopt the following conventions. Each M_i and M_F, for $F \subseteq I$, will be regarded as a submodule of M through the obvious identification; similarly, S_i and S_F will be regarded as subrings of S by extending elements of S_F trivially across $\oplus_{i \in I \setminus F} M_i$. ϵ_i and π_i will denote, respectively, the inclusion and projection maps with respect to M_i; similarly for ϵ_F and π_F, for $F \subseteq I$. (Thus $M_G \epsilon_F = 0$ whenever F and G are disjoint subsets of I.) For any $\alpha \in S$, we set $\alpha_{ij} = \epsilon_i \alpha \pi_j \in \text{Hom}_R(M_i, M_j)$; similarly, $\alpha_{FG} = \epsilon_F \alpha \pi_G$ and $\alpha_F = \epsilon_F \alpha \pi_F$ for any subsets $F, G \subseteq I$. Observe that $\alpha = \sum_{i,j \in I} \alpha_{ij}$, and that while this is an infinite sum, no problems of interpretation arise because for any $x \in M$, $x\alpha_{ij}$ is zero for almost all $i, j \in I$.

ON THE ENDOMORPHISM RING OF A DISCRETE MODULE   319

Before we begin the proof of the theorem, we require one more fact about discrete modules.

(vi) If $M = \oplus_{j \in J} M_j$ is a decomposition of a discrete module M into submodules M_j such that no two distinct M_j share an isomorphic direct summand, then this decomposition is locally semi-T-nilpotent.

Proof of (vi). We simplify notation by assuming that $x \in M_1$ and that homomorphisms $f_k \in \operatorname{Hom}_R(M_k, M_{k+1})$ are given; and we will show that $xf_1 f_2 \ldots f_n = 0$ for some n. For each $k \geq 1$, we may use (iv) to write $M_k = \oplus_{a \in A_k} M_a$ with each M_a a hollow module. For each $k \geq 1$, choose finite subsets $B_k \subseteq A_k$ with $x \in B_1$, $x f_1 \ldots f_k \in B_{k+1}$. Then, $x f_1 f_2 \ldots f_n = x f'_1 f'_2 \ldots f'_n$ where $f'_k = \epsilon_{B_k} f_i \pi_{B_{k+1}} = \sum_{a \in B_k, b \in B_{k+1}} \epsilon_a f_i \pi_b$. Hence, utilizing the König Graph Theorem, it remains to show that for every choice of $b_i \in B_i$, there exists an integer n with $x \pi_{b_1} \epsilon_{b_1} f_1 \pi_{b_2} \epsilon_{b_2} f_2 \pi_{b_3} \ldots \epsilon_{b_n} f_n \pi_{b_{n+1}} = 0$. Since, by hypothesis, M_{b_k} and $M_{b_{k+1}}$ are not isomorphic for any $k \geq 1$, each $\epsilon_{b_k} f_k \pi_{b_{k+1}}$ is a non-isomorphism, and it then follows from (v) that $x \pi_{b_1} \epsilon_{b_1} f_1 \pi_{b_2} \ldots \epsilon_{b_n} f_n \pi_{b_{n+1}} = 0$ for some n. □

Proof of the Theorem. To demonstrate (a), it suffices to show that the function $\alpha \xmapsto{\theta} \{\overline{\alpha_{ii}}\}_{i \in I}$ defines a ring homomorphism of S onto $\prod_{i \in I} S_i / J(S_i)$ with kernel $\theta = J(S)$. θ is clearly additive and surjective. To check that it is multiplicative, observe that for any $\alpha, \beta \in S$, $(\alpha\beta)_{ii} = \epsilon_i \alpha \beta \pi_i = \epsilon_i \alpha \pi_i \epsilon_i \beta \pi_i + \epsilon_i \alpha (1 - \pi_i \epsilon_i) \beta \pi_i = \alpha_{ii} \beta_{ii} + \epsilon_i \alpha (1 - \pi_i \epsilon_i) \beta \pi_i$, with $\epsilon_i \alpha (1 - \pi_i \epsilon_i) \beta \pi_i \in J(S_i)$ because $M_i \epsilon_i \alpha (1 - \pi_i \epsilon_i) \beta \pi_i \subseteq (\oplus_{j \in I \setminus \{i\}} M_j) \beta \pi_i \ll M_i$ by (iii). We next show that $\ker \theta = J(S)$.

For any $\alpha \in J(S)$, there exists $\beta \in S$ with $\alpha + \beta + \alpha \beta = 0$. Hence, from the calculation in the preceding paragraph, for each $i \in I$, $\alpha_{ii} + \beta_{ii} + \alpha_{ii} \beta_{ii} + \epsilon_i \alpha (1 - \pi_i \epsilon_i) \beta \pi_i = 0$ with $\epsilon_i \alpha (1 - \pi_i \epsilon_i) \beta \pi_i \in J(S_i)$. Thus α_{ii} is quasi-regular modulo $J(S_i)$ for each $i \in I$, and so $\alpha \in \ker \theta$ proving that $J(S) \subseteq \ker \theta$.

For the reverse inclusion we need a general result adapted from [2; Lemma 4.3.1].

Lemma. *Suppose that $M = \oplus_{j \in J} M_j$ is a decomposition of an arbitrary module M and that $\beta \in S = \operatorname{End}_R M$ is such that β_F has a left inverse in S_F for each finite subset $F \subseteq J$. Then for any $x \in M$, there exist $\mu_1, \mu_2, \cdots \in S$, $\gamma_1, \gamma_2, \cdots \in S$ and a sequence of finite subsets $F_1 \subseteq F_2 \subseteq \cdots \subseteq J$ such that for all $n \geq 1$, $x \mu_n \beta = x(1 + \gamma_1 \gamma_2 \ldots \gamma_n)$ with $x \in M_{F_1}$, each $\gamma_k \in S_{F_k} \beta_{F_k, F_{k+1} \setminus F_k} \subseteq \operatorname{Hom}_R(M_{F_k}, M_{F_{k+1} \setminus F_k})$ and $\gamma_k = 0$ if $F_k = F_{k+1}$.*

Proof of the Lemma. We proceed by induction on n. For $F \subseteq J$ set $F^c = J \setminus F$. Choose F_1 a finite subset of J with $x \in M_{F_1}$. Set $\beta_1 = \beta_{F_1} = \epsilon_{F_1} \beta \pi_{F_1} \in S_{F_1}$, $\beta'_1 = \epsilon_{F_1} \beta \pi_{F_1^c}$ and $\beta''_1 = \epsilon_{F_1^c} \beta$; then $\beta = \beta_1 + \beta'_1 + \beta''_1$. By hypothesis, there exists $\mu_1 \in \operatorname{End} M_{F_1}$ with $x = x \mu_1 \beta_1 = x \mu_1 (\beta - \beta'_1 - \beta''_1) = x \mu_1 (\beta - \beta'_1)$, because $M_{F_1} \beta''_1 = 0$. Hence $x \mu_1 \beta = x(1 + \mu_1 \beta'_1)$. Choose F_2 to be a finite subset of J which contains F_1 and with $x \mu_1 \beta \in M_{F_2}$. Then

$x\mu_1\beta_1' \in M_{F_2} \cap M_{F_1^c} = M_{F_2\backslash F_1}$ so, taking $\gamma_1 = \mu_1\beta_1'\pi_{F_2\backslash F_1} \in S_{F_1}\beta_{F_1,F_2\backslash F_1}$, we have $x\mu_1\beta = x(1+\gamma_1)$, which establishes the case $n = 1$. Observe that if $F_2 = F_1$, then $\gamma_1 = 0$.

Now assume that $n \geq 2$ and that the case $n-1$ has been established, so that $x\mu_{n-1}\beta = x(1+\gamma_1\ldots\gamma_{n-1})$ with $x\gamma_1\gamma_2\ldots\gamma_{n-1} \in M_{F_n\backslash F_{n-1}}$. As above, write $\beta = \beta_n + \beta_n' + \beta_n''$ where $\beta_n = \beta_{F_n} = \epsilon_{F_n}\beta\pi_{F_n}$, $\beta_n' = \epsilon_{F_n}\beta\pi_{F_n^c}$ and $\beta_n'' = \epsilon_{F_n^c}\beta$. Then, by hypothesis, there exists $\nu_n \in S_{F_n} = \text{End}_R M_{F_n}$ with $x\gamma_1\ldots\gamma_{n-1} = x\gamma_1\ldots\gamma_{n-1}\nu_n\beta_n = x\gamma_1\ldots\gamma_{n-1}\nu_n(\beta - \beta_n' - \beta_n'') = x\gamma_1\ldots\gamma_{n-1}\nu_n(\beta-\beta_n')$ because $M_{F_n}\beta_n'' = 0$. From the induction hypothesis, $x\mu_{n-1}\beta = x + x\gamma_1\ldots\gamma_{n-1} = x + x\gamma_1\ldots\gamma_{n-1}\nu_n(\beta - \beta_n')$ so that $x(\mu_{n-1} - \gamma_1\ldots\gamma_{n-1}\nu_n)\beta = x(1-\gamma_1\ldots\gamma_{n-1}\nu_n\beta_n')$. Set $\mu_n = \mu_{n-1} - \gamma_1\ldots\gamma_{n-1}\nu_n$ and choose F_{n+1} to be a finite subset of J which contains F_n and with $x\mu_n\beta \in M_{F_{n+1}}$. Then $x\gamma_1\ldots\gamma_{n-1}\nu_n\beta_n' \in M_{F_{n+1}} \cap M_{F_n^c} = M_{F_{n+1}\backslash F_n}$ so, taking $\gamma_n = -\nu_n\beta_n'\pi_{F_{n+1}\backslash F_n} \in S_{F_n}\beta_{F_n,F_{n+1}\backslash F_n}$, we have $x\mu_n = x(1+\gamma_1\ldots\gamma_n)$ with $\gamma_n = 0$ when $F_n = F_{n+1}$. This establishes the lemma. $\square$

Corollary. *If $M = \oplus_{j \in J} M_j$ is a locally semi-T-nilpotent decomposition of a module M and $\alpha \in S = \text{End}_R M$ satisfies $(\gamma\alpha)_{jj} \in J(S_j)$ for every $\gamma \in S$ and $j \in J$, then $\alpha \in J(S)$.*

Proof of the Corollary. Let $F \subseteq J$ be any finite subset of J, and for each $j \in F$, set $A_j = \{\sum_{i \in F}\alpha_{ij} | \alpha \in S \text{ and } \text{Hom}_R(M_j, M_i)\alpha_{ij} \subseteq J(S_j) \text{ for every } i \in F\}$. Then each A_j is a left ideal of S_F, in fact a quasi-regular left ideal of S_F because, as is readily checked, $\alpha_{jj} \in J(S_j)$ and $(1_F - \sum_{i \in F}\alpha_{ij})^{-1} = (1_{jj} - \alpha_{jj})^{-1} + \sum_{i \in F\backslash\{j\}}(1_{ii} + \alpha_{ij}(1_{jj} - \alpha_{jj})^{-1})$, with $(1_{jj} - \alpha_{jj})^{-1}$ the inverse of $1_{jj} - \alpha_{jj}$ in S_j.

Now suppose that $\alpha \in S$ satisfies the hypothesis that $(\gamma\alpha)_{jj} \in J(S_j)$ for every $\gamma \in S$ and $j \in J$. Then $\alpha_F = \sum_{i,j \in F}\alpha_{ij} \in \sum_{j \in F} A_j \subseteq J(S_F)$ for each finite subset $F \subseteq J$. Set $\beta = 1 - \alpha$ and let F_0 denote the empty set. Then $\beta_F = 1_F - \alpha_F$ is invertible in S_F for each finite subset $F \subseteq J$. With notation as in the lemma, we have for all $n \geq 1$, $x\mu_n\beta = x(1+\gamma_1\gamma_2\ldots\gamma_n)$ with $x \in M_{F_1}$, each $\gamma_k \in S_{F_k}\beta_{F_k,F_{k+1}\backslash F_k}$, and $\gamma_k = 0$ if $F_k = F_{k+1}$. In any event, $x\gamma_1\gamma_2\ldots\gamma_n = x\epsilon_{F_1}\gamma_1\epsilon_{F_2\backslash F_1}\gamma_2\ldots\epsilon_{F_n\backslash F_{n+1}}\gamma_n = x\gamma_1'\gamma_2'\ldots\gamma_n'$ where $\gamma_k' = \epsilon_{F_k\backslash F_{k-1}}\gamma_k \in \text{Hom}_R(M_{F_k\backslash F_{k-1}}, M_{F_{k+1}\backslash F_k})$.

We next show that $x\gamma_1\gamma_2\ldots\gamma_n = 0$ for some n, and for this it suffices, by another application of the König Graph Theorem, to show that for every choice of $j_k \in F_k\backslash F_{k-1}$, there exists an integer $n \geq 1$ such that $x\pi_{j_1}(\epsilon_{j_1}\gamma_1'\pi_{j_2})(\epsilon_{j_2}\gamma_2'\pi_{j_3})\ldots(\epsilon_{j_n}\gamma_n'\pi_{j_{n+1}}) = 0$. Since the decomposition $M = \oplus_{j \in J}M_j$ is locally semi-T-nilpotent, this can be accomplished by showing that for each $k \geq 1$, $\gamma_k'' = \epsilon_{j_k}\gamma_k'\pi_{j_{k+1}}$ is not an isomorphism of M_{j_k} onto $M_{j_{k+1}}$.

If, to the contrary, γ_k'' is an isomorphism for some $k \geq 1$, let $\eta_k \in \text{Hom}_R(M_{j_{k+1}}, M_{j_k})$ denote its inverse. Write $\gamma_k = \delta_k\beta_{F_k,F_{k+1}\backslash F_k}$ for some

$\delta_k \in S_{F_k}$. Then

$$\gamma'_k = \epsilon_{F_k \setminus F_{k-1}} \gamma_k = \epsilon_{F_k \setminus F_{k-1}} \delta_k \beta_{F_k, F_{k+1} \setminus F_k}$$
$$= \epsilon_{F_k \setminus F_{k-1}} \delta_k \epsilon_{F_k} (1 - \alpha) \pi_{F_{k+1} \setminus F_k} = -\epsilon_{F_k \setminus F_{k-1}} \delta_k \epsilon_{F_k} \alpha \pi_{F_{k+1} \setminus F_k},$$

so, using the hypothesis on α,

$$1_{j_{k+1} j_{k+1}} = \eta_k \gamma''_k = \eta_k \epsilon_{j_k} \gamma'_k \pi_{j_{k+1}}$$
$$= -\eta_k \epsilon_{j_k} \epsilon_{F_k \setminus F_{k-1}} \delta_k \epsilon_{F_k} \alpha \pi_{F_{k+1} \setminus F_k} \pi_{j_{k+1}}$$
$$= -\eta_k \epsilon_{j_k} \delta_k \epsilon_{F_k} \alpha \pi_{j_{k+1}}$$
$$= -((\pi_{j_{k+1}} \eta_k \epsilon_{j_k} \delta_k \epsilon_{F_k}) \alpha)_{j_{k+1} j_{k+1}} \in J(S_{j_{k+1}}),$$

which is impossible. Hence no γ''_k is an isomorphism and we conclude that there exists an integer $n \geq 1$ with $x \gamma_1 \gamma_2 \ldots \gamma_n = 0$.

Thus $x \mu_n \beta = x$ for some integer n, which proves that $\beta = 1 - \alpha$ is a surjection. Since β_F is a monomorphism for any finite subset $F \subseteq I$, β is a monomorphism and so $1 - \alpha$ is invertible in S. Furthermore, for any $\delta \in S$, $\delta \alpha$ also satisfies the hypothesis. Hence $1 - \delta \alpha$ is invertible for all $\delta \in S$, proving that $\alpha \in J(S)$. □

We are now in a position to complete the proof of (a). If $\alpha \in \ker \theta$ and $\gamma \in S$, then $\gamma \alpha \in \ker \theta$ and so $(\gamma \alpha)_{ii} \in J(S_i)$ for all $i \in I$. From (vi) and the corollary it follows that $\alpha \in J(S)$, as was desired.

To demonstrate (b), we proceed in a standard manner following the approach in [2]. We now assume that $M = \oplus_{j \in J} M_j$ is a discrete module which is a direct sum of isomorphic indecomposable submodules $M_j, j \in J$. We fix $M_0 = M_{j_0}$, $j_0 \in J$, together with isomorphisms $\mu_j : M_0 \to M_j$ for each $j \in J$. From (iv), we know that $S_0 = \mathrm{End}_R M_0$ is a local ring. Set $D = S_0 / J(S_0)$ and let $V = \oplus_{j \in J} D x_j$ be a left D-vector space with basis $\{x_j | j \in J\}$. Define $\psi : S = \mathrm{End}_R M \to \mathrm{End}_D V$ as follows: for $\alpha \in S$ and $i \in J$, $(\alpha) \psi : x_i \mapsto \sum_{j \in J} \overline{\mu_i \alpha_{ij} \mu_j^{-1}} x_j$ where $\overline{\mu_i \alpha_{ij} \mu_j^{-1}} \in D$ denotes the coset of $\mu_i \alpha_{ij} \mu_j^{-1} \in S_0$ modulo $J(S_0)$. Observe that α_{ij} is an isomorphism for at most finitely many $j \in J$ because for any $0 \neq x \in M_i$, $x \alpha_{ij} \neq 0$ for at most finitely many $j \in J$. Hence $\overline{\mu_i \alpha_{ij} \mu_j^{-1}} = 0$ for almost all $j \in J$, and $(\alpha) \psi$ therefore defines a linear transformation. It is straightforward to verify that ψ is a surjective ring homomorphism; this requires the fact that $\alpha_{ij} \beta_{jk}$ is an isomorphism if and only if α_{ij} and β_{jk} are isomorphisms. It now remains only to show that $J(S) = \ker \psi$.

If $\alpha \in J(S)$ then $M_0 \mu_i \alpha_{ij} \mu_j^{-1} \ll M_0$ for every $i, j \in J$, from which it follows that $\overline{\mu_i \alpha_{ij} \mu_j^{-1}} = 0$ for every $i, j \in J$, and hence that $\alpha \in \ker \psi$. For the reverse inclusion, assume that $\alpha \in \ker \psi$ and $\gamma \in S$. Then $\gamma \alpha \in \ker \psi$, so that $\mu_i (\gamma \alpha)_{ij} \mu_j^{-1} \in J(S_0)$ for any $i, j \in J$. Thus $(\gamma \alpha)_{ij}$ is not an isomorphism for any $i, j \in J$ and, in particular, $(\gamma \alpha)_{jj} \in J(S_j)$ for any $j \in J$. From (v) and the corollary, we conclude that $\alpha \in J(S)$. With this the proof of (b) and the theorem is completed. □

REFERENCES

1. M. Hasan Ali and J. Zelmanowitz, *Discrete implies continuous*, J. Algebra **183** (1996), 186–192.
2. F. Kasch, *Moduln mit LE-Zerlegung und Harada-Moduln*, lecture notes, Univ. of Munich, 1982.
3. S. Mohamed and B. Mueller, *Continuous and Discrete Modules*, Cambridge U. Press, 1990.

DEPARTMENT OF MATHEMATICS, UNIVERSITY OF CALIFORNIA, SANTA BARBARA, CALIFORNIA 93106

NONSINGULAR RINGS WITH FINITE TYPE DIMENSION

Yiqiang Zhou

ABSTRACT. Two modules are said to be orthogonal if they do not have nonzero isomorphic submodules. An atomic module is any nonzero module whose nonzero submodules are not orthogonal. A module is said to have type dimension n if it contains an essential submodule which is a direct sum of n pairwise orthogonal atomic submodules; If such a number n does not exist, we say the type dimension of this module is ∞. In this paper, we provide characterizations and examples of nonsingular rings with finite type dimension. A characterization theorem is proved for nonsingular rings whose nonzero right ideals contain nonzero atomic right ideals. Type dimension formulas are also obtained for polynomial rings, Laurent polynomial rings and formal triangular matrix rings.

Nonsingular rings with finite Goldie dimension have been studied by many authors (e.g., see [2, 4-14]) and they can be characterized by various nice properties. For example, a nonsingular ring has finite Goldie dimension iff its maximal quotient ring is a semisimple ring [10]. For more detail on the study of nonsingular rings of finite Goldie dimension, we refer to [7]. In this paper, we study the class of rings described by the title. The concept of type dimension was introduced in [15] in order to characterize the ring R for which every injective module is a direct sum of atomic modules. The difference between Goldie dimension and type dimension is obvious: One is the maximal number of nonzero right ideals which are independent and the other is that of nonzero right ideals which are pairwise orthogonal. The study of the rings under the title is motivated by the facts that there exist many nonsingular rings with finite type dimension but without finite Goldie dimension and, more importantly, those rings possess good nature. For example, we will see that a nonsingular ring has finite type dimension iff its maximal quotient ring is a finite direct sum of indecomposable self-injective regular rings. In this paper, after introducing some basic results on type dimension, we present various characterizations of nonsingular rings with finite type dimension; And then we prove that a nonsingular ring satisfies

---

1991 *Mathematics Subject Classification*. Primary 16D70, 16P99; Secondary 16D50, 16P60.

Key words and phrases. Nonsingular rings, type dimension, type submodules, atomic modules, self-injective regular rings.

Typeset by $\mathcal{A}_{\mathcal{M}}\mathcal{S}$-TEX

the property that every nonzero right ideal contains an atomic right ideal iff its maximal right quotient ring is a direct product of indecomposable right self-injective regular rings; Finally, type dimension formulas are obtained for polynomial rings, Laurent polynomial rings and formal triangular matrix rings. Analogues of those results for nonsingular rings with finite Goldie dimension are known and can be found in [7]. All rings R in this paper are associative with identity and all modules are unitary right R-modules. A nonsingular module is any module whose singular submodule is zero. The ring R is said to be nonsingular if R_R is a nonsingular module.

1. TYPE DIMENSION

We say two modules X and Y are *orthogonal* and write $X \perp Y$ if they have no nonzero isomorphic submodules. Two modules M and N are said to be *parallel*, written $M \| N$, if M is not orthogonal to any nonzero submodule of N and N is not orthogonal to any nonzero submodule of M. An *atomic module* is a nonzero module A such that any two nonzero submodules of A are parallel. The concept of atomic module was introduced by Dauns [3], where it was observed that every atom in the lattice of all saturated classes of R-modules is generated by an atomic module. For a submodule N of M, $N \leq_e M$ means that N is essential in M.

Definition 1.1. [15] *If there exist pairwise orthogonal atomic submodules $A_1, \cdots, A_n$ of a module M such that $A_1 \oplus \cdots \oplus A_n \leq_e M$, then such a number n is uniquely determined by M in the sense that if $N_1 \oplus \cdots \oplus N_m \leq_e M$ with $0 \neq N_i$ $(i = 1, \cdots, m)$ and $N_i \perp N_j$ $(i \neq j)$, then $m \leq n$. In this case, we call n the type dimension of M and write $t.dim(M) = n$. If such an n does not exist, we say the type dimension of M is ∞ and write $t.dim(M) = \infty$. If $M = 0$ we write $t.dim(M) = 0$.*

Lemma 1.2. *Let N be a submodule of a module M.*
1. $t.dim(N) \leq t.dim(M)$.
2. *If $M \| N$, then M has finite type dimension iff N has finite type dimension; And in this case, $t.dim(M) = t.dim(N)$. Conversely, if M has finite type dimension and $t.dim(M) = t.dim(N)$, then $M \| N$.*
3. *If $M = M_1 \oplus \cdots \oplus M_n$, then $t.dim(M) \leq t.dim(M_1) + \cdots + t.dim(M_n)$; The equality holds if $M_i \perp M_j$ whenever $i \neq j$.*
4. *If both N and M/N have finite type dimension, then M has finite type dimension and $t.dim(M) \leq t.dim(N) + t.dim(M/N)$.*
5. *If M has finite type dimension, then $Im(f) \| M$ for any monomorphism $f: M \longrightarrow M$.*

Proof. (1) − (3) follow directly from Definition 1.1, and (5) follows from (2).

(4). Let P be a complement of N in M. Then $N \oplus P \leq_e M$ and $P \hookrightarrow M/N$. So, $t.dim(M) = t.dim(N \oplus P) \leq t.dim(N) + t.dim(P) \leq t.dim(N) + t.dim(M/N)$. □

A ring R is said to be *indecomposable* if R can not be decomposed as a direct sum of two nonzero ideals. A module M is said to be *t-indecomposable* if M is not a direct sum of two nonzero orthogonal submodules. For any module M, we use $E(M)$ to indicate the injective hull of M.

Lemma 1.3. *Let M be an R-module.*
  1 *M is an atomic module iff $E(M)$ is t-indecomposable.*
  2 *If $E = E(M)$ is t-indecomposable, then $End(E_R)$ is indecomposable. The converse holds if M is nonsingular.*

Proof. Part (1) is obvious. For (2), suppose $S = End(E)$ is not indecomposable. Then there exists a central idempotent $e \in S$ such that $e \neq 0$ and $e \neq 1$. Then $E = e(E) \oplus (1-e)(E)$. We show that $e(E) \perp (1-e)(E)$. If not, then we have an R-module isomorphism $\pi : e(X) \longrightarrow (1-e)(Y) \neq 0$ for some $X \subseteq E$ and some $Y \subseteq E$. There exists an $f \in S$ such that f extends π. Thus, we have $(1-e)(Y) = f(e(X)) = f(e^2(X)) = e(f(e(X))) = e((1-e)(Y)) = 0$, a contradiction. So, $e(E) \perp (1-e)(E)$ and hence E is not t-indecomposable.

For the converse, let $E = E_1 \oplus E_2$ with each $E_i \neq 0$ and $E_1 \perp E_2$. Since E is nonsingular, we have $End(E) \cong End(E_1) \oplus End(E_2)$. Therefore, $End(E)$ is not indecomposable. □

A submodule N of M is said to be a *type submodule*, written $N \leq_t M$, if N satisfies the property that whenever $N \subseteq P \subseteq M$ with $N \| P$ it must be $N = P$. The concept of type submodules was discussed in [15]. Here, we are using an equivalent definition of type submodules (see [15, Lemma 1(b)]).

Lemma 1.4. *Let N be a submodule of a module M.*
  1 *If M has finite type dimension and N is a type submodule of M, then both N and M/N have finite type dimension and $t.dim(M) = t.dim(N) + t.dim(M/N)$.*
  2 *M has finite type dimension iff M has ACC on type submodules iff M has DCC on type submodules.*

Proof. (1). Let P be a complement of N in M. Then we have $N \oplus P \leq_e M$, $N \perp P$ and P is essentially embeddable in M/N. Therefore, $t.dim(M) = t.dim(N \oplus P) = t.dim(N) + t.dim(P) = t.dim(N) + t.dim(M/N)$.

(2). Suppose there exists a chain $M_1 \subset M_2 \subset \cdots$ with $M_i \leq_t M$ for all i. Let X_i be a complement of M_i in M_{i+1}. Then $X_i \neq 0$ and $X_i \perp M_i$ for each i. Thus, there is an infinite direct sum $\oplus_{i=1}^{\infty} X_i$ such that $X_i \perp X_j$ if $i \neq j$, implying $t.dim(M) = \infty$.

Suppose there exists a chain $M_1 \supset M_2 \supset \cdots$ with $M_i \leq_t M$ for all i. For each i, let P_i be a complement of M_{i+1} in M_i. Then $P_i \perp M_{i+1}$ and $0 \neq P_i \leq_t M$ (see [15]). Thus, we have an infinite direct sum of type submodules: $P_1 \oplus P_2 \oplus \cdots$. Let $X_1 = P_1$ and, for each $i > 1$, let X_i be a closure of $P_1 \oplus \cdots \oplus P_i$ in M such that $X_i \supseteq X_{i-1}$. Then $X_1 \subset X_2 \subset \cdots$ and each X_i is a type submodule of M (see [15, Lemma 1(g)]).

Suppose M contains a submodule which is an infinite direct sum $\oplus_{i=1}^{\infty} Y_i$ such that $Y_i \perp Y_j$ if $i \neq j$. For each i, let N_i be a maximal element in $\{P \subseteq M : Y_i \subseteq P \| Y_i\}$. Then each N_i is a type submodule of M and $N_i \perp N_j$ if $i \neq j$. So, we may assume each Y_i is a type submodule of M. Let $X_1 = M$ and, for each $i > 1$, let X_i be a closure of $\oplus_{j \geq i} Y_j$ in X_{i-1}. Then we have a chain $X_1 \supset X_2 \supset \cdots$ such that each X_i is a type submodule of M (see [15, Lemma 1(g)]). □

2. Nonsingular rings with finite type dimension

For the definition of the maximal right quotient rings, we refer to [7].

Lemma 2.1. *The following are equivalent for a nonsingular ring R:*
1. $t.dim(R) = n$.
2. *The maximal right quotient ring of R is a finite direct sum of n indecomposable right self-injective regular rings.*

Proof. (1) $\Rightarrow$ (2). It is well-known that the maximal right quotient ring of R is $Q = End(E(R))$ ($\cong E(R)$). Suppose that $I_1 \oplus \cdots \oplus I_n \leq_e R_R$, where each I_i is a nonzero atomic right ideal of R and $I_i \perp I_j$ (if $i \neq j$). Then $E(R) = E(I_1) \oplus \cdots \oplus E(I_n)$ and $Q \cong End(E(I_1)) \oplus \cdots \oplus End(E(I_n))$. By [7, 2.22] and Lemma 1.3, each $End(E(I_i))$ is a regular right self-injective indecomposable ring.

(2) $\Rightarrow$ (1). Suppose $Q = E(R) = E_1 \oplus \cdots \oplus E_n$ is a diret sum of regular right self-injective indecomposable rings. Then E_i is an atomic E_i-module by Lemma 1.3. Therefore, as Q modules, E_i is atomic and $E_i \perp E_j$ (if $i \neq j$). By [7, 2.7], as R-modules, each $R \cap E_i$ is atomic and $(R \cap E_i) \perp (R \cap E_j)$ (if $i \neq j$). Since $(R \cap E_1) \oplus \cdots \oplus (R \cap E_n) \leq_e R_R$, we have $t.dim(R) = n$. □

A module M is said to be weakly-injective (respectively, weakly R-injective) if for any finitely generated (respectively, cyclic) submodule Y of $E(M)$ there exists a submodule X of $E(M)$ such that $Y \subseteq X \cong M$ (see [1]). A TS-module is any module whose type submodules are summands [15].

Theorem 2.2. *The following are equivalent for a nonsingular ring R:*
1. R has finite type dimension.
2. The maximal right quotient ring of R is a finite direct sum of indecomposable right self-injective regular rings.
3. For every family $\{M_i : i \in I\}$ of pairwise orthogonal nonsingular modules, $\oplus_{i \in I} E(M_i)$ is injective.
4. For any family $\{E_i : i \in I\}$ of pairwise orthogonal nonsingular injective modules, $\oplus_i E_i$ is weakly injective.
5. For any family $\{E_i : i \in I\}$ of pairwise orthogonal nonsingular weakly injective modules, $\oplus_i E_i$ is weakly injective.
6. For any family $\{E_i : i \in I\}$ of pairwise orthogonal nonsingular weakly injective modules, $\oplus_i E_i$ is weakly R-injective.
7. Every cyclic (or finitely generated) nonsingular module has finite type dimension.
8. Every nonsingular TS-module is a direct sum of atomic modules.
9. Every nonsingular injective module is a direct sum of atomic modules.
10. Every nonsingular module contains a maximal injective type submodule.

Proof. (1) $\Leftrightarrow$ (2). By Lemma 2.1.

(1) $\Rightarrow$ (3). For a family $\{E_i : i \in I\}$ of pairwise orthogonal nonsingular injective modules, let $E = \oplus_i E_i$. Suppose $f : J \longrightarrow E$ is an R-homomorphism, where J is a right ideal of R. For each n, let $J_n = \{a \in J : f(a) \in \oplus_{i=1}^n E_i\}$. Then $J_1 \subseteq J_2 \subseteq \cdots \subseteq J_n \subseteq \cdots$. For each i, let Q_i be a maximal element in $\{L \subseteq R_R : J_i \subseteq L \| J_i\}$. Then Q_i is a type right ideal of R. By a well-known fact that any submodule of a nonsingular module M has a unique closure in M, we have $Q_1 \subseteq Q_2 \subseteq \cdots$. By Lemma 1.4, there exists a number n such that $Q_n = Q_{n+1} = \cdots$. Therefore, we have $J_n \| J_{n+1} \| \cdots$. Claim: there exists a number $s \geq n$ such that $J_s \leq_e J_{s+1} \leq_e \cdots$. In fact, if such a number s does not exist, then there exists a sequence $\{n \leq n_1 < n_2 < \cdots\}$ such that J_{n_i} is not essential in J_{n_i+1} for each i. Therefore, there exists $0 \neq K_i \subseteq J_{n_i+1}$ such that $J_{n_i} \cap K_i = 0$. Note that $K_i \hookrightarrow J_{n_i+1}/J_{n_i} \hookrightarrow E_{n_i+1}$. It follows that $K_i \perp K_j$ if $i \neq j$, and hence $t.dim(R) = \infty$. This is a contradiction. Therefore, such a number s exists. For any $t \geq s$, since $J_{t+1}/J_t \hookrightarrow E_{t+1}$ is nonsingular, we have $J_t = J_{t+1}$ for all $t \geq s$. Then $f(J) \subseteq \oplus_{i=1}^t E_i$. Therefore, there exists a homomorphism $g : R \longrightarrow E$ that extends f, showing that E is injective.

(3) $\Rightarrow$ (4) and (7) $\Rightarrow$ (1) and (8) $\Rightarrow$ (9). Obvious.

(4) $\Rightarrow$ (5) $\Rightarrow$ (6) $\Rightarrow$ (7). Similar to the proof of [1, Th].

(7) $\Rightarrow$ (8). Let M be a nonsingular TS-module. By [15, Prop.16], we only need to show that, for any family $\{X_\alpha : \alpha \in I\}$ of type submodules of M such that $X = \Sigma X_\alpha$ is direct and $\Sigma_{\alpha \in F} X_\alpha$ is a summand of M for any finite subset F of I, X is a summand of M. By [15, Lemma 1(g)], the closure X^c of X in M is a type submodule of M and hence a summand of M. So, it

suffices to show that $X = X^c$. If $X \neq X^c$ then, by an argument similar to the proof of [15, Prop.18], there exists a sequence $\{y_0, y_1, \cdots\}$ of elements in X^c such that $y_0^\perp \subset y_1^\perp \subset \cdots$ and $t.dim(\oplus_i R/y_i^\perp) = \infty$. Since each $t.dim(R/y_i^\perp) < \infty$, we may assume that there exist $\bar{0} \neq I_i/y_i^\perp \subseteq R/y_i^\perp$ such that $(I_i/y_i^\perp) \perp (I_j/y_j^\perp)$ if $i \neq j$. Since R is nonsingular, the nonsingularness of $R/y_i^\perp$ implies that $y_i^\perp$ is a complement right ideal of R. Thus, for each i, there exists $0 \neq J_i \subseteq I_i$ such that $J_i \cap y_i^\perp = 0$ and hence $J_i \hookrightarrow I_i/y_i^\perp$. Therefore, $J_i \perp J_j$ whenever $i \neq j$, implying $t.dim(R) = \infty$.

(9) $\Rightarrow$ (10) $\Rightarrow$ (3). Similar to the proof of "(c) $\Rightarrow$ (f) $\Rightarrow$ (b)" of [15, Th.22]. □

If R has finite type dimension, then R is a finite direct sum of indecomposable rings. But, a nonsingular indecomposable ring may not have finite type dimension as the following example shows.

Example 2.3. *Let $Q = \Pi_{i=1}^\infty R_i$ be a direct product of rings R_i with $R_i = \mathbf{Z}$ for each i and R be the subring generated by $\oplus_{i=1}^\infty 2R_i$ and 1_Q. Clearly, if $i \neq j$, then $2R_i \perp 2R_j$ as R-modules. So, $t.dim(R) = \infty$. It is easy to see that R is nonsingular and R has no nontrivial idempotents. Therefore, R is indecomposable.*

3. Nonsingular Rings With Enough Atomic Right Ideals

In this section, we will show that a nonsingular ring has the property that every nonzero right ideal contains an atomic right ideal iff its maximal right quotient ring is a direct product of indecomposable right self-injective regular rings. It is interesting to compare this to [7, 3.29] which states that, for a nonsingular ring R, every nonzero right ideal of R contains a uniform right ideal iff the maximal right quotient ring of R is a direct product of right full linear rings.

Lemma 3.1. *Let M be a module and N a submodule of M.*
  1 *Let $N \leq_t M$ and $N \subseteq X \subseteq M$. Then $X/N \leq_t M/N$ iff $X \leq_t M$; In particular, M/N is atomic iff N is a maximal type submodule of M.*
  2 *M has an atomic submodule iff M has a maximal type submodule; More precisely, N is a maximal type submodule of M iff any (or one) of its complements in M is atomic.*
  3 *If every nonzero submodule of M contains an atomic submodule, then the intersection of all maximal type submodules of M is equal to 0.*
  4 *If $M = X \oplus Y$ is nonsingular and $X \perp Y$, then $Hom_R(X, Y) = 0$.*
  5 *Let M be nonsingular and N_i ($i = 1, 2$) be maximal type submodules. If $N_1 \neq N_2$ then $(M/N_1) \perp (M/N_2)$.*

Proof. (1) Suppose X is not a type submodule of M. Then there exists $Y \subseteq M$ such that $X \subset Y$ and $X \| Y$. Thus, $(X/N) \subset (Y/N)$. For any

$\bar{0} \neq A/N \subseteq Y/N$, we have $N \perp B$ for some $0 \neq B \subseteq A$ since N is a type submodule of M. It follows from $X \| Y$ that $0 \neq C \cong D$ for some $C \subseteq X$ and $D \subseteq B$. Then $N \perp C$ and hence $\bar{0} \neq (C+N)/N \cong (D+N)/N$. Therefore, $(X/N) \| (Y/N)$ and so X/N is not a type submodule of M/N. Conversely, if X/N is not a type submodule of M/N, then there exists $Y/N \subseteq M/N$ such that $X/N \subset Y/N$ and $(X/N) \| (Y/N)$. We show that X is not a type submodule of M by showing $X \| Y$. For any $0 \neq A \subseteq Y$, we may assume $A \cap N = 0$. Then we have $\bar{0} \neq B/N \hookrightarrow A$ for some $B \subseteq X$. Since N is a complement submodule of M, there exists $0 \neq C \subseteq B$ such that $C \cap N = 0$. Thus, $C \hookrightarrow A$ and so $X \| Y$.

(2). It is routine.

(3). For any $0 \neq x \in M$, choose an atomic submodule A in xR. By Zorn's Lemma, there exists a submodule B of M maximal with respect to the property that $A \subseteq B$ and $A \| B$. It follows that B is a type submodule of M and is atomic. Let N be a complement of B in M. Then N is a maximal type submodule of M by (2). Note that $N \perp B$ and hence $A \perp N$, implying $x \notin N$.

(4). It is clear.

(5). For each i, let A_i be a complement of N_i in M. Then A_i are atomic by (2) and $A_i \| (M/N_i)$. So we only need to show $A_1 \perp A_2$. If not, then $A_1 \| A_2$. Note A_i are type submodules of M (see [15, Lemma 1]). It follows that $A_1 \cap A_2 \leq_e A_i$ for each i. Therefore, A_1 and A_2 are closures of $A_1 \cap A_2$ in M. But since M is nonsingular, it must be $A_1 = A_2$. This implies that $N_1 \| N_2$. Since N_1 and N_2 are type submodules, we have $N_1 \cap N_2 \leq_e N_i$ for each i. So, N_1 and N_2 are closures of $N_1 \cap N_2$ in M. Thus, $N_1 = N_2$. □

Theorem 3.2. *The following are equivalent for a nonsingular ring R:*
 1 *Every nonzero right ideal of R contains an atomic right ideal.*
 2 *The maximal right quotient ring of R is isomorphic to a direct product of indecomposable right self-injective regular rings.*

Proof. (1) $\Rightarrow$ (2). Let $\{E_t : t \in I\}$ be the set of all maximal type submodules of $E = E(R)$ and $E_* = \Pi_t(E/E_t)$. We first prove $End(E) \cong End(E_*)$. For any $f \in End(E)$, define $\phi(f)$ as follows: If $(x_t + E_t) \in E_*$ with $x_t \in E$ for each t, then we let $\phi(f)((x_t + E_t)) = (f(x_t) + E_t)$. We have $\phi(f) \in End(E_*)$ and ϕ gives a ring homomorphism. By Lemma 3.1(3), $\cap_t E_t = 0$. This implies that ϕ is one to one. In order to show ϕ is onto, for each t, we write $E = E_t \oplus E_t'$ and use π_t to indicate the canonical isomorphism from E_t' onto E/E_t (i.e., $\pi_t(x_t) = x_t + E_t$ with $x_t \in E_t'$) and $\pi = \oplus \pi_t$. For any $\theta \in End(E_*)$, we have $\theta(E/E_t) \subseteq E/E_t$ by Lemma 3.1(4,5). There exists a homomorphism $h \in End(E)$ such that h extends the map $\pi^{-1} \circ \theta \circ \pi$. It can easily be checked that $\phi(h) = \theta$. Therefore, $End(E) \cong End(E_*)$. Next, we note that $End(E)$ is the maximal right quotient ring of R and, because of Lemma 3.1(4), $End(E_*) \cong \Pi_t End(E/E_t) \cong \Pi_t End(E'_t)$. By Lemma 1.3 and [7, 2.22], each $End(E'_t)$ is an indecomposable right self-injective regular ring.

(2) $\Rightarrow$ (1). This can be proved by an idea similar to that in the proof "(2) $\Rightarrow$ (1)" of Lemma 2.1. □

Example 3.3. Let $F_1, F_2, \cdots$ be fields, $R = (\Pi F_n)/(\oplus F_n)$. Then $Z(R_R) = 0$ by [7, Ex.6,P94]. It is easy to show that every nonzero principal right ideal of R contains two nonzero right ideals which are orthogonal, and therefore R has no atomic right ideals. If let $S = R \oplus F$ for a field F, then some right ideals of S contain atomic right ideals but many do not.

4. Several type dimension formulas

We now present type dimension formulas for polynomial rings, Laurent polynomial rings and formal triangular rings. These formulas can be used to provide rings of finite type dimension.

Proposition 4.1. *We have the following:*
1. *Any (noncommutative) domain has type dimension 1.*
2. *Being a ring of type dimension n is a Morita invariant.*
3. *If $R = R_1 \oplus \cdots \oplus R_n$ is a direct sum of rings, then $t.dim(R_R) = t.dim(R_{1R_1}) + \cdots + t.dim(R_{nR_n})$.*

Proof. It is easy. □

Proposition 4.2. *Let $R = \begin{pmatrix} S & B \\ 0 & T \end{pmatrix}$ be the formal triangular matrix ring, where S, T are rings and B is an (S,T)-bimodule. Let $l(B) = \{s \in S : sB = 0\}$ and $r(B) = \{t \in T : Bt = 0\}$. Then*
1. $t.dim(R_R) = t.dim(l(B)_S) + t.dim((B \oplus T)_T)$.
2. $t.dim(_RR) = t.dim(_Tr(B)) + t.dim(_S(S \oplus B))$.

Proof. First note that, as right R-modules, $\begin{pmatrix} l(B) & 0 \\ 0 & 0 \end{pmatrix} \perp \begin{pmatrix} 0 & B \\ 0 & T \end{pmatrix}$ and $\begin{pmatrix} l(B) & 0 \\ 0 & 0 \end{pmatrix} \oplus \begin{pmatrix} 0 & B \\ 0 & T \end{pmatrix} \leq_e R_R$. Next, the map $f : X \longrightarrow f(X) = \begin{pmatrix} X & 0 \\ 0 & 0 \end{pmatrix}$ gives a bijection from the set of all right S-submodules of $l(B)$ onto the set of all right R-submodules of $\begin{pmatrix} l(B) & 0 \\ 0 & 0 \end{pmatrix}$ such that $f(X) \cong f(Y)$ iff $X \cong Y$. This implies that $t.dim(l(B)_S) = t.dim(\begin{pmatrix} l(B) & 0 \\ 0 & 0 \end{pmatrix}_R)$. Finally, the map $g : X \longrightarrow g(X) = \{\begin{pmatrix} 0 & b \\ 0 & t \end{pmatrix} : (b,t) \in X\}$ gives a bijection from the set of all right T-submodules of $B \oplus T$ onto the set of all right R-submodules of $\begin{pmatrix} 0 & B \\ 0 & T \end{pmatrix}$ such that $g(X) \cong g(Y)$ iff $X \cong Y$. It follows that $t.dim((B \oplus T)_T) = t.dim(\begin{pmatrix} 0 & B \\ 0 & T \end{pmatrix}_R)$. Now by Lemma 1.2(3), we have $t.dim(R) = t.dim(\begin{pmatrix} l(B) & 0 \\ 0 & 0 \end{pmatrix}) + t.dim(\begin{pmatrix} 0 & B \\ 0 & T \end{pmatrix}) = t.dim(l(B)_S) + t.dim((B \oplus T)_T)$. The proof of (2) is similar. □

Proposition 4.2 can be used to show the left-right asymmetry of the concept of type dimension.

Example 4.3.
1 Let $R = \begin{pmatrix} \mathbb{Z} & 2\mathbb{Z}_4 \\ 0 & \mathbb{Z}_4 \end{pmatrix}$. By Porposition 4.2, $t.dim(R_R) = 2$ and $t.dim(_RR) = 3$.
2 Let $Q = \Pi F_i$ with each F_i a field, S the subring of Q generated by $\oplus_i F_i$ and 1_Q, and $B = \oplus_i F_i$. Let $R = \begin{pmatrix} S & B \\ 0 & \mathbb{Z} \end{pmatrix}$. By Proposition 4.2, $t.dim(R_R) = 2$ and $t.dim(_RR) = \infty$.

Proposition 4.4. *Let R be a ring. Then $t.dim(R_R) = t.dim(S_S)$ if S is any one of the following:*
1 $R[x_1, x_2, \cdots, x_n]$.
2 $R[x_1, x_2, \cdots]$.
3 $R[x_1, \cdots, x_n, x_1^{-1}, \cdots, x_n^{-1}]$.
4 $R[\cdots, x_2, x_1, x_1^{-1}, x_2^{-1}, \cdots]$.

Proof. (1). We only need to prove the case of $n = 1$. First we show that if A is an atomic right ideal of R then $A[x]$ is an atomic right ideal of $R[x]$ by proving any two cyclic right ideals in $A[x]$ are parallel. Let f and g be any two nonzero elements in $A[x]$ with leading coefficients a and b respectively. By [7, 3.21], we may assume that the right annihilator of f (respectively g) in $R[x]$ is the same as the right annihilator of a (respectively b) in $R[x]$. Since A is atomic, there exist s and t in R such that $0 \neq asR \cong btR$ via $asr \longmapsto btr$. Then the map $\phi : fsR[x] \longrightarrow gtR[x]$ by $\phi(fsh) = gth$ ($h \in R[x]$) is an $R[x]$-module isomorphism. So, $A[x]$ is an atomic right ideal. Next, we show that if $A \perp B$ for right ideals A and B of R, then as right ideals of $R[x]$ we have $A[x] \perp B[x]$. Suppose $A[x] \perp B[x]$ does not hold. Then we have an $R[x]$-module isomorphism $\pi : fR[x] \longrightarrow gR[x] \neq 0$ by $\pi(fh) = gh$ for some $f \in A[x]$ and $g \in B[x]$. Let a and b be the leading coefficients of f and g respectively. By [7, 3.21], we may assume that, for some $c \in R$ with $gc \neq 0$, the right annihilator of f (respectively, gc) in $R[x]$ equals the annihilator of a (respectively, bc) in $R[x]$. Then the map $acR \longrightarrow bcR$ by $acr \longmapsto bcr$ is an R-module isomorphism, a contradiction. It follows that if $t.dim(R_R) = \infty$ then $t.dim(R[x]_{R[x]}) = \infty$. Now, if $t.dim(R) = n$ then there exist pairwise orthogonal atomic right ideals $A_1, \cdots, A_n$ of R such that $A_1 \oplus \cdots \oplus A_n \leq_e R_R$. Therefore, $A_1[x] \oplus \cdots \oplus A_n[x] = (A_1 \oplus \cdots \oplus A_n)[x] \leq_e R[x]_{R[x]}$. By the facts above, as right ideals of $R[x]$, $A_i[x]$ are atomic and $A_i[x] \perp A_j[x]$ when $i \neq j$. Therefore, $t.dim(R[x]_{R[x]}) = n$.

(2). The argument in the proof of (1) easily implies that if A is an atomic right ideal of R then $A[x_1, x_2, \cdots]$ is an atomic right ideal of $R[x_1, x_2, \cdots]$; and if $A \perp B$ in R_R then $A[x_1, x_2, \cdots] \perp B[x_1, x_2, \cdots]$ in S_S, where $S =$

$R[x_1, x_2, \cdots]$. Therefore, (2) follows from these and the easy fact that if $A_1 \oplus \cdots \oplus A_k \leq_e R_R$ then $A_1[x_1, x_2, \cdots] \oplus \cdots \oplus A_k[x_1, x_2, \cdots]$ is an essential right ideal of $R[x_1, x_2, \cdots]$.

(3) and (4). $R[x_1, \cdots, x_n, x_1^{-1}, \cdots, x_n^{-1}]$ and $R[\cdots, x_2, x_1, x_1^{-1}, x_2^{-1}, \cdots]$ are rings of right quotients (see the definition below) of $R[x_1, \cdots, x_n]$ and $R[x_1, x_2, \cdots]$ respectively, so (3) and (4) follow from (1) and (2) and the following Lemma 4.5. □

Let R be a subring of S and the identity of R is the identity of S. We call S a ring of right quotients of R if for every $0 \neq s_1 \in S$ and for every $s_2 \in S$, there exists $r \in R$ such that $s_1 r \neq 0$ and $s_2 r \in R$.

Lemma 4.5. *Let R be a ring and S a ring of right quotients of R. Then $t.dim(R_R) = t.dim(S_S)$.*

Proof. We first show that if $A \perp B$ in S_S then $(A \cap R) \perp (B \cap R)$ in R_R. This will imply that $t.dim(R_R) \geq t.dim(S_S)$. If $A \cap R$ and $B \cap R$ are not orthogonal, then there exist $a \in A \cap R$ and $b \in B \cap R$ such that $\phi : aR \longrightarrow bR \neq 0$ by $\phi(ar) = br$ is an isomorphism. Define $\Phi : aS \longrightarrow bS$ via $\Phi(as) = bs$ for all $s \in S$. Suppose $as = 0$ but $bs \neq 0$ for some $s \in S$. Since S is a ring of right quotients of R, there exists $r \in R$ such that $0 \neq bsr$ and $sr \in R$. Then we have $0 \neq \phi^{-1}(bsr) = asr$, implying $as \neq 0$. Thus, Φ is well-defined. Similarly, Φ is one to one. It is easy to see Φ is an R-module isomorphism, contradicting the fact that $A \perp B$.

Next, to prove $t.dim(R_R) \leq t.dim(S_S)$, we only need to show that if $aR \perp bR$ in R_R then $aS \perp bS$ in S_S. If aS and bS are not orthogonal then there exist $s_1, s_2 \in S$ such that $as_1 S \cong bs_2 S$ via $as_1 s \longmapsto bs_2 s$. Since S is a ring of right quotients of R, there exists $r_1 \in R$ such that $as_1 r_1 \neq 0$ and $s_1 r_1 \in R$. Then $bs_2 r_1 \neq 0$. Thus, there exists $r_2 \in R$ such that $bs_2 r_1 r_2 \neq 0$ and $s_2 r_1 r_2 \in R$. Therefore, we have an R-module isomorphism $a(s_1 r_1 r_2)R \cong b(s_2 r_1 r_2)R$ by $a(s_1 r_1 r_2)r \longmapsto b(s_2 r_1 r_2)r$, contradicting the fact that $aR \perp bR$. □

References

1. A.H. Al-Huzali, S.K. Jain and S.R. López-Permouth, *Rings whose cyclics have finite Goldie dimension*, J. Alg. **153** (1992), 37–40.
2. T.J. Cheatham, *Finite dimensional torsion free rings*, Pacific J. Math. **39** (1971), 113–118.
3. J. Dauns, *Module types*, Preprint, 1995.
4. P. Gabriel, *Des catégories abéliennes*, Bull. Soc. Math. France **90** (1962), 323–448.
5. A.W. Goldie, *The structure of prime rings under ascending chain conditions*, Proc. London Math. Soc. **8**(1958), 589–608.

6. K.R. Goodearl, *Singular torsion and the splitting properties*, Memoirs Amer. Math. Soc. **124**(1972).
7. K.R. Goodearl, *Ring Theory: Nonsingular Rings and Modules*, Marcel Dekker, Inc (1976).
8. J.J. Hutchinson, *Quotient full linear rings*, Proc. Amer. Math. Soc. **28**(1971), 375–378.
9. R.E. Johnson, *Quotient rings of rings with zero singular ideal*, Pacific J. Math. **11**(1960), 710–717.
10. F. Sandomierski, *Semisimple maximal quotient rings*, Trans. Amer. Math. Soc. **128**(1967), 112–120.
11. R.C. Shock, *Polynomial rings over finite dimensional rings*, Pacific J. Math. **42**(1972), 251–258.
12. M. Teply, *Some aspects of Goldie's torsion theory*, Pacific J. Math. **29**(1969), 447–459.
13. C.L.Walker and E.A.Walker, *Quotient categories and rings of quotient*, Rocky Mountain J. Math. **2**(1972), 513–555.
14. R.W. Wilkerson, *Finite dimensional group rings*, Proc. Amer. Math. Soc. **41**(1973), 10–16.
15. Y. Zhou, *Decomposing modules into direct sums of submodules with types*, Preprint, 1996.

DEPARTMENT OF MATHEMATICS AND STATISTICS, MEMORIAL UNIVERSITY OF NEWFOUNDLAND, ST. JOHN'S, NEWFOUNDLAND, CANADA A1C 5S7

Trends in Mathematics

Trends in Mathematics is a book series devoted to focused collections of articles arising from conferences, workshops or series of lectures.

Topics in a volume may concentrate on a particular area of mathematics, or may encompass a broad range of related subject matter. The purpose of this series is both progressive and archival, a context to make current developments available rapidly to the community as well as to embed them in a recognizable and accessible context.

Volumes of TIMS must be of high scientific quality. Articles without proofs, or which do not contain any significantly new results, are not appropriate. High quality survey papers, however, are welcome. Contributions must be submitted to peer review in a process that emulates the best journal procedures, and must be edited for correct use of language. As a rule, the language will be English, but selective exceptions may be made. Articles should conform to the highest standards of bibliographic reference and attribution.

The organizers or editors of each volume are expected to deliver manuscripts in a form that is essentially "ready for reproduction." It is preferable that papers be submitted in one of the various forms of TeX in order to achieve a uniform and readable appearance. Ideally, volumes should not exceed 350-400 pages in length.

Proposals to the Publisher are welcomed at either:

> Mathematics Department
> Birkhäuser Boston
> 675 Massachusetts Avenue
> Cambridge, MA 02139
> math@birkhauser.com

> Mathematics Department
> Birkhäuser Verlag AG
> PO Box 133
> CH-4010 Basel, Switzerland
> math@birkhauser.ch

Titles in the Series
ANDERSSON/LAPIDUS (eds). Progress in Inverse Spectral Geometry
JAIN/RIZVI (eds). Advances in Ring Theory

Forthcoming Title in the Series
ROSS/ANDERSON/LITVINOV/SINGH/SUNDER/WILDEBERGER (eds). International Conference on Harmonic Analysis